Laws of Men and Laws of Nature

Laws of Men and Laws of Nature

THE HISTORY OF SCIENTIFIC EXPERT TESTIMONY IN ENGLAND AND AMERICA

TAL GOLAN

HARVARD UNIVERSITY PRESS

Cambridge, Massachusetts
London, England

Library of Congress Cataloging-in-Publication Data
Golan, Tal.
    Laws of men and laws of nature : the history of scientific expert testimony in England
and America / Tal Golan.
        p. cm.
    Includes bibliographical references and index.
    ISBN 0-674-01286-0 (alk. paper)
    1. Evidence, Expert--England--History. 2. Evidence, Expert--United States--History. I.
Title.

KD7521.G65 2004
347.42'067--dc22

2004042215

For Dana

Contents

Acknowledgments

It would be impossible to properly thank the many historians, scholars, judges, scientists, lawyers, archivists and colleagues who contributed to this book by offering valuable materials, guidance, commentary and criticism. What follows is no more than a start. My editor Elizabeth Knoll has been a constant source of support and much-valued advice. David Lieberman and Thomas G. Barnes initiated me into the mysteries of Common Law and showed me how to handle its subtleties. Roger Hahn, Alex Pang, Sheldon Rothblatt, Carla Hesse, David Hollinger, Barbara Shapiro, Chris Kutz, Roger Smith, Jennifer Mnookin and the late Amos Funkenstein have spent many hours with me, discussing my ideas and providing me with invaluable guidance. My father-in-law, Leon Linskill, helped me to decipher obscured eighteenth-century materials and took great care in keeping my English proper. My dear friend Noah Efron has offered me unending camaraderie, energy and brilliance. I thank them all.

Institutional help was crucial to the completion of this book. UC Berkeley's history department and the Office for the History of Science and Technology provided me with the best education I could have hoped for. The inter-library loan unit at UC Berkeley admirably coped with the wide range of my requests. A two-year post-doctorate fellowship at the Dibner Institute for the History of Science and Technology, in MIT, provided me with access to the some of the finest libraries in the world and plenty of time to develop my ideas. The Rothschild Fund

gave me much-needed financial support and the Ben Gurion Research
Institute in Israel proved to be an excellent place to work on and finish
the book.

A slightly different version of Chapter 4 appeared as "Blood Will
Out: Distinguishing Humans from Animals and Scientists from Charlatans in the Nineteenth-Century Courtroom," *Historical Studies in the
Physical and Biological Sciences* (2000) 31:93–124. Some parts of Chapter
5 appeared as "The Authority of Shadows: The Legal Embrace of the
X-Ray," *Historical Reflections* (1998), 24:437–458. I thank both journals
for permission to use some of that material here.

Finally there is John Heilbron, to whom I owe more than words could
convey. His teachings are all over this book.

Laws of Men and Laws of Nature

Introduction

THIS IS A HISTORY OF SCIENTIFIC expert testimony in Common Law courts. Situated at the intersection of the two dominant institutions of science and law, scientific expert testimony has long been overlooked by both. Historians of science ignored it because they did not consider courts of law to be important sites of scientific activity before the twentieth century. Historians of law ignored it because they never considered science to be a significant factor in the development of judicial practices and jurisprudence related to evidence. As a result, there is relatively little scholarship about the history of the relations between the two most authoritative institutions in modern Western culture—science and law.

This book explores these relations. It describes the emergence of the law as a major patron of nineteenth-century science, a role it fulfilled in several ways. It was a patron in the most literal sense: expert testimony, arbitration, and counseling constituted a lucrative sideline activity for many men of science. In addition to funding scientists, the courts also underwrote scientific progress; the constant demand for additional and better scientific evidence spurred important discussions on central scientific issues such as standardization, accuracy, and reliability. Finally, the adversarial realities of the legal system provided a fertile breeding ground for an intensive nineteenth-century discourse on what it meant to be a man of science in an increasingly professionalized and industrialized world, and on the character of the scientific community—its function in society and the values by which it judged the work of its members.

Science and technology affected law no less than law affected science. New legal apparatuses evolved during the nineteenth century to deal with the rising tide of litigation involving complex scientific argumentation. Patent law evolved into a major mediator between the producers of scientific knowledge and those who adapted it to the various wants of society. The legal domain of regulation evolved to control the risks that scientific knowledge and its technological products created for public safety and the environment. Finally, and most important for this book, scientific and technological developments introduced, sometimes within days and weeks of their discovery or invention, novel forms of knowledge claims into the courts. These claims continually challenged judicial practices and inspired developments in the jurisprudence of evidence.

This leads us to the question of appearances. We are accustomed to thinking of science and law as two fundamentally distinct cultures. Science deals with nature, we are told, and law with society. Science organizes our knowledge of the world; law directs our actions in it. Science is an open-ended, impartial search for truth; the law is a normative process that ultimately seeks closure.[1] It is equally true, however, that science and law are mutually supporting belief systems and deeply connected social institutions heavily invested in each other. Scientific knowledge and techniques have played a growing role in the spread of justice in modern society. Institutions of the law have helped to clarify the character of legitimate scientific knowledge and practices and to readjust the social and institutional relations that their application required. We should not be surprised, therefore, to discover that the courts have not been neutral gatekeepers that simply exclude unreliable scientific testimony but rather active partners in the production and maintenance of credible scientific evidence. Similarly, we should not be surprised to find that science has been no mere supplicant to the law, but, again, an influential partner in the production and maintenance of credible legal theories and practices for fact-finding and proof. Here, then, lies the challenge of this book, to transcend the dichotomies of science and law. The result should be read not as a history of English patent law or American forensic science but as an exploration of key moments in the evolving relations between the expanding cultures of law and science on both sides of the Atlantic.

The account that follows is chronological, but its geography is more complicated. The book starts in late-eighteenth-century England with the birth of the modern partisan expert witness and with the legal embracing of Newtonian philosophy as a legitimate expertise. It ends in late-twentieth-century America, with the U.S. Supreme Court dipping into the murky waters of the philosophy of science in an effort to establish criteria that would allow the courts to distinguish between good and bad science. In between, this book explores important turning points in the practices, debates, and jurisprudence of expert testimony in England and America. Each chapter describes how an important set of scientific developments in engineering, chemistry, industry, public health, microscopy, experimental psychology, and other areas challenged the laws of evidence and the practices of expert testimony and produced new jurisprudence in patent law, nuisance law, expert medical and scientific testimony, the admissibility of visual images, and so on.

Mapping broadly the uncharted territories that lie between law and science, this book provides a much-needed historical perspective on the state of scientific expert testimony in Common Law courts today. The scientific controversies that accompanied many high-stakes legal cases that turned on scientific evidence—whether notorious criminal cases, such as the 1995 O. J. Simpson trial or the 1997 trial of Louise Woodward, or civil cases such as the tobacco, Bendectin, and breast-implant mass tort litigations—have been seen by many commentators as a sign of moral corruption. America's courts, they warn, are being swamped by "junk science" produced by opportunistic scientific experts at the behest of unscrupulous attorneys, who are able in the name of science to persuade credulous juries to acquit wealthy defendants or award millions of dollars each year to plaintiffs spuriously suing deep-pocketed corporations. Other critics blame the scientific disagreements on faulty legal procedures governing expert testimony. The adversarial nature of the legal process, they argue, promotes partisanship and prevents the appropriate resolution of the scientific issues presented in court. But whether they blame the experts or the law, almost all critics assume that the scientific disagreements in court are an aberration. If charlatanism and partisanship could somehow be swept aside, they believe, disagreement would diminish, and perhaps even disappear altogether.[2]

Not all scholars see the disagreement among the scientific witnesses as detrimental to either justice or science. Some maintain that the scuffles between scientists in court reflect the actual day-to-day workings of a healthy scientific community that constitutes an integral

part of society.[3] Thus, where the majority of commentators see decadence and unnecessary partisanship, these scholars see a normative
commitment by the adversarial legal system to develop two sides to
every story; where others cry judicial ineptitude, they celebrate the
superior ability of the adversarial procedures to disclose areas of uncertainty and inform the public of the interpretive conflicts and normative
biases left unacknowledged by the scientific community; and where
others emphasize the importance of reliable expert guidance for the
maintenance of a healthy judicial process in a society that grows exponentially in specialization and sophistication, they emphasize the
importance of deconstructing expert authority through the adversarial
process for the maintenance of a healthy political discourse.

Underlying this debate is one premise shared by all—that the malaise
of expert testimony is a sign of our times, the result of the growing
difficulties of the courts and the public in handling the increasing
complexity of modern science.[4] This assumption, we shall quickly see,
is mistaken. Far from being new, the putative problem of scientific
expert testimony in Common Law courts has a long and rich history.
Discontent with scientific expertise in the courts has existed as long as
there have been scientific expert witnesses, and by the mid-nineteenth
century, the debate over the meaning of these conflicts and the ways to
resolve them had all the features that today are blithely assumed to be
new. Understanding the long, twisted roots of today's conflicts will not
in itself resolve the vexed debate about the role of science in the courts.
If anything, it may suggest that current conflicts are more deeply
ingrained and hence less ameliorable than many pundits would like to
believe. But it will, at least, reveal that these conflicts are less a product
of human and institutional pathology than they are an illustration,
should we need one, of the complexity of the ongoing social negotiations needed to harmonize laws of men and laws of nature and to cut
truth and justice to human measure.

1

"Where There's Muck There's Brass": The Rise of the Modern Expert Witness

Lord Mansfield was a surprising man; ninety-nine times out of a hundred he was right in his opinion or decisions. And when he was wrong, ninety-nine men out of a hundred could not discover it. He was a wonderful man!

∼Remark attributed to Lord Chancellor Thurlow, quoted by James Oldham, *The Mansfield Manuscripts*

THIS CHAPTER RECOUNTS a late-eighteenth-century legal episode that serves in the legal literature as the origin story for the rise of expert testimony in the modern Anglo-American legal system. The case could have become an origin story because by the time it occurred both lawyers and men of science had obtained the same sort of authority they now exercise in their respective spheres. At the start of the eighteenth century, natural philosophy was but a bookish study of nature in general. By the end of the century, it had narrowed its focus to the inanimate world, supplemented learning from books with experiments, borrowed some mathematics, and showed indications of practical utility. Meanwhile, the lawyers were solidifying their control of the production of evidence and its deployment in the courtroom. At the beginning of the eighteenth century, the judiciary dominated criminal proceedings, and the accused represented themselves. Evidence was mostly adduced either by direct in-court altercation between accuser, accused and witnesses, or by the judge, who examined the parties and the witnesses himself. By the end of the century, the lawyers had reduced the trial judge to an umpire, took over the examination of witnesses, developed the techniques of cross-examination, established their right to argue points of law, and completely transformed the English legal system into the adversarial system as we know it today.[1]

The expert did not fit easily into this new adversarial environment. Traditionally, experts appeared in court either as a part of the jury or as court advisors. In both cases, their performance was initiated and controlled by the court, which assumed the impartiality of the experts. But during the eighteenth century, as the court gradually assumed a neutral position, as the litigants started to summon their own experts to represent them before the jury, and as adversarial ideology was given free reign, a new place had to be found for the expert. The incipient conflict came to a head in 1782, in the civil case of *Folkes v. Chadd*. In this case, also known as the Wells Harbor case, litigants summoned to court several sorts of "men of science," to testify before the jury as to what had caused the decay of a certain harbor on the Norfolk coast of England. The testimony of one of these experts, a prominent Newtonian philosopher, was disallowed because of the lawyers' objection that his philosophical explanations were a "matter of opinion, which could be no foundation for the verdict of the jury." On appeal, Lord Mansfield, chief justice of the King's Bench, found the silencing of the philosopher to be an error and granted a new trial on the ground that the philosopher's theory "was very proper evidence."[2]

Lord Mansfield's opinion in the Wells Harbor case has served in legal literature as the principal precedent that shaped the most dominant option of using experts' knowledge in the modern Anglo-American courtroom—that of calling experts to testify before the jury as partisan witnesses. It has been unanimously declared "the foundation of the rules governing expert evidence." Some went even further and considered it "the court's seal of approval on the whole adversarial apparatus including contending experts, hypothetical questions, and jury evaluation."[3] Still, in spite of its prominent status, the Wells Harbor case has received almost no attention from historians of either law or science. All sources refer either to each other or to the original legal report of *Folkes v. Chadd*, which was published in 1831, half a century after the events of the case.[4]

This opening chapter offers a close and detailed analysis of the Wells Harbor case. When its legal, social, economic, engineering, and scientific backgrounds are reconstructed, the Wells Harbor litigation indeed emerges as an important historical junction in the evolving relations between science and law—but for reasons other than those so far suggested by historians. In *Folkes v. Chadd* Lord Mansfield was not intent on inaugurating a new practice of calling experts as partisan witnesses, nor on solving the difficulties that await such a practice in the adversarial courtroom. Rather, Mansfield was trying to clarify the legal status of a nascent species of expert—Newtonian philosophers, who expressed in court theories whose position on the legal continuum between fact and speculation was yet unsettled. Mansfield's solution, which shaped the practice of modern expert testimony for the next two centuries, maintained that the law should not give preference to one kind of science over another and required that all kinds of science be heard in open court.

In developing the many dimensions of *Folkes v. Chadd*, this opening chapter has a three-part agenda. First, it outlines the long career of the expert in Common Law courtrooms from its early medieval origins to the late eighteenth century, thereby providing the background for the rest of the book. Second, recounting the contest of authority and expertise among the various sorts of men of science—handled not retrospectively, by historians or philosophers, but contemporaneously by the rough epistemology of the legal process, the chapter offers a unique picture of late-eighteenth-century English science, one that is not washed out by the bright light of the Newtonian sun. Third, pointing to the gulf between the facts and the thinking that went into *Folkes v. Chadd*, and its subsequent reputation, the chapter suggests a reformulation of the conventional narrative describing the rise of modern expert evidence. The modern expert witness was indeed the creation of the late eighteenth century. However, far from being deliberately molded as a judicial solution to the problem of partisan expertise in the adversarial courtroom, the expert was conceived as a necessary exception, the only source of information the new system could not rationalize under its evolving doctrines. And such the expert would stay—a freak in the new adversarial world, an incompatible and inharmonious, yet indispensable and influential, figure in the modern adversarial courtroom.

The Decay of Wells Harbor

The town of Wells is situated on the north coast of the county of Norfolk, England. Its inhabitants like to call their town Wells-Next-the-Sea, but it has been centuries now since the sea slipped away from the town, leaving only its name behind. To get to the sea, one has to cross a strip of extensive salt marshes, wide sand dunes, and long shingle spits. Created by the ceaseless sedimentation of the wild North Sea, the strip ranges from three-quarters of a mile to three miles in breadth and stretches for many miles eastward and westward of Wells. The strip is elevated above the neap tides, and the sea makes its appearance on it only in an incomprehensibly intricate net of creeks, branching and subbranching to infinity. Some of the largest creeks are big and deep enough to serve as safe harbors. Such is the harbor of the small town of Wells. It extends sinuously for more than three miles through the sandy and shingle beach and through the low and flat marshes until it reaches the quay of the town.[5]

The north coast of Norfolk was notorious for being one of the most dangerous and most fatal to sailors in all Britain, and Wells Harbor has been from time immemorial a safe haven for ships that routinely dared the wild North Sea on the busy route between London and the northern coasts of Great Britain. As a gate to an agricultural county, Wells was also a seaport that handled considerable imports of coal and exports of corn and malt to London and the Continent during the seventeenth century. It was thought of such importance that a Parliamentary Act was passed in 1663 that allowed the town to tax the goods imported there in order to enlarge the quay and cover its maintenance.[6] For the maintenance of the harbor itself, having no river or any other inland fresh water source, Wells had always relied on the strength of the ebbing tide to scour the rich silt that the violent tides and winds constantly deposited at its bottom. While filling the harbor's channel, the influx of the tide filled all the other creeks and gullies, and in spring the tides overflowed the entire salt marshes, creating a natural reservoir that covered thousands of acres. With the ebbing of the sea, much of this water ultimately collected in the central channel of Wells Harbor, providing sufficient scouring to maintain its depth and safety. At the beginning of the eighteenth century the deeper part of the channel, called the pool, where the ships anchored, was reported to be so deep that even at low tide two or three tiers of vessels could lie afloat and swing around.[7]

Situated near the "Good Sands" region, the reputed birthplace of the Agricultural Revolution, during the eighteenth century Wells Harbor became the second largest harbor for the exportation of malt within the kingdom, second only to Great Yarmouth, in East Norfolk. The prosperity of the region was due neither to its sandy soil nor to its dry climate, but rather to the successful application of new farming methods imported from the continent for the purpose of increasing revenues. Well-placed to enjoy the growing demands of the urban markets both at home and across the North Sea, a new breed of capitalistic farmers evolved in Norfolk. They were led by landowners who realized that large profits could be made and were eager to experiment with new methods of farming that would produce an ever-larger surplus for sale. By the end of the eighteenth century, Norfolk's harbors were shipping more grain than all the rest of England combined, and its husbandry came to be known worldwide as the Norfolk system.[8]

Salt-marsh silt forms very fertile soil once the salt has been washed out. For this reason many salt marshes along the eastern coast of England have been reclaimed for farmland. Reclamation along the north coast of Norfolk, where fertile land is exceedingly scarce, has been going on since Roman times. Reclamation reached its culmination in the late seventeenth and early eighteenth centuries, during which time nearly eight thousand acres of coastal marshland were embanked and reclaimed from the North Sea. Around Wells, Sir Thomas Coke, first Earl of Leicester, and Sir Charles Turner, Lord of Wells Manor, two of the biggest local landlords and leaders of the new farming movement, reclaimed close to eight hundred acres of salt marshes on both sides of the harbor's channel between 1719 and 1721. The embankments erected to prevent the tidal water from flowing through and over the reclaimed marshland greatly weakened the body of backwater available for scouring the harbor. The effect was quickly noticed. Just a few years later, in 1725, according to an eyewitness account, "the said harbour of Wells, its channel and pool, have very sensibly decayed, as have done all the channels that has been anyways deprived of their ancient stock of back waters." Soon the parts of the harbor furthest from the sea became clogged to such a degree that the quay became inaccessible to shipping and the greater part of the cargo had to be carried to and from the town by lighters.[9]

Wells Harbor was not the only Norfolk harbor that suffered from the massive reclamation of coastal marshlands. The fresh water harbors at King's Lynn and Wisbech went into decay after the monumental seventeenth-century reclamation project of the vast marshland known as the Great Bedford Level shut out the tides that kept the mouth of their rivers open. The Sussex port of Rye was so silted up after the reclamation of its bordering marshes that a new entrance to the harbor had to be cut in 1724, and a large stone sluice had to be built to keep it clean. "If private men, to get a little land, may be guilty of such encroachments," a Norfolk pamphlet from 1724 expressed local frustration, "all our ports may be ruined in time, in the same manner that Lynn, Wisbech, and Rye have been."[10]

To remedy the growing problem of their harbor, a group of Wells merchants agreed to finance a sluice that would scour the harbor and keep it open. It is not clear whom they hired to do the job. Turning to foreign help in drainage and other hydraulic matters was not unusual in early-eighteenth-century England, which still depended for its engineering "even more than we did for our pictures and music, on foreigners."[11] Most of these hydraulic experts were recruited from the neighboring Low Countries, where much practical knowledge had already been accumulated on such matters. The experts hired by Wells merchants erected dams and cut new passages so as to redirect some of the creeks and gullies branching from the harbor's main channel into a large creek opposite the quay, the mouth of which was artificially contracted. The flooding waters lying in these various creeks created a reservoir that discharged itself forcefully into the main channel through the narrow opening of the large creek. Freestone Sluice, as it was called, started to operate in 1748, and for a while it successfully scoured away the mud that disturbed the upper parts of the harbor. With time, however, the mouth of the sluice widened and its effect was gradually impaired. In addition, in 1758, Sir John Turner, heir of Lord Charles Turner, embanked and reclaimed another 172 acres of salt marsh, among them a 66-acre site called Wharham Slade that had previously supplied Freestone Sluice with much of its water. This further reduced the reflux of the ebbing waters in general and the scouring powers of the sluice in particular. Within a few years, the combined effects of the deteriorating sluice and the new embankment brought the harbor to a worse state than ever.

Fearing for the loss of their harbor, local merchants, ship owners, and inhabitants entered into a voluntary subscription and secured a large loan. With these resources a bigger sluice was built upon a new site in the remaining unembanked part of the salt marshes in 1765. In 1768, the town succeeded in obtaining a Parliamentary Act that allowed it to increase the duty on incoming cargo and to use it to pay its encumbering debt. The Act further ordered the appointment of a board of commissioners for the harbor with the powers "to make such bye-laws, rules, orders, and regulation, as should be found necessary for the purposes of . . . the improving, preserving, and maintaining the harbor, quay, and other works belonging thereto."[12] The newly appointed board constituted a rough model of the harbor's uneasy politics. On one side sat the town merchants and the ship owners whose livelihood depended on the harbor. On the other side sat Sir John Turner and Thomas William Coke, the two local landlords who owned the reclaimed marshlands adjoining the harbor. The underlying tensions soon surfaced as new predicaments began to multiply.

One of the great advantages of Wells Harbor had been the northwest direction of its channel, which coincided with the flow of the rising tide and allowed vessels to come from the sea through the channel and into the pool with relative ease. During the 1760s and 1770s, the mouth of the channel had changed its orientation, moving considerably toward the east so that it ended up facing northeast by east. As a result, the tide, which flowed from northwest to southeast, drove the vessels across the channel and into the eastern bank. This meant that no vessels could enter the harbor safely without a strong leading wind, and many ships had been lost, to the great disadvantage of the port. In addition in 1777, the new sluice that had successfully cleared the harbor for a while was found to have been nearly destroyed by a sea worm that had attacked the timber with which it was built. The commissioners sought advice from a local engineer named Wooler, who advised them to build an entirely new sluice, made of stone so that the worms could not touch it. Because of the expense (over £2000), the advice was rejected. The previous debt was not yet fully paid, and the commissioners knew that the duty under the present Act was a great burden on the merchants and could not be further raised without losing trade from the port. Unable to raise the money for a new sluice and confident that the embankments on the marshlands were the first and principal cause of all of Wells Harbor's troubles, the commissioners of the harbor considered their legal options for bringing the embankments down.

Enter Law

The decision to go to court was not taken lightly by the commissioners. To start with, Turner and Coke sat on the board and could offer formidable opposition. Second, the legal action could turn into a protracted and costly business that would force the commissioners to proceed through two systems of justice. As a general rule, late-eighteenth-century Common Law took a hard-nosed attitude toward private improvements that hindered navigation in public waterways. These were considered a public nuisance, an offense against the King's subjects, indictable as a misdemeanor. Still, not every structure erected in tidal water was, *ipso facto*, a public nuisance, and not every recognized nuisance was to be eliminated, especially if the damage could be compensated.[13]

Whether or not a nuisance existed was considered a question of fact to be decided by a jury in a Common Law court. In the still largely rural and agricultural society of the eighteenth century, the standard that the jury followed in such cases was the absolute right to free use and enjoyment of land. According to this conservative guiding rule, which had remained remarkably constant since the Middle Ages, any conduct producing unreasonable interference with the use of a neighbor's property constituted a legal nuisance.[14] Therefore, the commissioners stood a good chance of winning their cause in court. Alas, Common Law courts were only empowered to grant damages. Thus, even if successful, the commissioners could still find themselves forced to seek an order for abatement from the Chancery Court, a notoriously time-consuming and costly procedure.

The commissioners sought the advice of Mr. Nash Grose, a veteran Serjeant-at-Law who enjoyed a prosperous practice in the central court of Common Pleas. The experienced Grose advocated a cautious approach. "All the embankments," he pointed out, "were Nuisance at the time they were made; but some of them have existed so long that there will be some Difficulty in getting them removed." If the commissioners wanted to pursue the matter in court, Grose advised, they should use the money raised following the Act of 1768 to indict "those persons who have made or continued the last of the Embankments which are prejudicial to the Harbour." However, Grose recommended that before making such an attempt, the commissioners write to Sir John Turner, "stating the inconveniences occasioned by the embankments and begging him to remove them . . . [For] he may find out some plan to remove the nuisance complained without doing much injury to the land embanked, which would be the most desirable object on both sides."[15]

Sir John was a recognized leader of the "the Norfolk science of agriculture," which turned Norfolk into the principal site of what contemporaries called "the agricultural revolution." This so-called revolution hinged on the alteration of the traditional patterns of communal land ownership. This was done throughout the eighteenth century by a flood of Parliamentary Enclosure Acts that wiped out the traditional rights of commonage by requiring that private land be fenced off from common land. Enclosure, not unlike the contemporaneous conversion to the metric system across the Channel, offered the best way forward from the confusion of traditional practices to the orderly methods of the new scientific spirit. Fencing made systematic cultivation possible. It kept one's livestock in, kept other people's livestock out, and liberated the landowners from the need to consult others about what they should do. Backed by the Enclosure Act, landowners were able to take advantage of the gradual collapse of the old village economy to perform massive purchases of land intermixed with or adjacent to their property. Their objective was to create large consolidated estates that would allow for the concentration of capital, rationalization of labor, and mechanization of production. In this emerging protocapitalistic environment, the science of agriculture became "the most useful science a gentleman can obtain," tending, as it was claimed, "to the increase of both private property and public benefit."[16]

According to Arthur Young, the great publicist of scientific farming during the late eighteenth century, "it would seem that the farmers of Wells neighborhood owe more to Sir John Turner than to anyone else." On his estate Sir John experimented with husbandry, forestry, and crop rotation. The embankment he erected in 1758 was no doubt part of his systematic reclamation of the fertile marshland, which otherwise served only as a sheepwalk and as a meager common for cattle.[17] Such improvements required prudent administration and sizeable resources. The large-scale operation, the ceaseless conversion and reconstruction, and the legal expenses involved in the redistribution of the land necessitated large investments, which were tied up for a long time before they showed a return. Some landowners eventually got hefty returns from their improved estates—no one more so than Turner's famous neighbor, Coke of Norfolk. Coke inherited his estate in 1776, at the age of 22, and within the next forty years spent more than half a million pounds upon its improvement, thereby almost tripling its yearly rental value from £12,332 to £31,050.[18] Other landowners, however, piled up mainly debts. Such seemed to be the case with Sir John Turner. Old, indolent, and heavily in debt for mortgages on his estate, his response to the commissioners' request to take down his embankment was one of delay and evasion. Frustrated, the commissioners finally decided to seek justice in the courts.

Twice a year, around January–February and July–August, all twelve judges of the three central courts—King's Bench, Common Pleas, and Exchequer—would leave their comfortable Westminster seats for two or three weeks to spread royal justice among the counties of England. These judicial tours were called assizes. The counties were grouped into six assize circuits, and a pair of judges, each from a different court, rode each circuit, trying before local juries the cases originating there. One judge presided over the criminal cases and the other over the civil ones.[19] The commissioners took their cause to the assizes for the county of Norfolk, held in Norwich. There, a grand jury found for an indictment against Sir John, declaring his 1758 embankment a public nuisance and ordering it to be taken down. Obstinate, Turner still refused to abate and his lawyers continued to temporize. Then, in June 1780, just before the commissioners were ready to proceed with their legal action in the summer assizes, Turner died. Alarmed by the delay that Turner's death might engender and by a new inspection showing the continuing deterioration of their harbor, the commissioners resolved not to wait any longer and to employ men in order to cut open the embankment. The decision was reached despite the violent objection from the other great landlord on the board, Thomas Coke, who warned the commissioners that any similar attempt to destroy his embankments would be defended by him "as if his house was attacked."[20]

Turner's heirs were his two daughters, and his two sons-in-law, Sir Martin Browne Folkes and Robert Hales, jointly managed the insolvent estate. Notified of the commissioners' intentions, they were advised against physically resisting the cutting. Instead, they filed a bill in Chancery Court requesting that it direct the commissioners to refrain from any activity that would injure their property. "We are willing," they declared," to try it by any Action they think fit, or the Court shall approve, whether we are entitled, or Sir John had not a Right to make this Erection and to continue it, or whether they have any Right to complain of it as a Nuisance . . . We desire only that irreparable Mischief may not be done to the bank till it is tried." Approving, the Chancery Court issued an injunction against the commissioners and ordered the dispute to be brought before the court of the King's Bench.[21]

The main features of the eighteenth-century procedures of Common Law had already been fixed by the beginning of the fourteenth century, when every legal action derived its final authority from a specific royal mandate, called a writ. Each writ specified the reasons for its issuance and the agencies enjoined for the specified actions to be taken. This highly formal system of writs served also as the main classificatory scheme for Common Law, and eighteenth century legal procedures were still organized around it. By that time, however, this genuinely medieval mechanism had been transformed into a labyrinth of technicality through which none but the most skillful or most fortunate could find their way. Eighteenth-century suitors were required to select, at their peril, the appropriate writ for their cause of action, and then strictly follow its ancient procedures. These typically required the parties to frame a single narrow question of fact to be decided by the trial jury. This was done during a pretrial stage called *pleading.* The pleadings were launched by the plaintiffs' written declaration, explaining their claim in strict conformity with the original terms of their chosen writ. Upon receiving the opening declaration, the defendants would choose either to *demur* (attack it in point of law) or to *plead* (oppose it in point of fact). By choosing to demur they would be admitting the plaintiffs' version of the facts but denying that it disclosed a legal cause of action. By choosing to plead they could either make *dilatory pleas* (opposing the legal action on technicalities such as choice of jurisdiction or writ) or *peremptory pleas* (impugning the right of action itself). If making peremptory pleas, they could choose to go either by *traverse* (denying the plaintiffs' statement of facts *in toto)* or by *confession and avoidance* (admitting the statement of facts but adding others that contravened it). Pleas could then be followed in succession by a replication, a rejoinder, a surrejoinder, a rebutter, a surrebutter, and so forth *ad nauseam.*[22]

Exercising this black-letter art of pleading, the legal representatives of the Turner estate and the commissioners of Wells Harbor finally agreed that the method of litigation be this: Folkes and Hales would bring an action of trespass against the commissioners as if they had actually cut their embankment down. This would place the burden of proof on the commissioners, who would try to justify cutting down the embankment by showing that it was a nuisance "which any of the King's Subject has a right to abate." The narrow question put before the jury to decide would therefore be whether the mischief that the bank did to the harbor was a justification for the cutting. If the trial jury decided that the cutting was unjustified, then the injunction would stand and the commissioners would have to compensate Folkes and Hales for their expenses. If the jury decided that the cutting was justified, then Folkes and Hales would have to compensate the commissioners for their injuries and probably also destroy the embankment. The trial itself, where the jury would hear the facts of the case, was set to take place in August 1781, at the summer assizes for the county of Norfolk in Norwich.[23]

Enter Science

Sir Martin Browne Folkes was a descendant of highly scientific lineage. His grandfather was Sir William Browne, president of the Royal College of Physicians. His uncle and godfather was Sir Martin Folkes, Newton's handpicked successor as president of the Royal Society of London. Named after both famous relatives, Sir Martin Browne Folkes owned a large estate near Lynn, Norfolk, which he no doubt farmed scientifically. We should not be surprised therefore that in preparing for the coming trial he enlisted the services of science in the shape of Robert Mylne, F.R.S. Mylne agreed to come to Wells, study the harbor and its surroundings, and submit a report as to whether and in what ways the embankment erected in 1758 affected the harbor.

Ten years earlier, in 1771, Robert Mylne had been among the seven founding members of the Society of Civil Engineering, whose formation best symbolized the emergence of the new English profession of civil engineering during the late eighteenth century. On the Continent, the profession of civil engineering was already well established, and official academies had long taught the preparatory sciences and arts necessary for able professionals to comprehend the technicalities of the projects they took part in. In England such establishments did not exist. There works of civil engineering such as fen drainage, construction of canals, mills, bridges, turnpikes, and improvement of harbors and rivers had long been undertaken, if not by foreigners, then by a welter of craftsmen, millwrights, stone masters, surveyors, and instrument makers. These craftsmen made up for their lack of formal education with much technical ingenuity, but they never saw themselves as part of a larger vocation.

This state of things was changing fast during the latter part of the eighteenth century in response to the rapid increase in the number and scale of public and industrial works throughout the country. The overwhelming demand for expert planning and professional supervision turned what used to be at best a haphazard occupation into a promising and respected career. By 1781, at least half a dozen well-known engineers had their own independent consulting practices, while many others had regular employment managing the construction of canals, lighthouses, bridges, and harbors; supervising fen and coal mine drainage; and designing water mills, steam engines, and other machinery required by the fast-growing industry.[24]

Mylne came from a distinguished Scottish family that had provided master masons to the Scottish Crown since the sixteenth century. He was born in Edinburgh, where his father, in addition to being the city surveyor, had an extensive architectural practice. In 1754, at the age of 21, Robert went to Rome, where he studied architecture for four years. In 1759, he returned to England just in time to win the prestigious competition for the design of the Blackfriars Bridge on the Thames. His controversial bridge introduced for the first time into England the elliptic arch, and its successful completion in 1769 made Mylne famous. Known as an architect-engineer, a common Continental liaison but unique in England, he soon developed a thriving private architectural-engineering-surveying practice. His specialty remained bridges, but his designs were varied. Among other things, he was appointed architect of St. Paul's Cathedral and chief engineer of the great New River Company, which supplied water to London. A recognized leader of this new and rising order of professional men who called themselves civil engineers, Mylne was often invited to Parliament to testify before the various committees of the House of Commons that dealt with the improvement of internal navigation so necessary for the growing economy. Occasionally, he also served as an arbitrator for the King's Bench, settling legal disputes the royal judges thought fell within his professional expertise. Securing Mylne as his expert, Sir Martin Browne Folkes retained therefore the services of an eminent professional who was considered as authoritative in his own field of civil engineering as any doctor or lawyer could be in their respective professions.[25]

In October 1780, Mylne traveled to Wells to study the problems of the harbor. The report on his findings was long overdue, but when it finally arrived in late May 1781, it contained all that the lawyers for Turner's estate could have hoped for. Mylne's conclusion was clear: the troubles of the harbor could not be attributed to the said embankment or to any other embankment. What had caused the decay of the harbor were the vast quantities of materials discharged at the immense western estuaries by six rivers—the Ouse, Nene, Witham, Trent, Wharfe, and Swale—and deposited along the north coast of Norfolk by the rising tides and the strong winds of the North Sea. "The Sea is embanking, of itself, the whole Coast, and the time will probably come," Mylne predicted, "when this Harbor will diminish to a creek and that again, by slow Degrees, to a solid land."[26]

Mylne made sure to support his opinion with careful calculations and factual observations, and to illustrate it by two maps carefully made under his direction by local surveyors. According to Mylne's calculations, more than sixteen hundred acres of unembanked marshes and another 150 acres of creeks and inlets existed on both sides of the harbor. What could the sixty-four acres of embanked marshland that were the object of the dispute do, he asked, that these unembanked 1750 acres could not? If anybody was to blame for the decay of the harbor, it was those who had designed the sluices. Only contracting the mouth of the creeks, instead of incorporating a true stop gate to retain the backwater to a full head before letting it out, the two sluices failed to exploit the full scouring impact of their retained backwater. With the coming trial in mind, Mylne concluded his report by indicating to his employers his willingness to help disprove the accusations against them from the witness stand:[27] "Human Nature is too apt to form Systems and draw Conclusions, without sufficient Observation and Scrutiny. Too often the Search is made for an evil in our neighbor's property, when, perhaps, it is to be found at home. The facts herein stated, I am ready to testify when required; and the deductions made from them, I hope will be found to be well founded in the Opinions of impartial Men."

Experts in the Courtroom

The English jury system had long acknowledged the importance of experts like Mylne in cases such as that of Wells Harbor, where the disputed facts were such that the jury lacked sufficient knowledge to draw an informed decision. "If matters arise in our law which concern other sciences or faculties," declared an English judge in 1554, more than two centuries before the Wells Harbor case, "we commonly apply for the aid of that science or faculty, which it concerns. Which is an honorable and commendable thing in our law. For thereby it appears that we don't despise all other sciences but our own, but we approve of them and encourage them, as things worthy of commendation."[28] Over the centuries, the court had developed two procedural options to deploy such men of science, who, from their special training and experience, could instruct the court and the jury in regard to the disputed facts. The first option was to call them as jurors. The second option was for the court to nominate them as consultants whose advice the court or the jury could adopt as they pleased. There was also a third option, which was for the parties to call them as witnesses testifying on their behalf. However, there was no special procedure to define witnesses as experts as was the case with court experts or those serving in expert juries. In the absence of such a procedure they were regarded and treated merely as witnesses.

Expert Juries

Summoning to the jury people with special knowledge concerning the disputed facts of the particular case was originally but a natural extension of the medieval community-based jury, whose basic principle was that "those were to be summoned who could best tell the fact, the *veritatem rei.*" Drawn from the vicinity where the case arose, and chosen because of their direct knowledge of the facts of the case and of the reputations and intentions of the parties involved, the medieval jurors functioned as witnesses, investigators, and decision makers all at once. They went around informing themselves before and during the trial and based their decision on what they personally knew and what they had learned "through the words of their fathers and through such words of persons whom they are bound to trust as worthy." On occasion, however, the jurors needed specialized knowledge in order to do justice. In these cases, special expert juries were assembled under specific writs.[29]

Trade disputes were probably the most frequent cause for summoning expert juries. Juries of goldsmiths, booksellers, wine merchants, attorneys, and fishmongers, to name but a few, were often summoned to investigate and decide whether a specific guild's trade regulations had been violated. Such cases, we are informed, show that "the practice was well established in the fourteenth century of having the issue actually decided by people especially qualified." A second class of "expert juries" came to be defined under the writ *de medietate linguae* (of the half tongue), which secured the right of a foreigner to request a trial where at least some of the jurors were from the defendant's own country and were able to understand his or her point of view and explain it to the rest of the jury. The traditional right for a *de medietate* jury was gradually extended also to Jews (but not gypsies), Welshmen (but not Scots), clerics, university scholars, merchants, and other guild members.[30]

A third class of expert juries included all-female juries summoned as experts in cases involving sexual assault, pregnancy, and childbirth. Of these, the most common was the jury of matrons, which was usually impaneled in criminal trials and consisted of married women or widows experienced in childbirth. Another, less common jury of this class came to be defined under the twelfth-century writ *de ventre inspiciendo* (to inspect the belly). Under this writ the court would impanel a jury of experienced women for the limited purpose of establishing disputed facts of pregnancy. Unlike the *de medietate* jury, which functioned both as an investigator and as a decision maker, the *de ventre* jury functioned only as an investigative body whose conclusions were presented as advice to the court.[31]

Court Experts

As with the jury of experts, the second option for deploying experts in the courtroom—that of an expert nominated by the court—dates back to the early days of English Common Law. Historians have unearthed many cases in which the court chose to consult directly with independent experts. The earliest reference we have is from 1299, when physicians and surgeons in London were called to advise the court on the medical value of the flesh of wolves. Medical advice was most often sought in cases of malpractice or when the nature of wounds was an issue. The advice of grammarians was also in demand in cases where the disputed facts concerned language. We know of cases from 1429, 1494, and 1554 in which the court referred to the practice of consulting grammarians on the meaning of Latin pleas and commercial instruments as an old and well-established practice. In all of these cases, the experts apparently were summoned to advise the court rather than to provide evidence to be evaluated by the jury. In 1619, for example, two physicians advised the court that a wife might bear a legitimate child "forty weeks and nine days" after the death of her husband. The conclusion must have satisfied the court, because it instructed the trial jury to use it as *datum* in their final conclusion.[32]

In juryless courts, such as the Patent and Admiralty Courts, experts sat alongside the judge providing him with their advice. The custom was said to have been adopted during the fifteenth century from the international court of the Councils of the Sea, which in earlier times sat in Barcelona and settled disputes among members of the Merchants' and Mariners' Guilds. The judges of this international court conferred with the authorities of the guilds and in case of conflicting advice took the independent advice of professional navigators. Based on this custom, the Admiralty Court consulted regularly with the elders of the Corporation of Trinity House, the famous club of sea captains that was chartered by Henry VIII in 1514 and performed many official marine functions such as licensing and supervision of lighthouses. Juryless trials had their own advantages, and the Admiralty Court prospered in the early modern period by offering its clients speed and predictability. By the eighteenth century, the King's Bench had also adopted the practice of consulting the Trinity masters in maritime cases.[33]

The eighteenth century saw the culmination of the use of expert knowledge under traditional procedures of both the expert jury and the court expert. At the height of its popularity, the practice extended from factual decisions to actual rulings upon points of law. In 1730, a Parlia-

mentary statute united all kinds of special juries under the new procedure of the "struck jury," which allowed the parties to strike names from
an unusually large panel of prospective jurors. In the second half of the
eighteenth century, this powerful procedure was used by Lord Mansfield to train a corpus of merchant-jurors to act as a permanent liaison
between law and commerce. The advice of his merchant juries
accounted in large measure for Mansfield's monumental success in creating a coherent merchant law within Common Law. Lord Holt in
1703, Chief Justice Lee in 1753, and Lord Hardwicke in 1755 chose the
other option and nominated merchants as court experts with whom
they would consult before ruling in trade cases. Lord Mansfield also
used this procedure in insurance cases.[34]

Experts as Witnesses

Experts appeared also as witnesses, called by the parties in the case to
support their cause in court. This practice had become increasingly
common since Tudor, and later Stuart, England, which experienced an
enormous expansion in the volume of civil litigation. "The English
people," according to one historical account, "were never before
nor since so litigious and law minded as during the reigns of Queen
Elizabeth I and her two Stuart successors."[35] The growing population,
a buoyant land market, and the expanding national and international
commerce were all factors that brought more business to the courts.
These same forces also dissolved the medieval system of self-informed
juries by disintegrating the static communal organization that supported it. By the sixteenth century, the selection of trial jurors was
already determined at least as much by status and administrative experience as by geographical proximity to the location where the offense
had occurred.

Once the jurors ceased to be truly local, they must have found it
impossible to acquire relevant knowledge of the trial's issues before they
came to court. Other means were required to present them with the
relevant facts in court. On the Crown side in criminal proceedings, a
system of magistracy developed, which took over the production of
evidence and its presentation in court. Officials involved in the process
of the arrest—justices of the peace, constables, coroners, and
bailiffs—began to play an increasingly active role in gathering evidence
to be used in court.[36] On the civil side, however, no such official mechanism existed, and it was the interested parties who became the principal

producers of evidence presented in court.[37] No longer private to the jury, evidence, properly displayed in public, gained such importance during the sixteenth century that "if none come in to give evidence, although the malefactor hath confessed the crime to the Justice of Peace . . . the Twelve men will acquit the prisoner."[38]

As the environment of civil proceedings became more diversified, legal questions involving special skills came up more frequently, and witnesses were increasingly called by the parties in the case to testify before the jury as to particular facts within their expertise. Physicians and surgeons testified in insurance and will cases; surveyors, in property cases; merchants, concerning the particular customs and norms of trade; tradesmen, concerning the quality of particular goods; ship builders, concerning the state and construction of vessels; other artisans, concerning their respective subjects of mechanical skill. Still, in spite of their growing presence in the courts, experts testifying as witnesses were not regarded as a distinct legal entity. There was no special procedure that would define them as experts like that applied to their colleagues who were court experts or those serving on expert juries. Moreover, eighteenth-century courts did not see it as improper for lay witnesses to testify as to their opinion, if this was based on their direct knowledge of the facts of the case. Thus, the testifying experts, who usually had personally observed the facts of the case and testified as to their conclusion, were not differentiated from other lay witnesses, who often did the same.[39]

The First Trial

In early August 1781, the hearing concerning the facts of the Wells Harbor case took place at the Norfolk assizes in Norwich. The hearing lasted two days. Starting at seven in the morning of the first day, the commissioners' lawyers marched a long line of pilots, mariners, and other eyewitnesses to the witness stand to testify from their personal experience to the rapid deterioration of the harbor following the embankment of 1758. It was only late on the second day that Folkes' lawyers finally produced Mylne, as their last witness, to persuade the jury that their embankment had done no harm to the harbor. No objection was made and Mylne was allowed to testify from the witness stand as to what, in his opinion, had caused the decay of the harbor, and as to his judgment that the embankment in question had no part in this. Mylne explained to the jury that, because they were unfamiliar with the true principles of nature, the experts for the defense had been misled by their own perceptions. The filling up of the harbor they had witnessed was but a mere link in the temporal chain of causes imperceptible to lay observation, and this chain of causes would have led to the silting up of the harbor whether or not any embankment had been made. The authoritative testimony of this famous engineer, a fellow of the Royal Society, who had made a special study of the situation that had led to the trial, made a strong impression on the jurors, who, we are told, "relying on the weight of Mr. Mylne's abilities of knowledge, and not having the least doubt of the truth of his evidence, found a verdict for the plaintiffs."[40]

The commissioners of Wells Harbor were outraged by the arrogant attitude of the metropolitan expert. How could Mylne, a foreigner to the county, who lacked intimate knowledge of the facts of the case, claim, on the basis of the shortest inspection, to recognize forces at work unobserved by their own experienced experts, who had spent their entire lives at the harbor? Their lawyers requested a new trial on the grounds that they "were surprised by the doctrine and reasoning of Mr. Mylne." This was the standard legal argument for a new trial made by a party who felt cheated by a case falsely made at a trial, which it had no reason to expect and therefore could not come prepared to answer.[41]

Having no systematic appellate procedures, the granting of a new trial upon a finding of an improper verdict was a procedural device of critical importance in Common Law. First applied in the middle of the seventeenth century, its increased use gave the royal judges of the central courts an opportunity to correct hasty decisions given at the assizes. It also helped the judges to bridge two uneasy and ever-shifting fault lines—the one dividing the responsibilities of judge and jury, and the one separating the highly formal and largely outdated procedural machinery from the practical needs of the vibrant environment that existed outside the courtroom. In the first instance, granting a new trial allowed the judges to eliminate lenient verdicts that clearly went against the weight of the evidence, without punishing the juries and disturbing the fragile political balance between judge and jury. In the second instance, it allowed them to mold the highly formalized and outmoded rules of procedures so that cases could be decided on their own merit.[42]

When a question of law arose at the assizes in a civil case, the practice was to refer the question back to London, to the full bench in the court in which the case originated. Thus, it was the four royal judges of the King's Bench, headed by Chief Justice Lord Mansfield, who convened to discuss the commissioners' request for the new trial. Having looked into the evidence given at the trial, they agreed that the commissioners should have had the opportunity to counter Mylne's claims with those of their own experts. "In matters of science," they dictated, "the reasoning of men of science can only be answered by men of science." A new trial was therefore granted, and to avoid additional surprises in this important litigation, which "has influenced the whole county of Norfolk, and perhaps the whole country may be affected by it," the judges directed the parties to exchange between them in writing, before the new trial, the opinions of the experts whom they intended to produce in court "so that both sides might be prepared to answer them."[43]

The Race for the Experts

The second round of the Wells Harbor case, all seemed to agree, was going to be decided upon the opinions of men of science. Thus, in preparing for the new trial, the commissioners of Wells Harbor went out to recruit the best expert they could find, preferably an expert who would be more famous and more authoritative than their adversaries' champion, Mylne. Not many could have qualified for that job. In fact, in the whole of the kingdom there existed only one person who clearly fit the description—John Smeaton, civil engineer, F.R.S.[44]

John Smeaton was the driving force behind the formation of the Society of Civil Engineering in 1771. He was the first person in England to sign his reports as a "civil engineer," and it was his lead that other engineers, such as Mylne, followed when they came to consider themselves as professionals, much like doctors and lawyers. Like Mylne, Smeaton also served as an arbitrator for the royal courts of the King's Bench and Common Pleas. And, like Mylne, he was as much at home at the construction site as he was in the Royal Society and in Parliament, where he was invited to discuss the improvement of rivers and harbors and the building of bridges and docks, and where "no person was heard with more attention, nor had any one ever more confidence placed in his testimony."[45]

Smeaton started as a maker of scientific instruments before turning to engineering for his livelihood in 1753, at the age of 29. In 1759, he captured the imagination of the public and the admiration of his fellow engineers with his successful design of a lighthouse on the Eddystone reef outside Plymouth Harbor, one of the most difficult sites imaginable. By 1781, Smeaton had developed a career as a consulting engineer that was unrivaled in its volume and diversity of work. His numerous designs included water mills and windmills, bridges, harbors, river and canal navigation constructions, steam engines, and fen drainage projects. Among the myriad engineering projects he was involved in, harbors were his specialty. By 1781, Smeaton had been consulted on more than thirty different harbors in England and Scotland.[46]

In addition to his flourishing engineering practice, Smeaton also developed a prominent scientific career. In 1753, he was elected to the Royal Society, recommended by Lord Cavendish for "his great skill in the theory and practice of mechaniks . . . as also [for] being well versed in the knowledge of the mathematiks and natural Philosophy."[47] By 1781, Smeaton had already contributed about fifteen papers to the Society's *Philosophical Transactions.* One of them, "An Experimental Enquiry Concerning the Natural Powers of Water and Wind to turn Mills, and other Machines, depending on a circular Motion," in which he described his experiments with waterwheels and windmills, was awarded the Copley Medal, establishing his scientific reputation as one of the Royal Society's most valued members.

Thus, in 1781, John Smeaton was at the top of his profession, respected by all for his extensive practice as well as for his intellectual powers, integrity, and sense of vocation. As the leading authority on harbors in the country, he was largely responsible for the recent successful rescue of Ramsgate Harbor, one of England's largest harbors, from the vast amounts of sand that threatened to choke it completely.[48] In short, he was the best expert the commissioners of Wells Harbor could have hoped for. How disappointed they must have been, therefore, when they requested Smeaton's services and were informed that his services had already been secured by their foes.

Folkes' lawyers found Mylne a difficult man to work with.[49] Moreover, his arrogant performance at the first trial had backfired. In the heat of the trial, he had testified that the rich materials discharged at the western estuaries were brought into the harbor due to the action of the the western and northern tides meeting across the harbor's mouth. This was a hasty speculation that was seized upon after the trial by the commissioners to undermine his credibility. It could be proved, the commissioners maintained, that the western and northern tides acted in contrary directions and moved away from, not toward, Wells Harbor; thus, they could not have possibly affected the harbor.[50] Folkes' lawyers preferred, therefore, not to rely on Mylne for the second trial, where cross-examination concerning his previous testimony could severely undermine his credibility and their case. Instead, they turned to

Smeaton, their request preceding that of the commissioners' lawyers by a few days.

Smeaton preferred to arbitrate between the parties rather than to serve one of them as a partisan witness. In reply to the commissioners' request for his services, he suggested that they contact Folkes' lawyers and ask them to "lay the matter before me jointly as a judge instead of a witness; for otherwise you see I stand engaged to go upon the matter at the request & expenses of the other party and must be brought in as evidence on their side of the question, which in effect probably deprives me of that particular information I might secure from you." Informing Folkes' party of his correspondence with the commissioners, he repeated his suggestion that "it might be for the advantage for both parties to proceed upon the business jointly for there seemed an opening, in case they [the commissioners] thought it proper, for an application to you that I might determine the matter as an arbitrator; in consequence of which I would become possessor of the whole scope of the argument on both sides."[51]

The commissioners declined. Slighted once by highbrow science, they preferred to put their trust in the hands of a local jury of their peers. Still, they were not going to be caught off-guard again. Having been late in the race for the best expert, they opted instead for quantity and hired no less than four senior engineers—John Grundy, Joseph Nickalls, Thomas Hogard, and Joseph Hodskinson—to represent them in the coming trial. Like Smeaton and Mylne, Grundy and Nickalls were among the seven founding members of the Society of Civil Engineering, while Thomas Hogard had joined the Society a year later, in 1772. John Grundy was an experienced engineer who had an extensive practice and was well known in Norfolk. His main specialty was the improvement of river navigation and the drainage of adjacent low lands. He was also a personal friend of Smeaton, who had a high regard for his abilities. Joseph Nickalls, a millwright by training, served as an appointed engineer to the Thames commissioners, representing their cause in Parliament against loud opposition from promoters of competing canal schemes. Thomas Hogard specialized in fen drainage and served as a commissioner for several fen drainage projects in Lincolnshire. Joseph Hodskinson was vice-president of the Society of Civil Engineering and one of the most respected land surveyors in England. There was no doubt the commissioners of Wells Harbor were preparing themselves well for the coming scientific battle.[52]

Smeaton's Report

Since its inception in the mid–seventeenth century, the practice of demanding new trials had been an effective delaying tactic. Lawyers had been using it to put off undesirable verdicts for years. That changed after Lord Mansfield was sworn in in 1756 as chief justice of the Court of King's Bench. Emphasizing expeditious case handling, Mansfield introduced procedural reforms that shortened the wait between trials. The second jury hearing in the Wells Harbor litigation was, therefore, promptly set for the following summer term of the Norfolk assizes, in July 1782. Preparation time was short, especially since the parties were required to print and exchange their experts' reports before the trial. Thus, in mid-March, Smeaton spent three snowy days in Wells, marching along the sea walls, measuring tidal depths from a moored boat, and riding furiously along the harbor channel and along the shore at low tide. Then, he returned to London to study further the history of the harbor, read the evidence produced in the first trial, and write his report.[53]

Smeaton's reports were famous for their clarity and organization, often starting with a philosophical discussion of the general principles involved and advancing methodically to the practical solution he derived from them. The exhaustive report that Smeaton finally submitted to his employers, on May 4, 1782, was no different. "To have a clear and comprehensive view of the cause of [the] decay," Smeaton launched his report, "it will be necessary to shew the natural causes by which the port of Wells has been formed." Thus, Smeaton commenced with a discussion of the general principles that govern the creation and decay of tidal harbors by the natural forces of the sea and the weather.

There was a time, Smeaton theorized, "when nothing more than naked sand lay against the bare coast upon which the town of Wells now stands." At that stage, the tide flowed and ebbed uniformly as a continuous sheet of water. However, as the steady deposition of tidal silt increased the breadth and the height of the sand, tidal waters were eventually left behind and started to cut gullies on their way back to the sea. These gullies were soon enlarged, by the scouring action of the backwater, into a series of intricate tidal creeks, the lesser ones conducting water into the larger. Steadily raised by the sedimentation, the land reached a height where it was no longer being flooded by the neap tides. The tidal silt that accumulated during the spring tides, dried and hardened and soon became suitable for the growth of grassy vegetation, which further held the sedimentation in place and increased the rate at which the land was raised. At the same time that the marshes increased in height, they also increased in breadth, and greater bodies of water were left on them. The gullies and the creeks multiplied, and the larger ones increased in depth and size. "If all are ultimately collected into one, as has been the case with the channel of Wells Harbor," Smeaton wrote, "the scour would be sufficient to maintain a channel through which vessels might be brought from the sea and thus a useful Harbour would be formed, which would increase in depth and utility by the continuance of the forming powers, but yet, only to a certain degree."[54]

As the marsh was elevated, the depth of the water left on it at high tide constantly diminished. (See Figure 1.) At first, this had no perceivable effect, since most of the backwater flowed back directly to the sea. It was only the lowest half a foot or so that needed the gullies in order to find its way back to the sea. But with the continual elevation of the marsh, a height would eventually be reached from which the scour would start to diminish. At first, it would be only a few times a month that the gullies would be deprived of the scouring action of the backwater. However, the elevation process would continue to raise the maturing marsh above the reach of higher and higher tides, causing the water inlets to choke up and dwindle. Ultimately, the process would stop at the height of the extreme high water, but long before that the gullies, then the creeks, and finally the main channel would be choked. The parts that were further from the sea would be the first to land up until "progressively, from the extremities towards the sea, the gullies, creeks, fleets, and main channel will become solid land."[55]

Unable to support his dynamic theory of tidal harbors with positive evidence, Smeaton attacked its alternative with carefully constructed quantitative argumentation. He showed by calculations that even if one assumed that Turner's embankment contributed to the problems of the harbor, its contribution was negligible. He computed the unembanked marshland that was lost due to elevation to be around sixteen hundred acres and the relevant marshland embanked before 1758 to be 572 acres. Dividing the total by the sixty-four acres that were lost due to the 1758 embankment, Smeaton concluded that the 1758 embankment could have been responsible for no more then $\frac{1}{34}$ of the overall problem. Constructing a second quantitative argument, Smeaton pointed out that since these sixty-four embanked acres were highly elevated, they were overflowed no more than four or five times a year and only as the result of the most extreme spring tides. Thus, Smeaton argued,

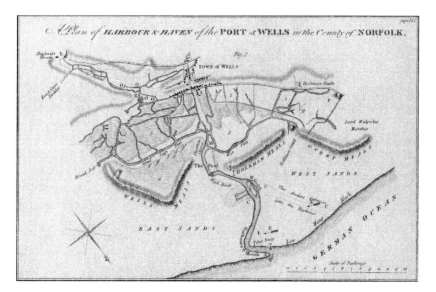

Figure 1: A Plan of the Port of Wells, taken from John Smeaton, "Report on Wells Harbour," 1782.

1 – Holkham Marsh, embanked in 1719 by Sir Thomas Coke, first Earl of Leicester
2 – Wells West Marsh, embanked in 1719 by Sir Charles Turner, Lord of Wells Manor
3 – West Salt Marsh
4 – Lodge Marsh
5 – North Marsh
6 – Church Marsh
7 – Wharham Slade

A – the entrance to the harbour
B – The course of the old channel
C – the west side of the ancient entrance into the harbour
D – Freestone's Sluice
E – the direction of the tide over the coast
G – the present navigable channel
H – the pool
I – the quay
K – the bank made by John Turner in 1758
O.O. O.O. – the bank raised in 1719 by Sir Charles Turner for enclosing the Church Marsh

even if one generously endowed the gushing backwater from these sixty-four high acres with a scouring power equal to that of all of the sluice, operating only five times a year their benefit could only be as $\frac{1}{70}$ of that of the sluice that operated 365 good tides a year.

"I find myself," Smeaton concluded, "forced by fair induction to infer . . . that the embankment of the slade marshes in the year of 1758 could not in any sensible degree, capable of measure or estimation, have contributed to the landing up [of the harbour]." The true story of Wells Harbor was "the progressional operation of nature, which originally formed the harbor of Wells and brought it to maturity, has also occasioned it to grow more and more into a state of decay, and will finally close it up, and convert to a firm ground, fit for arable purpose, and those of pasturage, the very spot where ships have rode at anchor." The deterioration of Wells Harbor, Smeaton closed his report, could have been counteracted by the hand of man. After all, "the greatest part of the sea-ports in Flanders and Holland are kept open, and under circumstances more unfavorable than the port of Wells." That, however, required money and expertise. Alas, the commissioners of Wells Harbor had chosen to build cheap and faulty sluices, and later blamed other people for their troubles.[56]

Men of Science

Smeaton, and Mylne before him, wrote as natural philosophers, confident not only of the existence of natural laws but also of man's ability to discern them, however subtle their workings. Their theory of tidal harbors relied on the concept of a slow and continuous change in the earth's shape brought about by permanent and uniform laws of nature operating over a vast amount of time. Ordaining the strength and the direction of the tides and the winds, these laws were fundamentally laws of motion—the oldest and most respectable of all natural laws. The forces described in these laws alone had steadily built up and worn down the earth. Smeaton and Mylne had no use for universal floods or other cataclysms as the means to explain phenomena or accelerate their change. In a word, they were uniformitarians well before Hutton's famous *Theory of the Earth* established the conventions of uniformitarianism in 1795. Hutton could have been quoting Smeaton and Mylne in writing that "nothing but the most philosophic eye, by reasoning upon chain of facts, is able to discover [true causes]."[57] Equipped with such a philosophic eye, Smeaton and Mylne were able to dismiss the mounting human testimony as to the rapid deterioration of the harbor after the embankment was erected in 1758. On their side were the hidden laws of nature, which ordained powers that no part of the earth's surface was immune to.

By 1782, half a century after Newton's death, Newtonian science had conquered British natural philosophy. However, Hutton the geologist and Smeaton the geographer were Newtonians also in a stricter sense.

Newton's interest in geography had caused him to publish two Latin editions of Bernhard Varenius' *Geographia Naturalis*, in 1672 and 1681. Breaking with the Scholastic geographical tradition, which considered the works of man and the earth itself to be the creations of divine providence, Varenius treated geography as a branch of mixed [applied] mathematics that dealt with the quantitative features of the earth and its parts. Although based on Cartesian principles, Varenius' *Geographia* remained a staple with Newtonian disciples, who brought out two more editions. The earlier, in 1711, by James Jurin, one of Newton's lieutenants during his presidency of the Royal Society, converted Varenius' geography into Newtonianism by supplementing it with notes that replaced Varenius's Cartesian explanations with Newtonian ones. The second, in 1733, rendered the Jurin edition into English, with many more Newtonian additions, and was recommended as the most useful book on its subject by James Hodgson, F.R.S., master of the Royal Mathematical School at Christ's Hospital, an institution in which Newton had taken an interest. The English edition proved to be remarkably influential and was updated three times, the last in 1765.

Varenius' *Geographia* gave a program of physical geography based on uniformitarian principles and included a dynamic theory that treated the process of the formation and decay of sandbanks and coasts. It pointed out the effects of winds and currents, the methods by which sand and silt build up beaches, and the almost inevitable creation, by natural causes alone, of bars and banks that destroy even the "most rich and flourishing trading towns." It is not unlikely that Smeaton, who read widely on subjects of interest to him, gained much of his theory from reading Varenius in one of the English editions.[58]

However, while Newton's science had conquered British natural philosophy by 1782, it did not conquer all of science. Nor could it have done so, for other sciences existed outside natural philosophy. The generous usage of the eighteenth century defined science as any product of long experience and systematic observation. As the *Oxford Dictionary* put it, in a definition now designated as obsolete, a man of science was a person "who possesses knowledge in any department of learning, or trained skill in any art or craft." We have already met with a few of these sciences. One was the Norfolk science of agriculture. "Unquestionably the first of all sciences, as it nurses and supports the rest," this science was practiced by a highly influential group of landlords and sponsored by such luminaries as Sir Joseph Banks, president of the Royal Society, and "Farmer George," King of

England.[59] Another science was practiced by the commissioners' experts—Hodskinson, Grundy, Hogard, and Nickalls. Like Mylne and Smeaton, they too belonged to the rising order of professionals called "civil engineers." Unlike Mylne and Smeaton, however, they were practical men of science, skilled practitioners, who concerned themselves with the observable and the measurable alone.

The Defendants' Reports

After waiting for better weather, the defense team—Joseph Hodskinson, John Grundy, and Thomas Hogard—set out to Wells in early April and spent ten days there, inspecting the harbor and its surroundings. They did not bother themselves with the general laws governing the dynamics of tidal harbors—that would have been to play the part of the natural philosopher. They were civil engineers, practical men whose business was to estimate the relative effectiveness of the backwater deprived by Turner's embankment. To that end, they concerned themselves with the mappings, measurements, and calculations related to the current state of the land; its existing areas, enclosed or not; the locations and orientations of the creeks; the length and the depth of the channel; the directions of the winds and the tides; and so forth.[60]

Back in London, Hodskinson promptly delivered a detailed report to his employers. While Smeaton launched his report with a long theoretical lecture that created the basis for his expert opinion, Hodskinson grounded his report on the description of his interviews with the local pilots and harbor masters, and the observations and measurements he had made to satisfy himself that the information given was truthful. The interviews convinced him that within local memory the coast had not grown in breadth or height; on the contrary, the sea had gained ground in many places along the coast. In addition, he found out that the shifting of the harbor's channel toward the east had occurred after 1758 and that it could not possibly have been affected by the silt discharged at the western estuaries, because the tides in Wells came from a different direction. "From these points," Hodskinson wrote, "I concluded that the Causes of the Obstructions [to the harbour] were not to be sought in the main Ocean; and therefore I turned my inquiries internally."[61]

Turning inland, Hodskinson began to calculate. Following is a summary of his copious calculations:

The original weight of the waters available for scouring the harbor was:	1,050,754 tons
The weight of the backwater eliminated by the embankments of 1719 was:	368,242 tons
The body of backwater eliminated by the 1758 embankment was:	214,122 tons
Thus, the body of water available for scouring the harbor has been cut by more than a half to:	468,390 tons

Not all drops in backwater are alike, however. In spite of eliminating the larger body of backwater, Hodskinson considered the embankments erected in 1719 to be less injurious to the harbor than later embankments, because the former were located close to the sea and the backwater they eliminated used to flow to the sea "at the first quarter's ebb," before the channel was sufficiently empty to benefit from their scouring powers. By the same line of reasoning, he judged the embankment erected in 1758 to be most detrimental to the harbor. First, because it was located further inland, behind the harbor, the backwater it deprived the harbor of had scoured the upper parts of the harbor's channel, from the quay downward, where it was most needed. Second, due to the great length of its course and the elevation of Wharham Slade, its backwater rushed through the channel with great momentum. Third, being farther away from the sea, its backwater used to flow to the sea during "the last quarter's ebb," when the channel was sufficiently empty to best benefit from its forceful action. "Upon the whole," Hodskinson concluded, "I am of opinion that the present bad and ruinous state of the harbor is to be in a great measure, if not wholly, imputed to the embankment of Wharham Slade in 1758; and that if the tide of the sea is permitted to flow and reflow in its ancient course and manner over the Wharham Slade, and if the harbour's mouth and channel should be first cleansed and restored to its former position, the navigation will be supported and maintained in a safe, useful and commodious state by the natural operation of the tidal waters thereon."[62]

John Grundy and Thomas Hogard did not submit reports of their own. "Having accompanied Mr. Hodskinson in his survey of the harbour of Wells, in pursuance of your directions," they wrote to their employers, the commissioners, "we join and entirely concur in his report." Joseph Nickalls did submit a separate report, but it was structuredmuch like that of Hodskinson and was probably coordinated by him. Appending a list of tidal harbors kept open and deep exclusively by the flux and reflux of the sea, Nickalls' conclusion also was that "unless the embankment of Wharham Slade is removed, and the tidal waters permitted to flow and reflow therein, as they used heretofore to do, this harbour must be destroyed."[63]

The Rise of the Adversarial Spirit

Norfolk men were famous for their litigiousness, and the assizes on the Norfolk Circuit were notorious for their length.[64] "The inhabitants," wrote one seventeenth-century commentator, "are so well-skilled in matters of the law as many times even the baser sort at the plough-tail will argue *pro et contra* cases in law, whose cunning and subtiltie hath replenished the shire with more lawyers than any shire whatsoever though far greater."[65] Thus, come late July 1782, both parties appeared in Norwich at the summer assizes, accompanied by their lawyers and by what had grown by then into an elite group of the new British profession of civil engineering. Each side was equipped with the printed reports of its rivals, ready to do battle before the trial judge and his special jury.

The notion of doing battle in the courtroom has its roots in medieval criminal law, according to which the perpetrators of a felony committed a twofold offense—against their victims and against their king. Thus, they could have been prosecuted in two distinctive ways—by a private suit of the victim or by a suit of the Crown by indictment. Proof by combat, the paradigmatic procedure of proof in a private suit, provided an early and most dramatic example of the active role that can be played in court by the litigants. However, it was rarely pursued. With the rising authority of the Crown, indictment became the norm. In this mode of prosecution, the litigants were left "to what a set of strangers might say, witnesses selected by a public officer" and were permitted only to state their respective contentions.[66] The litigants seized control over the production and presentation of evidence only during the eighteenth century. With this change, ironically, the ancient spirit of proof by combat returned to dominate criminal proceedings. The new warriors brought in to do battle for the litigants were the lawyers. They may have substituted black robes for armor and law books for weapons, but their ideology was no less adversarial. The truth, they still maintained, was that which was still standing on its feet after the battle was done.

From the decline of the priesthood after the Reformation until the

nineteenth-century elevation of the medical profession, lawyers constituted perhaps the most prominent profession in English culture. A legal
career offered young men of talent the best chance for wealth, power,
and social standing, and the number of professional practitioners of the
Common Law mushroomed accordingly. Records suggest that by the
early seventeenth century there were no fewer than 500 barristers and
serjeants practicing in Westminster Hall, as compared to 50 a century
earlier. Still, until the early eighteenth century, lawyers were generally
kept out of criminal trials.[67] Defendants were not allowed to have
counsel, and although a prosecution counsel was allowed, his court
appearances were rare. The judge dominated the proceedings and he
examined the parties and the witnesses himself. In this environment,
which kept criminal proceedings quick and simple, the dogma was that
knowing the truth, defendants were their own best advocates, and that
the judge was there to help them pronounce it and to ensure them a fair
trial.[68]

By the 1730s, defense counsels had begun to participate in regular
criminal proceedings. Still not allowed to address the jury, they were
permitted to gather and adduce evidence and examine and cross-
examine witnesses. It is not yet clear what the reasons were for this
change, which was considered by the noted judge and legal historian
James Stephen to be "the most remarkable change" that took place in
English criminal law ever. Currently, the leading theory is that the
appearance of the defense lawyer was, at least partially, a response to the
expanding prosecutorial capacities of the Crown. It was feared that new
prosecutorial practices, such as the Crown witnesses system, and the
growing activities of professional thief catchers, encouraged false testimonies and brought about the condemnation of innocent men. But
whatever the reasons were, by the second part of the eighteenth century, according to Stephen, "a practice sprung up, by which counsels
were allowed to do everything for prisoners accused of felony except
addressing the jury."

Outside the courtroom, lawyers' participation gave new significance
to pretrial activities such as preparing records and seeking out witnesses. Inside the courtroom, the active participation of lawyers slowly
reshaped the processes of criminal litigation. They increasingly took
over the examination of witnesses, developed the techniques of cross-
examination to perfection, and established the right to argue points of
law.[69] As the notion that the parties could be held responsible for
developing their own proof by presenting persuasive evidence in court
took root, judicial involvement in the processes of litigation dimin-

ished. Consequently, the parties gained the right to choose the evidence to be considered. The articulation of prosecution and defense cases and the burdens of the production of evidence and proof followed, and the courtroom was soon polarized. In this emerging adversarial environment, free reign was given to partisan views. By the end of the eighteenth century, the courtroom had splintered into invisible but well-entrenched territories consisting of the judge, the jury, the litigants, and the lawyers, and highly intricate rules of evidence were developed in order to mediate the newly dispersed authority of the law among these elements.[70]

William Best, a leading mid–nineteenth-century authority on legal evidence, looking back to the beginning of his subject, wrote that "the necessary consequence of [allowing defense counsel in criminal cases] was that the objections to the admissibility of evidence were much more frequently taken, the attention of the judges was more directed to the subject of evidence, their judgment was better considered, and their decision better remembered." Indeed, prohibited from speaking directly to the jury, the lawyers often fought their battles over the content and the presentation of the evidence before the jury through evidentiary objections. The evolving use of evidentiary objections in the strategic effort to win the client's case was not limited to criminal proceedings. Many of the lawyers involved in criminal proceedings also practiced civil litigation, and as they moved back and forth between the two arenas, they deployed on the civil side the same skills and techniques they had formulated in its criminal counterpart. Civil proceedings were less prone to evidentiary objection, since they relied, first and foremost, on written evidence, and only to a lesser degree on oral testimony supplied by party-selected witnesses. Nevertheless, the 1780s saw the rising adversarial spirit manifesting itself in both criminal and civil litigation.[71]

The Second Trial

The second round of the Wells Harbor litigation began on July 25, 1782, before Henry Gould, chief justice of the Royal Court of Common Pleas, and a special jury. Two eminent counsels ran the trial. Leading the legal team for Folkes and Hales, unchanged from the previous trial, was Henry Partridge, a well-known barrister with strong Norfolk connections. This time, the commissioners of Wells Harbor brought their own heavy legal artillery in the person of George

Hardinge, barrister of the Middle Temple and solicitor-general to Queen Charlotte. Hardinge's planned line of defense was to attack the notion that imperceptible causes for the decay of Wells Harbor recognized by experts at a distance were to be preferred to the factual observations made by experienced witnesses on the spot. "When the Jury are informed of the Causes, which are perceptible and proved upon Oath," the defense noted upon reading Smeaton's report, "it is hoped that they will not attend to Causes which were imperceptible and never existed, but in the head of Mylne [and Smeaton]."[72]

The reports of the experts on both sides were made available to the jury a week before the trial. However, when the trial began Hardinge chose not to call upon his experts to give oral testimony and be cross-examined. Instead, he summoned a long line of eyewitnesses to testify to the rapid deterioration of their harbor. He also called upon mariners and navigators to testify that the western and northern tides acted in contrary directions and did not meet across the harbor's entrance. Partridge, on the other hand, planned to repeat his previously successful strategy and to summon his famous scientific expert to the witness stand to persuade the jury that Turner's embankment was not responsible for the Wells Harbor troubles. However, when Partridge tried to do so, Hardinge objected. The illustrious Smeaton, he contended, should not be permitted to speak to the jury because his testimony concerning the hidden causes of nature "was matter of opinion, which could be no foundation for the verdict of the jury, which was to be built entirely on facts, and not on opinions."[73]

The requirement of Common Law to limit oral testimony to that of witnesses who possessed personal knowledge based on factual perception was very old and had its roots in medieval law that demanded that witnesses speak only "what they see and hear." However, within the context of the medieval community-based proceedings, such constraints had little, if any, meaning, and hearsay testimony was received without question. Objections to hearsay evidence started to grow in number during the sixteenth and seventeenth centuries as the legal fact-finding procedures turned increasingly public.[74] Still, well into the eighteenth century, the court was content with allowing such objections to go to the weight of the testimony rather than to its admissibility. With the rise of the new adversarial system, however, objections to hearsay evidence grew in strength and sophistication and were applied in a constantly broadening range of cases. By the 1780s, the traditional hearsay rule had already given birth to two distinct legal doctrines: the hearsay doctrine, which attempted to limit testimony to information based solely on personal observation, and the opinion doctrine, which sought to control the form in which witnesses communicated their perceptions to the jury, requiring them not to use inferences when the subject matter was susceptible to factual statements. "Henceforth," John Wigmore, the legal scholar, summarized the status of the new opinion doctrine at the end of the eighteenth century, "the only question can be how far there are to be specific exceptions to it." It was with this question in mind that Chief Justice Gould addressed Hardinge's objection to Smeaton's testimony in the Wells Harbor trial.[75]

The answer to this question was not simple by any means, and by 1782 it had given rise to inconsistent decisions. The leading rationale of the day, set out by Lord Mansfield in 1766, was that opinion testimony was not to be admitted when, "if rightly formed, it could only be drawn from the same premises from which the court and jury were to determine the cause; and therefore it is improper and irrelevant in the mouth of a witness."[76] This rationale left ample opportunity for the admissibility in evidence of all kinds of opinion testimony that was based on knowledge relevant to the case but not accessible to the jury. It even allowed for the opinions of lay witnesses to be given, if such opinions were relevant and based on direct experience. As for experts—they were allowed to testify to their opinion from facts that fell within their expertise as long as they had observed them directly. Occasionally, the judges also allowed them to testify to their conclusions from the evidence of other witnesses, because their experience enabled them to draw conclusions that neither the jury nor the judge could draw.

Smeaton's case seemed straightforward. He was a well-known expert, and the facts of the Wells Harbor case, which he had observed directly, fell well within his field of expertise, thus constituting a proper object for his expert opinion. Nevertheless, the defense counsel, Hardinge, had firm legal ground to stand upon when he maintained that Smeaton's opinion did not rest on the facts of the case but on speculation that had no place in court. Smeaton, Hardinge argued, had gone beyond the established legal range of expertise. For centuries, experts had been summoned to court to give their opinions about matters that were not common knowledge but that were nonetheless concrete and tangible—matters such as norms of trade, the construction of ships or wagons, meaning of Latin words, foreign laws, the nature of wounds, the traits of witches, and so forth. These expert opinions were regarded as being based on empirical observations, readily traceable to the particular training and experience of the expert pronouncing them.

Smeaton, on the other hand, propounded in court a high-minded hypothesis about some natural processes, imperceptible to anyone but himself, which allowed him to shift the blame for the undisputed decay of the harbor from the obvious human hand of his employers to the hidden hand of nature. But what kind of training or experience could have qualified a person like Smeaton in 1782 to be an authority on such matters? And what kind of legal reasoning could acknowledge such latent causes as a proper foundation for the verdict of the jury, which was to be built entirely on facts?

The 72-year-old Chief Justice Gould was known for the strictness of his approach to his law.[77] His logic, the logic of Common Law, was Baconian. It disdained abstract explanations, suspected elegantly constructed theories, and stressed the necessity for direct observational data in processes of proof. Hypotheses were but preliminary means of directing further observation and measurements. Experts in the courtroom were summoned for their privileged knowledge, but were nevertheless expected to base their conclusions on careful observations and calculations. Yet, Smeaton's evidence was based on a hypothetical natural process that could have taken centuries and could not be measured, tested, or otherwise verified. Thus, Chief Justice Gould accepted Hardinge's argument that Smeaton's evidence indeed "could be no foundation for the verdict of the jury" and did not permit Smeaton to address the jury from the witness stand. With Smeaton and his imponderable science out of the way, Hardinge continued to win the day, by the end of which the jury gave a verdict for the defense, allowing the commissioners of the harbor to cut the embankment that choked up their harbor. Folkes and Hales immediately asked for a new trial on the grounds that their expert had been improperly silenced.

Mansfield's Decision

The law of evidence developed slowly and erratically during the eighteenth century. The inherited medieval procedural machinery was becoming more and more subtle and rigid. Unable to change the established procedures, the courts had overlaid them through the years with a mass of conventional practices, which had, in effect, substituted a wholly new system. The confusion was further magnified by the fact that most of the rulings upon evidence were made at the assizes and not reported, except through notes taken by the counsels, which were carelessly done and often contained mistakes. Motions for new trials, which often turned on the application of the evolving principles of evidence gave the royal judges an excellent opportunity therefore, to clarify problematic issues of law and articulate the principles behind them in a precise and detailed manner so that they could serve as unambiguous authority in the future.[78]

Thus, on November 21, 1782, when Lord Mansfield, probably the most influential judicial figure of the eighteenth century, delivered the King's Bench's judgment concerning the new appeal in the Wells Harbor litigation, he was delivering more than a mere opinion on whether the expert opinion given by John Smeaton was proper evidence or not. He was using the opportunity to clarify the true issue, which, as he himself liked to say, "was winnowed from the chaff of the trial." His decision is worthy of a lengthy quotation:[79]

> The facts of this case are not disputed. In 1758 the bank was erected, and soon afterwards the harbor went to decay. The question is, to what has this decay been owing? The defendant says, to this bank. Why? Because it prevents the back-water. That is matter of opinion. The whole case is a question of opinion from the facts agreed upon. Nobody can swear that it was the cause; nobody thought that it would produce this mischief when the bank was erected. The commissioners themselves look on for above twenty years . . . It is a matter of judgment what has hurt the harbor. The plaintiff says that the bank was not the occasion of it . . . Mr. Smeaton is called. A confusion now arises from a misapplication of terms. It is objected that Mr. Smeaton is going to speak, not as to facts, but as to opinion. That opinion, however, is deduced from facts which are not disputed—the situation of banks, the course of tides and winds, and the shifting of sands. His opinion, deduced from all these facts, is, that, mathematically speaking, the bank may contribute to the mischief, but not sensibly. Mr. Smeaton understands the construction of harbors, the causes of their destruction, and how remedied. In matters of science no other witnesses can be called. An instance frequently occurs in actions of unskillfully navigating ships. The question then depends upon the evidence of those who understand such things; and when such questions come before me, I always send for some brethren of Trinity House. I cannot believe that when the question is, whether a defect arises from a natural or an artificial cause, the opinions of men of science are not to be received . . . I have myself received the opinion of Mr. Smeaton respecting mills as a matter of opinion.

The cause of the decay of the harbor is also a matter of science, and still more so whether the removal of the bank can be beneficial. On this such men as Mr. Smeaton alone can judge. Therefore we are of opinion that his judgment, formed on facts, was very proper evidence.

Winnowing the Wheat from the Chaff in Folkes v. Chadd

Mansfield's decision in "the great case of *Folkes v. Chadd*" has long served as the origin story for the rise of partisan expert testimony in the modern Anglo-American legal system. The original report on *Folkes v. Chadd*, published in 1831, half a century after the events of the case, was the first document to register the importance of the case. The author of the report was Henry Rosqoe, an experienced barrister, who based his reports on the records of various leading judges. According to Rosqoe, the early nineteenth-century courts regarded Mansfield's decision as "the principal case on the admissibility of matter of opinion." "Professional men," Rosqoe further elaborated on what was to be learned from Mansfield's decision, "when examined on the subject of their art or science, are of necessity allowed to state their opinion."[80]

The next significant reference to Mansfield's decision came 60 years later from Professor James Thayer of Harvard's law school. Mansfield's decision, Thayer taught in his course on evidence, represented the onset of judicial recognition in the modern practice of party-called expertise. "For a long time," he wrote when introducing the case in his textbook, "experts were thought of in the old way, as being helpers of the court . . . But at last the modern conception came in, which regards the expert as testifying, like other witnesses, directly to the jury."[81]

One difficulty with Thayer's reading of Mansfield's decision lies in the fact that the practice of party-called expertise was not novel in 1782. As early as 1678, in a murder case, some of the most eminent physicians in the kingdom, including Sir Hans Sloan, the future president of the Royal Society, were called by both sides to testify as to the causes of certain symptoms observed in the autopsy, and to the general proposition that a man could die of wounds without a fever.[82] In the eighteenth century, party-called expertise was also documented in civil proceedings, noticeably in the growing area of patent litigation. The practice recurred in relation to the growing textile trade during the 1760s and 1770s.[83] Tax litigation and nuisance litigation also saw the deployment of party-called expertise within an intense adversarial spirit.[84] Having presided over many patent trials, Lord Mansfield himself was quite familiar with the use of party-called expertise. In the 1760s, for example, he presided over an important patent case that involved the London optician Peter Dollond, F.R.S. and a group of other London opticians, in a struggle for the patent rights for the design of the refracting telescope. The case brought to the witness stand a long line of experts, who testified as to the optical principles, previous designs, and trade secrets involved. In the summer of 1781, concomitantly with the Wells Harbor litigation, Lord Mansfield presided over another important case involving expert testimony, one that revolved around Richard Arkwright's attempt to enforce his monopoly on the carding machine in the textile business. The attempt failed but a second one, in 1784–1785, succeeded, largely thanks to the testimony of scientific figures such as William Herschel, Erasmus Darwin, James Watt, Robert Mylne, and J. A. de Luc, who vouched under oath as to the validity of the patent's principle and specifications.[85]

Thus, the practice of relying on party-called expertise was not novel in 1782. One could still maintain that its adversarial context was, though, and that Mansfield's decision was the first to recognize and legitimize the practice in this new context. However, little in Mansfield's decision could support such a claim. Mansfield's decision displays a complete disregard of Smeaton's appearance as a partisan witness. In fact, Mansfield's opinion treated Smeaton as if he were a court expert. "When such questions come before me," Mansfield reasoned, "I always send for some brethren of Trinity House." The Trinity House was a famous club of retired sea captains, and its brethren functioned as arbitrators and official court experts in cases arising out of events on the high seas. Clearly, the deployment of court-nominated expertise is not the precedent one would choose if one were intent on inaugurating a new practice of calling experts as partisan witnesses selected and paid for by the parties.[86]

John Henry Wigmore, Thayer's student and the leading early twentieth-century authority on legal evidence, also placed the origin of modern Anglo-American expert testimony in late-eighteenth-century England. Wigmore was aware, however, that the practice of relying on party-called expertise was not new to the period. He also realized that late eighteenth-century expert witnesses, who had observed the facts of the case and testified to their conclusions, were not yet differentiated from lay witnesses, who were also allowed to testify to their opinion, if this was based on their intimate knowledge of the facts of the case.

Wigmore maintained, therefore, that the distinctiveness of the modern expert witness sprang not from the license to testify to opinion (which was still shared with lay witnesses), but from the expert's exclusive privilege to pronounce an opinion whether he had observed the facts of the case directly or not. It was this distinction, according to Wigmore, that made its first successful appearance in Mansfield's' decision in *Folkes v. Chadd*:

> Here was a man [Smeaton], who had never seen the place, had no "facts" to add, and was going to give . . . his opinion upon the general question in doubt, the cause of the decay. Why should he do this? Why waste time in listening to numbers of such persons when the twelve men in the box have been specially selected for the very purpose of having their opinion serve as decisive? There would be only one reason for listening to such outside opinions, namely, that the witness was such a person that the jury would be really aided by his opinion.[87]

Thus, Wigmore claimed, Mansfield's decision epitomized "the general recognition by the end of the 1700s, that there was a class of persons, i.e., those skilled in matters of science, who, though they personally knew nothing about the circumstances of the particular case, might yet, perhaps by way of exception, give their opinion on the matter."[88]

Wigmore, we know by now, had his facts wrong. Smeaton not only had seen the place but also had written a detailed report on his findings that had been accepted by the court as primary evidence and was delivered to each juryman a week ahead of the trial. The source of Wigmore's mistake was probably the original report of *Folkes v. Chadd*, which failed to mention that Smeaton had based his opinion upon his own inspection of the facts of the case. The report further claimed that Mansfield's decision was "followed and confirmed by a variety of similar decisions," and specified as its leading example a case from 1790 in which a second royal judge, Lord Kenyon, "admitted the evidence of a ship-builder on a question of sea worthiness, though he had not been present at the survey." Alas, Lord Kenyon's decision could hardly have been based on Mansfield's ruling in *Folkes v. Chadd*, since Smeaton possessed first-hand knowledge of the facts of the case. Indeed, in 1785, in an insurance case, Mansfield excluded the testimony of ship builders on the same question of seaworthiness for "not having examined the ship themselves." Thus, although Wigmore's claim that there was general recognition of a new class of experts by the end of the eighteenth century may still be true, it nevertheless cannot be based on Mansfield's decision in *Folkes v. Chadd*.[89]

The iconic status of Mansfield's decision as the origin story for the rise of partisan expert testimony is therefore an exercise in *post hoc* rationalization, which is commonly made in a legal system that legitimizes its present decisions by reference to precedents. Often, it is only the later consensus of the legal profession that turns a series of largely isolated judge-made solutions to particular problems at different times into a coherent rationale.[90] Still, if Mansfield's decision was neither about inaugurating a new practice of calling experts as partisan witnesses before the jury, as Thayer maintained, nor about allowing experts to pronounce opinions without being personally familiar with the facts of the case, as Wigmore maintained, then what was it about? What was the issue that Lord Mansfield tried to winnow from the chaff of the protracted litigation?

Having reconstructed the facts of the case, we are now able to propose an answer to this question. Lord Mansfield, just like Chief-Justice Gould before him, was trying to decide the merits of Hardinge's objection, which pitted men of science in the old sense (that is, men of great and tested experience) against men of science in the new sense (that is, Newtonian philosophers who based their opinions on their privileged knowledge of the laws of nature). Thus, Mansfield's decision delivered the King's Bench's authoritative interpretation of the implications of the nascent opinion doctrine for handling the opionion of experts like Smeaton and Mylne, proto-scientists who functioned like skilled professionals but cogitated like natural philosophers.

The Wells Harbor litigation, therefore, did constitute an important historical moment in the deployment of expert knowledge in the courtroom, but for reasons different from those so far suggested. It was a junction where the expanding late eighteenth-century cultures of law and science finally crossed paths. The lawyers had been solidifying their control over the production and presentation of evidence in the legal courtroom. Meanwhile, natural philosophy had shown signs of becoming a competent branch of applicable knowledge. As a noted Newtonian, Smeaton was a conspicuous example of the growing importance of natural philosophy as a worthwhile, even useful, pursuit. His philosophical studies of waterwheels, the most important source of energy during the early stages of the industrial revolution, revolutionized their usage and improved their performance by more than 30 percent. Thus, although late-eighteenth-century Englishmen may have still considered the pronouncements of natural philosophers to be less than facts, they were nevertheless already more than mere opinions. That was also the case with the causes behind matters such as the decay of harbors, which traditionally had been matters to be addressed by the experience of the craftsmen and artisans who had built and used them. By 1782, as Mansfield made certain to clarify in his decision, these were already a "matter of science," upon which "men as Mr. Smeaton alone can judge."[91]

Wigmore was, therefore, right when he considered Mansfield's decision as illustrating the growing legal recognition by the end of the eighteenth century that there was a new class of persons, skilled in matters of science, who could give their opinion, even if it was not based directly on the traditional trustworthiness of the senses. However, this lack of positive first-hand evidence was not merely a contingent deficiency occasioned by the experts' failure to personally inspect the facts of the case. Rather, it was an inevitable consequence of the knowledge these new men of science brought to the courts, knowledge that was often based on the imponderables of nature, which "nothing but the most philosophic eye, by reasoning upon chain of facts, is able to discover."[92]

Hardinge's objection to Smeaton's testimony forced the legal profession to reflect on the epistemological status of this philosophical reasoning and on the status of its bearers in the courtroom. The conservative Chief-Justice Gould chose to remain within the guarded line delineated by the evolving rules of evidence and to exclude Smeaton's theory for not being clearly reducible to hard and concrete evidence. This formalist approach, which denied the court the services of the most respected expert on the issue upon which the whole litigation turned, made no sense to Mansfield. "I cannot believe," he remarked, "that when the question is, whether a defect arises from a natural or an artificial cause, the opinions of men of science are not to be received." Thus, Mansfield declared the opinion of men of science an exception to the opinion doctrine. Unwilling to distinguish one science from the other, Mansfield measured professional reputation instead. If the proposed witness was known as an expert on the matter before the court, Mansfield prescribed, his opinion, formed on facts, was proper evidence.[93]

The Third Trial

Folkes' lawyers got what they wanted. The third trial was set for August 1783, at the Norwich summer assizes, and this time all men of science were going to be heard as witnesses in open court. Still, they were worried. Public opinion in Norfolk was unmistakably on the commissioners' side. After the second trial, the local community had entertained a hope for an end to the litigation and a recovery for their harbor. Therefore, Folkes' successful appeal precipitated much resentment, which found Smeaton and his fanciful science an easy target. An anonymous letter published in the *Norwich Mercury* trumpeted against "the great, the skilful, the celebrated, the omniscient Mr Smeaton," and against his "speculative doctrines, scientific opinions, and visionary conjectures." Mansfield's ruling for a new trial only to allow men of science to be fully heard seemed absurd to the writer. The facts of the case were plain for everyone to see, and the third trial would differ from the second only in that the new jury would have to go through the trouble of hearing the experts instead of just reading their reports. "Suffice it to say," he comforted himself, "that they are now to be examined as witnesses in open Court, where they will be liable to a nice and judicious cross Examination, if their Opinions shall be thought worthy of it."[94]

Smeaton reassured his anxious employers that his report would hold in cross-examination. No one, he wrote them, not even himself, "can take it to pieces and show it in reality to be infirm and invalid." Nevertheless, Smeaton was worried too. The commissioners had just added his former brilliant apprentice, William Jessop, to their scientific roster. "I scarcely know an engineer remaining of any name in the harbour branch who is not engaged on one side or the other," he wrote his employers. He urged them to make sure that at least Mylne would appear next to him in the coming trial, "not only for the sake of justice in the determination of the cause, but for my own sake as an individual, as I wish not to stand single and unsupported against a legion."[95]

The third trial was indeed a vituperative affair. When Mylne stepped onto the witness stand, Hardinge launched a menacing cross-examination, attacking Mylne's character, accusing him of perjury, and threatening to prosecute him for it. Smeaton was treated better; still his testimony, which lasted four hours, failed to sway the jurors, who by the end of the day agreed that the embankment was injurious to the harbor and gave a verdict for the defendants. Lord Ashurst, the presiding judge, directed the jury to inform him also as to whether the removal of the embankment would tend to restore the harbor or not, but the jury failed to agree on this issue. Seizing on this disagreement, Folkes' lawyers appealed again. Since the jury could not decide whether the removal of the embankment would be likely to restore the harbor, they argued, the cutting of the embankment could not be justified. Hence, they should have the verdict or at least the right for yet a fourth trial.[96]

Thus, come late November 1783, the judges of the King's Bench convened once again to discuss the new twist in the Wells Harbor litigation. The question laid before the jury, the judges concluded, was whether the embankment was injurious to the harbor, on which the jury answered in the positive. The additional issue, whether the removal of the embankment would tend to restore the harbor, was not for the jury to decide but for the Court of Chancery to decide, if the commissioners were to seek from it an order for abatement. Thus, the judges rejected the plaintiffs' appeal and declared the verdict for the commissioners final. The 1781 injunction granted by Chancery was dissolved, and the commissioners of Wells Harbor were allowed costs and expenses, £562 altogether, to be paid by Folkes and Hales. "That melancholy scene, it is hoped, will now be closed," the local newspapers reported. The true culprits, they moralized in summation, were not the parties who fought for their property, but "those engineers, whose speculative opinions and unfounded conjectures have served no other purpose but to mislead their employers, protract the suit, and occasion expense to both parties."[97]

The legal battle was over but the tension between men of law and men of science was just beginning to take shape. Robert Mylne was much offended by the treatment he had received from the lawyers. Early in 1784, he demanded from Hardinge a public apology that "will do away the foul aspersions you attempted to fix on my reputation, not as an artist, but as an honest man." Hardinge refused to apologize. "The zeal of an advocate often carries him too far," he replied, "but I trust that I am not more faulty than others in the same line with myself." Mylne refused to back down and threatened to take the matter into the courts. Hardinge asked an influential friend, Lord Camelford, to mediate. "Few gentlemen of the [legal] profession, I believe, account themselves accountable for their pleadings out of court," Lord Camelford wrote Mylne. "If they were it would be difficult for them indeed to do justice to their clients without offending their antagonist." Mylne was not persuaded. "The stage of the advocates is precluded from every one else;" he replied, "and they take this advantage to say and act, that, which they are ashamed of and would blush to say and do privately." After much additional mediation and negotiation, Hardinge grudgingly agreed to produce a letter of apology, which Mylne promptly returned as unsatisfactory. A second letter, expressing explicit regret for the allegations made during the Wells Harbor litigation, "which could not be justified by your conduct or character," faired better. Mylne accepted it and distributed it widely within all relevant social circles.[98]

While men of law and science skirmished in the wake of *Folkes v. Chadd*, Folkes and Hales accepted defeat and removed the embankment erected in 1758 so that the tide could flow over the Wharham Slade and restore the harbor into its ancient golden state. That, however, did not happen. For a long time, the fate of the struggling harbor remained doubtful. In 1785, Folkes and Hales sold Turner's estate to Thomas William Coke, the neighboring landowner, for £57,750. The loss of the drained Wharham Slade marsh lowered the price by about £2000. The large purchase, made just before the sharp rise in land values of the war years, was a good one and helped to establish Coke's legendary reputation as the leading scientific farmer in the kingdom.[99]

Twenty-five years later, in January 1808, Coke requested the commissioners' permission to again reclaim the fertile Wharham Slade. In return, he promised to finance a whole new sluice with a true stop gate that would be able to collect enough tidal water to successfully scour the harbor. Conceding that such a project would have "great superiority over the slade in its present state," the commissioners of Wells Harbor admitted that, "for a long time past the utility of the said marsh, as a receptacle for back water, has been very inconsiderable, it being now ascertained that the water in common spring tides does not run to the level of its surface, and that the marsh is in reality never overflowed so as to send forth a body of efficient scouring water, unless in cases of forced or raging tides, which very rarely occur." The commissioners approved Coke's proposition, and he went on to cut off Wharham Slade marsh from the harbor's creek. Alas, he never made good on his promise to build a new sluice. The attempts to improve the shrinking harbor continued throughout most of the nineteenth century, but to no avail. In 1881, an exact century after *Folkes v. Chadd*, Wells Harbor was officially demoted to the status of a creek, losing both its independence and its commissioners.[100]

Smeaton was thus proved to be right about the insignificance of the Wharham Slade for the scouring of the Wells Harbor. Was Smeaton's entire theory right? Was the decay of the Harbor part of the "progressional operation of nature"? The answer to this question remains uncertain—not for lack of historical information but because even today, more than two centuries later, science still cannot provide us with a definite answer. Too much is dependent on contingent factors such as the specific distribution of the vegetation and the various gradients of the marshland.[101] Thus, although much of Smeaton's theory is accepted today as true, his expert opinion remains tentative, at best. Not knowing which party to the Wells Harbor litigation was correct may help us to appreciate better how, in the late eighteenth century, the lawyers could have set science against itself: the science of the natural philosopher against the science of experience, the science of cut-and-try against the science of principles; and applied mathematics indifferently on all sides.

The Narrative of the Rise of Modern Expert Testimony

The debate between the "Goulds," who maintain that the law should exclude from the courtroom certain expert opinions for not being scientific enough, and the "Mansfields," who maintain that the law has no way to give preference to one kind of science over another and requires that all kinds of science be heard in open court, is as alive today as it was in the late eighteenth century. As with the scientific dispute over Wells Harbor, not knowing who is right—the Goulds or the Mansfields—may help us to better appreciate the subtleties of the historical moment during the late eighteenth century when Common Law first came to face this problem in the context of the new adversarial system.

Traditionally, experts, whether part of the jury or court advisors, were summoned and controlled by the court, which conferred on these experts a large degree of impartiality. During the late eighteenth century, as the interested parties began to summon their own experts to represent them before the jury, a demanding problem emerged: how to ensure that in this adversarial environment lay jurors would still have access to reliable expert guidance when they needed it. Legal lore has it that Lord Mansfield's decision in *Folkes v. Chadd* shaped the modern solution to this problem, that is, "the whole adversarial apparatus including contending experts, hypothetical questions, and jury evaluation."[102] However, the analysis of Mansfield's decision finds little awareness, let alone angst, about the issues of Smeaton's appearance as a partisan witness. The analysis of other major late–eighteenth and early–nineteenth-century rulings on Common Law provides similar results. While the judges were busy with the delicate act of balancing the demands of the increasingly defined rules of evidence with the growing supply of expert testimony, they seem far less concerned that the practice of calling experts as partisan witnesses was expanding and that the position of the experts in court was shifting from an independent and impartial status to total partiality to the side that hired them.

Could it be that the experienced royal judges overlooked the difficulties that might await the deployment of partisan expertise in the new adversarial courtroom? The absence of judicial anxiety is all the more remarkable if we take into account that there was ample judicial dismay about lay witnesses for hire. The late eighteenth century was a period in which the slightest interest in the result of the trial rendered a witness unreliable. Persons were not allowed to testify in cases in which they had financial interest. Husbands and wives were forbidden from testifying for or against each other. Even the parties to the lawsuit themselves, by the same reasoning, were not allowed to testify.[103] Why then the partisan scientific expert?

The answer, I would like to suggest, is that late-eighteenth-century judges counted on men of science to give, by ties of honor, unbiased opinions on matters beyond the ken of the jurors. Men of science had long adopted the gentlemanly code of honor as a necessary condition for the reliability of the scientific discourse. Gentlemen were bound to credit the word of their fellows. The status of the gentleman—his economic independence, his freedom of action, the moral discipline he imposed upon himself—all guaranteed the credibility of his word.[104] This social contract worked both ways. Nothing ruined gentlemanly status quicker than dishonesty. "Twenty faults are sooner to be forgiven," John Locke gave notice in his 1690 guide to the education of English gentlemen, "than the straining of truth, to cover any one by an excuse:"

> [Lying] a quality so wholly inconsistent with the name and character of a gentleman, that nobody of any credit can bear the imputation of a lye; a mark that is judged the utmost disgrace, which debases a man to the lowest degree of a shameful manners, and ranks him with the most contemptible part of mankind, and the abhorred rascality, and it is not to endure by anyone, who would converse with people of condition, or have any esteem or reputation in the world.[105]

The seventeenth-century founders of the Royal Society deemed it essential that among its members "the Farr greater Number are Gentlemen, free, and unconfine'd." During the eighteenth century, the Royal Society continued to strengthen its status as a body of disinterested gentlemen, who impartially investigated nature and toiled for no end but the improvement of public good. The royal judges, therefore, were not worried about the behavior of men of science and trusted that their expert testimony would correspond to their true opinion.[106]

The circumstances of *Folkes v. Chadd* provided a good example of this judicial trust. Mylne and Smeaton were elite members of the Royal Society, the only engineers elected to the its Council.[107] Lord Mansfield had previously benefited from their expert services and respected them enormously. "Everybody knows the character of Mr. Mylne for ability," he commented during deliberation of the first appeal. "He is a man very considerable in his profession, a good architect, a good mathematician, & a good engineer. There is no doubt he deserves the character that has been giving to him." During deliberation of the second appeal he conferred similar praises, "on the consumate skills and integrity of Mr. Smeaton who had on several occasions been examined in causes of a similar nature tried before his lordship." Clearly, Mansfield was not worried that experts like Mylne and Smeaton would testify dishonorably. They were men of honor, and their integrity guaranteed the truthfulness of their stories.[108]

In retrospect, one can only appreciate the irony in this late-eighteenth-century judicial leniency towards the new partisan role that men of science took as witnesses in the modern adversarial courtroom. This leniency seems to carry the mark of the aloofness of the eighteenth-century judiciary, who dominated the courtroom to such an extent that they could not imagine it otherwise—that a time might come when their judicial powers would no longer suffice to control the play of partisan expertise in the courtroom. They were soon proved wrong. In the early nineteenth century, the tremendous expansion of science and technology into industry and other public sectors quickly established the scientific expert witness as a pivotal figure in the courtroom and turned partisan expert testimony into an acrimonious and persistent thorn in the side of Common Law. We turn to these developments in the second chapter.

2

The Common Liar, the Damned Liar, and the Scientific Expert: The Growing Problem of Expert Testimony

A pitiable specimen is that poor man of science, pilloried up in the witness box, and pelted by the flippant ignorance of his examiner! What a contrast between the diffident caution of the true knowledge, and the bold assurance, the chuckling confidence, the vain-glorious self-satisfaction, and mock triumphant delight of his questioner!

~ "Cornelius O'Dowd upon Men and Women, and Other Things in General," *Blackwood's Edinburgh Magazine*, September 1864

WRITTEN IN THE EARLY 1720S and published post-humously in 1754, Lord Gilbert's seminal work *Law of Evidence* stated that "the first, therefore, and most signal Rule in Relation to Evidence is this—that a Man must produce the utmost Evidence that the nature of the Fact is capable of. For the Design of the Law is to come to legal Demonstration in Matters of Right, and there can be no Demonstration of a Fact without the Best Evidence that the nature of the thing is capable of."[1] By the end of the eighteenth century, it was clear to the legal profession that in a growing number of cases the "Best Evidence that the nature of the thing is capable of" could be produced by science and science alone. Indeed, the 1782 contest of expertise over the implications of the tide for Wells Harbor, and Lord Mansfield's subsequent decision to allow opinion testimonies by men of science in the courtroom, were signs of the time—the industrial revolution, which brought about a rising tide of legal cases involving technological and scientific arguments. Thus, during the early nineteenth century, among the crowd of experts who were allowed into the witness stand we can already find an increasing number of men of science, including chemists, microscopists, geologists, engineers, mechanists, and so forth. These experts untangled for the court and the jury the complexities of the growing number of cases involving science. They appraised the disputed claims with their experimental techniques and offered their knowledge of the principles of nature, which the jurors then could apply to the facts before them.

In 1795, the fourth edition of Gilbert's *Law of Evidence* became the first legal text on evidence to devote a distinct discussion to the matter of expert testimony. "The proof from the Attestation of Persons on their Personal Knowledge, we may properly, with the French Lawyers, call proof by Experts," wrote the editor, an English barrister, Capel Lofft, in a new section he added and titled "Of proof by Experts." "In proportion as Experience and Science advances, the uncertainty and danger from this kind of proof diminishes," he reasoned. Still, he cautioned his readers:

> Formerly, when the Mother of an illegitimate Child was indicted for the Murder of it, if the Lunges, being immersed in Water would float, it was held Proof on which Surgeons might justify an opinion that the Child was born alive; the inflated lunges, in consequences of the Air which had been drawn into them having being rendered specifically lighter than Water. But this presumption is now held insufficient, for that the Air included in the vesicles of the Lunges from other causes may be adequate to the production of this effect in the lunges of a still-born Child.[2]

Lofft's moral was clear. Caution must be exercised with this rising genus of witnesses who were given exceptional license to give their personal opinions in court. Alas, vigilance was not the order of the day, and scientific witnesses were soon generally allowed, on the strength of their knowledge of the principles of nature, to give their opinions even on facts they had no personal knowledge of. "Though witnesses can in general speak only as to *facts*, yet in questions of science, persons versed in the subject, may deliver their *opinions* upon oath, on the case proved by other witnesses," explained Serjeant Thomas Peak in his 1801 *Compendium of Law on Evidence*. "For though not a particular fact, it is still general information which the rest of mankind stand in need of, to enable them to form an accurate judgment on the subject of dispute."[3]

The growing judicial recognition of their status as a special class of witnesses was not all good news for men of science. It may have underlined their critical importance for the judicial process, but at the same time it also perpetuated their marginalization within this process. In moving across professional and institutional boundaries, from the exclusivity of their lecture theaters, workshops, laboratories, and societies to the public courtroom, men of science hoped to present there laws that were not controlled by human whim. Instead, they found themselves manipulated as mere tools in the hands of the lawyers. As part of the jury or as advisors to the court, they were independent and active participants in the legal decision-making process. As witnesses, they found themselves isolated in the witness box, away from the decision-making processes. Browbeaten and set against each other, they found that their standard strategies for generating credibility and agreement were unsuitable for the adversarial heat of the courtroom. The result was a continuous parade of leading men of science zealously contradicting each other from the witness stand, a parade that cast serious doubts on their integrity and on their science in the eyes of the legal profession and the public. Thus, while the volume of expert testimony was constantly increasing throughout the nineteenth century, the respect paid to it by the courts and the public was constantly diminishing. In 1782, Lord Mansfield resolved that, "in matters of science, the reasoning of men of science can only be answered by men of science." Eighty years later, by 1862, the judicial conviction was already that scientific testimony, "which ought to be the most decisive and convincing of them all, is of all the most suspicious and unsatisfactory."[4] The distance traveled by science and law in their relations, from late-eighteenth-century optimism to mid-nineteenth-century disenchantment and mistrust, is the focus of this chapter.

Insurance Litigation: *Severn, King and Company v. Imperial Insurance Company*

An early example of the importance of scientific testimony in the courtroom and the difficulties it created was shown in an important series of insurance cases that followed a spectacular fire on November 10, 1819, which devastated one of the largest sugar factories in London, owned by Severn, King and Company.[5]

The early decades of the nineteenth century were sweet times for the British cane sugar industry. The demand for sugar was constantly rising, and the sugar industry found itself in a growing and competitive market in which each manufacturer looked for opportunities to improve his methods of production. The processes of sugar refining required heating the sugar solutions to about 250°F. The traditional techniques involved heating the sugar solutions in large pans over an open fire. This method was inconvenient, inefficient, and dangerous. If the fire was not brisk enough the sugar boiled too slowly. If the fire was too strong the bottom part of the sugar spoiled. Constant care was needed to prevent the highly combustible solutions from boiling over onto the dry wooden floors already saturated with sugar. Even if not spilled, the boiling solutions were still capable of producing flammable gas that was apt to explode. The dangers of fire were therefore constant. At least 20 sugar factories had burned down within the first 20 years of the nineteenth century.[6]

During the late 1810s, several attempts to improve the processes of sugar refining had been tried. There was a clear need to avoid the open fire. Some attempted to avoid it by passing steam through the sugar. But because the heat of ordinary steam was not sufficient to boil the sugar, it was necessary to use a high-pressure engine, which was also very dangerous. E. C. Howard, a well-known technologist, patented a process in which the sugar was steamed in a special pan, *in vacuo*. The process was not widely used, however, because it was so expensive to install. Henry Wilson, a civil engineer who had made his name in chemical manufacturing, offered a different solution. In Wilson's process, whale oil was heated in a remote boiler. Once a thermometer ascertained that the oil had reached the desired temperature, a cast-iron pump circulated the oil through copper coils immersed in the sugar pan and back to the boiler to be reheated. Heated to about 350°F, the circulating whale oil kept the sugar solutions around the desired temperature of 250°F. According to Wilson, who took out three patents on his process in 1816, 1817, and 1818, his temperature-controlled process offered a great improvement in efficiency and safety. The process, Wilson argued, eliminated the danger of flammable gases, since sugar solutions were known to emit them only above 350°F and whale oil only above 600°F. In addition, the controlled temperatures kept the sugar from being spoiled by overheating and from boiling over. Even if it did boil over, there was no danger of fire, since there was no open fire under the pan.[7]

Severn, King and Company had licensed Wilson's process and started to use it three months before the fire occurred. After the fire, the four insurance companies that had insured the factory refused to honor their policies on the grounds that Wilson's process, the use of which had not been reported to them, introduced an increased risk of fire that voided the terms of the policies. Severn, King and Company sued to recover their loss, and in the series of trials that followed, the relative difference in safety between the old and the new processes of sugar refining became the main legal bone of contention. Since the only significant difference between these processes was the substitution of hot whale oil for an open fire, an unprecedented crowd of England's finest chemists and chemical technologists was summoned by the litigants to represent them in the £70,000 debate concerning the little-known characteristics of whale oil and its behavior under the frequent and prolonged application of intense heat.

On April 11, 1820, the first trial started in London, in the Court of Common Pleas before Chief Justice Lord Dallas and a special jury. In his opening speech, the chief counsel for the plaintiff, General Serjeant John Singleton Copley, promised the judge and the jury that he would prove to them three things: First, Wilson's process offered great improvements in the safety of sugar refining. Second, for the oil to reach 600°F, when it would start to emit flammable gases, would have taken at least several hours of hard firing, which was not the case according to both eyewitnesses and the timetable of the night of the fire. Third, even if such a temperature had been reached or flammable gases somehow produced in the vessel where the oil was heated, they would have gone up the steam pipe created to discharge the vapor from the water in the heating oil.

In his turn, James Scarlet, the chief counsel for the defendants, the directors of the Imperial Insurance Company, also promised that he would supply the court with unequivocal proof of three things: First, Wilson's process was extremely dangerous because the repeated heating and cooling altered the nature of the whale oil, making it increasingly volatile and apt to explode. Second, the temperature of used whale oil could rise much more quickly than expected, and used oil produced highly flammable gases at much lower temperatures than those specified for new oil. Third, these flammable gases have a greater specific gravity than atmospheric air, thus they would not escape through the venting pipe but rather accumulate close to the ground.

During the first day of the trial, an unprecedented line of celebrated experts stepped onto the witness stand for the plaintiffs and swore under oath that the new process was infinitely "less hazardous" than the old one. During the second day, an equally long and impressive roster of defense experts testified under oath that they were certain that the old process did not have "the slightest danger" and that the new process was "extremely hazardous." Thus, William Allen, F.R.S., a lecturer on chemistry at Guy's Hospital since 1802, stated for the plaintiffs that "he was convinced that if any person had left the whole [of Wilson's] apparatus to its fate, and paid no attention to it, danger would not have occasioned, since the gas, if any were formed, would be gradually carried off by the tube." On the defense side, however, Arthur Aikin, secretary to the Society for the Encouragement of the Arts and the author of the *Dictionary of Chemistry*, who had also written all the chemistry articles for Rees' *Cyclopedia* until 1807, thought that "there must be great danger attending its use," since "there was always the risk that the boiler would give way under the pressure of the volatile [heating] oil. It was a dangerous and unmanageable fluid; and the more frequently it was subjected to the action of fire, the more volatile and inflammatory it became."[8]

Charles Sylvester, chemist, who succeeded Aikin as the contributor of the chemistry articles to Rees' *Cyclopedia* from 1807 onwards, and Thomas Barry, a well-known manufacturing chemist, testified for the plaintiffs that even "if in consequence of any negligence gas was generated in the retort, it would not be sufficient in quantity to do any mischief." On the defense side however, Richard Phillips, a professor of chemistry at the Royal Military College, a lecturer on chemistry at the London Institution, and chairman of the London Chemical Society, testified that "there was great danger of an inflammable gas communicating with the external air." Friedrich Accum, the librarian of the Royal Institution, a lecturer on chemistry at the Surrey Institution, and chief engineer of the Chartered Gas Light and Coke Company, who was reputed to be the founder of the English coal gas industry, swore for the plaintiffs that "all inflammable gases were lighter than common air." The defense, however, delivered Michael Faraday, Sir Humphry Davy's former assistant and the chemical operator at the Royal Institution, who testified that in various experiments on whale oil undertaken especially for the trial, "he found that it emitted vapor denser than the atmosphere."[9]

True to their tradition, most men of science who stepped onto the witness stand made sure that, like Faraday, they also could back up their opinions with results from experiments, many of them performed especially for the trial. The experiments carried out by the plaintiffs' experts were rather simple. They used regular laboratory devices and did not attempt to reproduce either the apparatus or the original conditions of its operation at the burnt factory. The plaintiffs' strategy reflected the empirical nature of early nineteenth-century industrial knowledge, when the use of controlled experiments for the study of technological processes had not yet been firmly established. This strategy also reflected the time-honored practice of expert testimony summarized in the legal maxim *"Cuilibet in sua arte credendum est,"* (every one must be believed in his own art). For centuries, experts had stepped onto the witness stand to inform judges and juries about factual matters that were not part of common knowledge, things such as guild norms, nautical conventions, foreign laws, and witching techniques. If they carried out a demonstration, it was usually designed to illustrate a point or enhance credibility, not to carry the full weight of their argument. The notion of reproducing a given event, artificially, under definite conditions and with controllable parameters, was absent from such demonstrations.[10]

General Serjeant Copley, the leading counsel for the plaintiffs, knew what all lawyers knew—that lay juries could not usually follow elaborate technical arguments and that these only gave the rival counsel an opportunity during cross-examination to deconstruct the testimony of the best experts. The hard lesson learned through the centuries of medical and other expert testimony was that experts should avoid technical explanations as much as possible and let their own credibility be the main support of their testimony. Following this strategy, the plaintiffs' experts functioned in court as if they were character witnesses providing *"bona fide"* testimony for Wilson's process, or as medical experts, who, after taking temperatures concurred that Wilson's process was a highly salubrious treatment for one of sugar refining's chronic maladies.[11]

Thus, William Brande, secretary of the Royal Society, Sir Humphry Davy's successor as professor of chemistry at the Royal Institution and author of the textbook of the day, the *Manual of Chemistry*, reported from the witness stand that he had made some experiments for the plaintiffs and that "he could not produce inflammable gas from oil under a temperature less than 600 degrees." Samuel Parkes, a manufacturing chemist whose much reputed *Chemical Catechism*, written for the education of his daughter, was in its tenth edition by the time of the trial, also reported on his experiments for the plaintiffs and agreed with Brande that 600 degrees "was the lowest temperature at which permanent inflammable gas was produced from oil." John Thomas Cooper, a manufacturer of chemical and laboratory apparatus and a highly regarded lecturer on chemistry at the Russell Institution, also made some experiments for the plaintiffs in the presence of Henry Coxwell, secretary of the Committee of Chemistry in the Society of Arts, and Bryan Donkin, chairman of the Committee of Mechanism of the same Society and vice president of the Institution of Civil Engineers. Once on the witness stand, they all reported that it took them more than two and a half hours to raise the temperature of the oil from 350°F to 600°F and swore that they "had never procured permanent inflammable gas under 610 degrees." All other experts for the plaintiffs who had performed experiments reported the same. They had all measured the temperatures at which new and used whale oil began to emit flammable gas and had found them to be no less than 600°F for new oil and 10 to 15 degrees lower than that for used oil.[12]

Serjeant Copley made sure to back up his scientific luminaries with a multitude of technologists. He called witnesses such as Anthony Robinson and James Harris, sugar refiners whose combined forty years of experience and unblemished reputations qualified them to assert under oath that the old method "required infinite vigilance to prevent the recurrence of danger," while "the present plan required much less vigilance and attention to prevent accidents than the old one." Altogether, the plaintiffs' formidable array of experts presented in court a uniform and authoritative theme, according to which the safety of the process pursued by the plaintiffs was greatly superior to the old-fashioned process they had abandoned. All these experts were examined, cross-examined, and reexamined until 6 P.M., when Chief Justice Dallas, exhausted, declared it was "useless to call any more witnesses," and closed the first day of the trial.[13]

The second day was dedicated to the articulation of the defense case. The leading defense counsel, James Scarlet, perhaps the most successful jury-trial lawyer of his time, was in a difficult position. The burden of proof was on him since it was the defense that averred the increased risk of Wilson's process. However, the process was new, which meant that Scarlet could not present witnesses who could testify from their experience as to its alleged dangers. Instead, Scarlet decided to champion a dynamic theory that, much like that of Mylne's and Smeaton's in 1782, allowed him to rearrange the facts of the case, which on their face pointed the other way. His theory was that the repeated heating and cooling of whale oil eventually caused the production of highly combustible products at relatively low temperatures. His scientific experts, however, were able to offer him no simple explanation, indeed not even a complex one, to sustain his argument. Left with little choice, Scarlet decided to base his defense on the results of a series of experiments especially designed to show the dangers of Wilson's process.

Scarlet's innovative strategy was built around two simultaneous moves: discrediting the contemporary state of chemistry, while emphasizing the utility of experimentation as a reliable guide to true knowledge. If only "the scientific gentlemen" representing the plaintiffs "had conducted the same train of experiments as his witnesses have done," Scarlet told the jury in his opening statement during the morning of the second day of the trial, he had no doubt they would have stated very different opinions. The London Times quoted Scarlet's speech in detail:

It was not by one or two experiments that the questions in chymistry could be decided, for that science underwent continual fluctuation, and its principles required more patient investigation than those of any other science. In his [Scarlet's] younger days what was called the phlogistic theory was very prevalent, in consequence of Dr. Priestley's writings; and the late celebrated Dr. Milner certainly gave very brilliant lectures on the subject which he [Scarlet] attended. He probably should have been also led away by this theory, had he not happened to form an intimacy with that eminent man, the late Mr. Tenant . . . [who] told him not to be led away by this theory, for the French were changing the whole nomenclature of the science, and all this phlogistic theory would in consequence be soon overturned. On hearing this, he had desisted from the study, and the result had verified the prediction of Mr. Tenant. Since that time Sir Humphrey [sic] Davy had worked another revolution in the name of chymical substances, and perhaps the present generation might be fated to witness another change. This variety of classification and frequent substitution of one defective nomenclature for another," Scarlet enlightened the jury, "arose from the propensity of men to generalize from particular appearances. It appears to him that Mr. Wilson was meddling with a subject which he did not understand, when he spoke of the properties of oil."[14]

Scarlet's arrogance epitomized the central position that the legal counsel came to possess in the courtroom. Indeed, by the nineteenth century, the art of manipulating evidence before the jury had become as significant as the evidence itself and was admired or criticized accordingly.[15] Scarlet's novel strategy also anticipated the new position that the scientific expert would come to occupy in the nineteenth-century adversarial courtroom. No longer an impartial figure whose function was to explain the bearing of his specialized knowledge on the facts of the trial, but a quasi-advocate, a vehicle through which the barrister could establish the technical foundation of his case and expose the weaknesses of the other side's case. Thus, John Taylor and John Martineau, chemists and engineers who owned a large sugar factory and held some important patents on sugar manufacturing, designed a series of experiments for the defense with the sole intent of exposing the dangers of Wilson's process.

Samuel Wilkinson, foreman in Taylor and Martineau's factory, who carried out the actual experimentation, described these experiments to the jury in detail. In the first experiment, which "wished to ascertain whether common oil mix with oil previously boiled, would produce inflammable vapor of low temperature," he heated twenty to thirty gallons of oil, one-third used and two-thirds new, in a closed boiler equipped with a 4-foot venting tube. When the mixture reached 100° F, Wilkinson found that vapor began to rise from the tube. At about 280° F, Wilkinson applied a light to the top of the tube and the vapor took fire with sudden gusts and caused an explosion in the boiler. In a second experiment, Wilkinson deposited thirty-three gallons of new whale oil in the boiler and subjected it to heat for twelve days, eleven hours each day, allowing it to cool at night. Each day he measured the temperature at which the oil began to emit flammable vapors. Except for the first and the third days, Wilkinson told the jury, he found each day flammable vapors at a temperature range between 310°F and 500°F. Wilson reported similar results from a third experiment, which was similar to the second, except that it lasted twenty-three days instead of twelve and the boiler was heated for twelve hours rather than eleven.

Dr. John Bostock, F.R.S., a lecturer on chemistry at Guy's Hospital, who had witnessed some of Wilkinson's experiments, described to the jury one in which suddenly "the temperature of the oil was raised in twenty minutes from 360°F to 460°F, at which point it threw out some highly inflammable vapors." The same was done by Alexander Garden, a chemist who had just recently isolated naphthalene from coal; by John

Children, F.R.S., a close associate of Davy, who often did joint experiments with him; by John Taylor and John Martineau; and by others. They had all witnessed Wilkinson's experiments, and each described gusts of fire, combustive vapors, and sudden explosions.[16]

The presiding judge, Chief Justice Lord Dallas, expressed his utter frustration at the conflicting scientific evidence in his charge to the jury:

> They [the jurors] had heard the evidence, he [Dallas] would not say of the most intelligent, but of as intelligent men in chymical and scientific pursuits as were to be found in this country or in Europe. He had himself read the works of some of them, had derived pleasure from their labors, and entertained the greatest respect for their talents and information. But they had, nevertheless, left the Court in a state of utter uncertainty; and the two days during which the results of their experiments had been brought into comparison, were days, not of triumph, but of humiliation to science.[17]

Dallas advised the jury to throw "the contradictory results of experiments" out the window and stated his disgust at the partisanship that had been displayed during the trial. "It must be a matter of general regret," he said, "to find the respectable witnesses to whom he was alluding drawn up, not on one side, and for the maintenance of the same truths, but, as it were, in martial and hostile array against each other." It took the scientific experts two long days to testify to their conclusions. It took the jury only half an hour to find a full verdict for the plaintiffs. Indeed, even in those days juries did not like insurance companies, especially when they attempted to avoid payment.

The defendants moved for a new trial on the ground that the verdict was against the weight of the evidence. In the following Easter term, the judges of the Court of Common Pleas convened and approved the request of the defense for a new trial. The Court's opinion was not registered, but we have a fair indication of its spirit because the new trial was suspended "till one of the other causes should also have been tried, and the result of certain proposed experiments affecting the point in dispute be made known." Informed by their superior, Chief Justice Dallas, about the confusing nature of the evidence, the judges, it seems, were still hoping that with further experimentation, the chemists and their science would be able to clarify the evidentiary mess and offer the jurors a better basis on which to draw an informed conclusion in this important litigation. Such clarification was important because the case involved not only large sums of money but also the general practice and principles by which fire insurance was regulated.[18]

The Second Round: *Severn, King and Company v. Phoenix Insurance*

The stakes were raised for the second trial, this time against the Phoenix Insurance Company. Both parties were well aware that the results of the trial were going to affect not only the first case, but also two other suits that were still waiting in the wings. Sticking to their basic arguments, both parties focused their attention on producing more convincing experimental data. This was especially true of the plaintiffs, who were unsettled by Dallas's acerbic remarks in the first trial, by the successful motion for a new trial, and by the Court's expectation for new and improved experimental data. Keeping their victorious scientific roster, the plaintiffs nevertheless reinforced it with new stars such as Thomas Thomson, professor of chemistry at the University of Glasgow, editor of *Annals of Philosophy*, and author of the comprehensive and influential *System of Chemistry*; Dr. John Davy, F.R.S., Sir Humphry's younger brother; John Dalton, president of the Manchester Literary and Philosophical Society, whose scientific integrity was irreproachable and his work on gases well known (in great part, thanks to Dr. Thomson); and a score of other well-respected men of science, sugar manufacturers, and oil refiners.

This time, instead of their casual experimentation of the first trial, the plaintiffs carried out a careful and elaborate program of experimentation designed to carry the full weight of their argument in court. The testimony of their leading expert, Dr. Thomson, demonstrated their new strategy. The defense had used real-life instead of laboratory apparatus? Thomson's experiments reproduced both the apparatus of Wilson's process and its original conditions of operation. The defendants had heated about thirty gallons of oil? Thomson heated 100 gallons in a boiler identical to the one that was destroyed in the fire with a similar 16-foot venting-pipe. The defendants had heated their oil for twenty-three days? Thomson heated it for forty-one days. "I have tried experiments on whale oil," he testified, "and I have not been able to satisfy myself that it emits gas at so low a temperature as 640 degrees; certainly not lower." Moreover, Thomson told the jury, even if gas were produced, it would still not burn unless it be at least $\frac{1}{12}$ the quantity of

atmospheric air. Thus, even "if the whole of the oil used at Severn and
Co. were to be turned into gas, it would [still] be impossible to produce
combustion, considering the state of the premises." On top of all that,
he added, it was also "impossible that at 360 degrees any vapor could
pass at the mouth of the leaden tube which is 16 feet from the vessel. No
inflammable vapor, even at the heat of 600 degrees, could pass from the
oil vessel through the [16-foot venting] tube. It would become oil and
fall down again before it reached near that height." Thomson's conclusion was clear. "I think that it is impossible," he told the jury, "that any
danger could occur in the vessel if there was a fire twenty miles long
under it." Thomson even explained away the defense's findings. "At
340°F," he told the jury, "oil gives out an aqueous matter—a steam
which condenses at the top of the vessel. From there it falls back into
the oil; and as the water is heavier than oil, it sinks; in its way down it is
expanded again by the heat and makes crackling noises as if the oil were
boiling. There is, however, no danger in this."[19]

Following Thomson's vigorous testimony came a long line of scientific witnesses for the plaintiffs. Stepping onto the witness stand, they
all swore that, "in their judgment, the new process was infinitely safer
than the old," and then harmonized with one or more of Thomson's
arguments. Dr. John Davy knew little about oil and had made no
experiments himself. He nevertheless brought to the witness stand the
authority of the name Davy when he testified that he had witnessed
some of the experiments and concurred, "as far as his personal knowledge of the facts extended, with the answers given by Dr. Thompson."
Other witnesses, however, had made experiments to ascertain the
inflammability of oil, and they all found that it required a degree of heat
far beyond 360 degrees to produce an inflammable gas. Thomas
Brande, for example, told the jury "he once thought that oil boiled at
220 deg., but afterwards found that it was the water escaping and falling
again." Taking some oil, which had been exposed for 29 days to heat
from 400°F to 500°F, and then heating it to 575°F and above, Brande
was able to produce vapor and gas, but the vapor was not inflammable,
and the gas, being carbonic, extinguished the fire. Brande reiterated
Thomson's point that, even if some vapor was created, it would still be
highly unlikely for it to escape the tall pipe without condensing and
falling back. Accum, who like Thomson had experimented with apparatus similar to that of Severn and Co., designed an experiment to prove
this last point. He introduced valves at different heights along the pipe
to check whether vapor or gas had been formed, but condensed

before reaching the orifice. He failed to find any. He also failed to produce any flammable vapor or gas in spite of heating the oil for seventeen days at 360°F.[20]

Samuel Parkes' testimony illustrated the great length to which the plaintiffs' experts had gone in their efforts to render their experimental data credible. Parkes used a third apparatus identical to that destroyed in the fire.[21] The boiler was equipped with two thermometers, and Parkes employed two technicians to test for flammable vapors and keep a detailed logbook of the temperatures "not day by day, but hour by hour." The large apparatus was kept in the plaintiffs' factory, in a chamber fitted with two different locks. Each technician had one of the keys "to prevent any individual from entering by himself." In addition to the technicians, Parkes also devised a flammable vapor detector by placing a lamp in the boiler's chamber with a little piece of tow hanging next to it. If flammable gas escaped when no one was in the room, he explained to the jurors, it would still be detected because the lamp would ignite it, and the tow would be consumed. Parkes kept Wilson's process going for forty-five days at 360°F, during which time he became so confident that no gas was generated, that "on more than one occasion he took out the plug from the vessel itself . . . lighted a large sheet of paper by the lamp and put it into the vessel." Had he been mistaken and "inflammable gases had been produced," Parkes recounted the drama before the jury, "he must have been blown to pieces." The two technicians were called to corroborate Parkes' testimony, and even the oil dealer, James Philips, was produced to testify as to the oil used in the experiments.[22]

The defense's experimental program was not as comprehensive as that of the plaintiffs. Nevertheless, its experts still trusted their experimental findings to carry them through their day in court. They did not use an apparatus similar to the one destroyed in the fire. Instead, Serjeant Scarlet, again chief counsel for the defense and ingenious as ever, attempted to augment the credibility of his experts and their experimental data in a different way. The defense experts, Scarlet told the jury in his opening statement, took Chief Justice Dallas's remarks at the first trial, "that these conflicting opinions were not the triumph of science but its humiliation," very seriously. Consequently, "they endeavored ever since to induce the scientific gentlemen who had been examined on the other side to attend at their experiments; but, for what cause he could not say, this invitation was not attended to. He regretted this exceedingly, but he did not cast blame anywhere." In spite of the plaintiffs' experimental extravaganza, Scarlet told the jury, they should consider "that all the opinions they had heard were formed from experiments on oil used in a particular manner, and heated to a particular degree." He, however, "would clearly establish that this oil was a most unmanageable and uncertain agent, and therefore was so dangerous as to create imminent risk, and a greater one than that against which the plaintiffs insured."[23]

Michael Faraday served as the leading scientific witness for the defense. He testified that having previously been heated twenty-two times, the oil gave off flammable gas at 360°F when heated in a glass retort and at 410°F when heated in a large boiler. It arrived at that temperature in ten minutes and after another ten minutes, the temperature was already 460°F and the oil boiled over. He was obliged, with the help of Taylor and Wilkinson, who had witnessed his experiments, to throw water on it to put out the fire, but even after that, the "ebullition continued for a considerable time." In another experiment, the steam-pipe was filled with vapor in sixteen seconds, and ultimately took fire, and a wet barrel had to be inverted over the apparatus. Describing additional experiments that produced equally blazing results, Faraday told the jury "he felt perfectly confident that oil, used as it was in the process of the plaintiffs, would be rendered volatile and liable to be very rapidly heated."[24]

Faraday thought very little of the plaintiffs' experimental program. As he told the jury, he "did not consider that it was a satisfactory experiment to bring the oil, with great caution to 360 degrees, and then leave it there . . . So heated, it was safe. But left on a large fire at that temperature it was dangerous, from the rapidity with which it heats afterwards." Faraday's experiments were designed to prove this point and to prove it efficiently. Thus, they ignored what Faraday considered to be contingent or irrelevant factors. This, he quickly found out, was a mistake. Legal nature abhors shortcuts. Legal facts are all unique, at least in the sense that they all happened once, and once only. Conclusions about the facts of the case, which are made by the jury, are always contingent. They do not carry precedential powers and have to be decided anew in each and every case. Scientific facts, on the other hand, are always ordinary, at least in the sense that the uniform laws of nature regulate them. Accordingly, the scientific instinct is to recreate them at will and to reduce contingent factors in an effort to develop generalized cases. The legal instinct, however, is to be suspicious of extrapolation from artificially created facts to the original events of the case and tends to demand that the experimental circumstances be as much as possible the same as those in the case at hand.

The plaintiffs' counsels made good use of this antinomy between science and law in their cross-examination of Faraday. Their examination demonstrated how easy it was for a skilled counsel to deconstruct scientific expert testimony. Faraday was made to admit that his experiments had been conducted on a comparatively small-scale apparatus; that the longest venting-pipe he had used was only 4 feet; that thermometers in general were highly unreliable at high temperatures; that since oil was a very bad conductor of heat, thermometers placed in different areas of the boiler may differ significantly; that he did not use a pump in his experiments even though when a pump is in motion the circulated oil becomes lower in temperature; that in his experiments his object was to have the oil boiled as soon as possible and that in doing so he did not observe the original proportion "between the surface of the fire and the quantity of the oil." Pressed, Faraday attempted to make the point that "it was not always necessary in experiments for matters of science to have the apparatus as large as they are in nature, or the ordinary process of art," but his argument, broken up and mocked by the lawyers, lacked sway.[25]

Alexander Tilloch, the editor of the *Philosophical Magazine*, was the only new scientific reinforcement brought in by the defense. He had not carried out his own experiments and only testified to those he had witnessed. The rest of the defense's scientific experts, including Bostock, Aikin, Phillips, Garden, and Children, had little to add to their testimonies from the previous trial. They had all attended Faraday's experiments and concurred with his opinions. Many of them had also conducted their own experiments and reported flammable vapors and sudden dramatic jumps in the temperature of the heated oil. The real danger with the oil, Bostock explained to the jury, was that not much was known on the subject, not enough, at least, "to use the oil in a sugar-house, without having made previous experiments." Aikin reiterated this point. The "oil was the more hazardous as it was not necessarily dangerous," since workers "might be much deceived by it."[26]

Chief Justice Dallas was not as philosophical as Bostock and Aikin. His summation of the evidence was brief. "After the long and patient, I had almost said painful, attention you bestowed upon this case," he told the jury, "I feel it a duty not to detain you one moment longer than is absolutely necessary." His distress with the scientific evidence was unmistakable:

A vast body of evidence had been laid before the jury; medical men, chymical men, eminent men in every department of science, had been examined in the course of the trial; but what was the lamentable result? The jury had heard of opinion opposed to opinion, judgment to judgment, theory to theory, and what was still more extraordinary, they had seen the same experiments producing opposite results. Who should decide this mighty controversy? He [Dallas] professed himself unable to give an opinion. He was not unacquainted with scientific subjects, but the little he knew only convinced him how much was beyond the reach of his knowledge . . . This he [Dallas] would say of science in its present state, that all that belonged to the theory was doubtful, and that all that rested on experiment was new.[27]

This time, it took the jury an hour and forty-five minutes before they gave, again, full verdict for the plaintiffs. "The new process," the verdict read to the decided satisfaction of the courtroom crowd, "is less dangerous than the old," and its introduction did not "require notice to be given to the insurance office." The defendants, the directors of the Phoenix Insurance Company were directed to honor their policy and to compensate the plaintiffs for their costs for the trials. In sending their bills to the court's clerk, the plaintiffs included the expenses incurred in process of the experiments and allowed also for the trouble and loss of time of their chemical experts. Although not listed, these costs must have been considerable. Parkes alone sent a bill for £213, and Thomson, whose experiments were no less expensive, came to London three times and had to hire someone to teach for him at Glasgow. And there was a long line of other scientific experts waiting to be compensated for their expenses and loss of time.[28]

It was not clear, however, what legal basis there was for the compensation of men of science. Expenses incurred in the process of furnishing evidence in matters of fact were always compensated. The current expenses, though, were incurred in the process of furnishing evidence in matters of opinion, not fact, and no precedent existed on how these should be treated. As for loss of time, the general rule was that witnesses were entitled to no compensation for loss of time, unless they were medical men or attorneys, who constituted an exception to the rule. Thus, when the insurance company received these bills, it refused to pay them.

On November 12, 1821, both sides argued their views before the judges of the Court of Common Pleas, which convened to decide the issues of cost allowances. On the first issue—whether the experimental costs should be allowed for—the plaintiffs' counsel argued that since the process of boiling sugar by the application of heated oil was a new discovery, "it was impossible for the most scientific witnesses to speak on the subject, without having had recourse to actual experiment; so that the results of these experiments were a necessary part of the evidence adduced." With regard to the second issue—the allowance for the time and trouble of scientific witnesses—counsel argued that his scientific experts should be counted as professionals like the lawyers and the doctors and thus must be considered to fall within the spirit of the exception. The defense counsel disagreed on both points:

> Witnesses have no allowance for making a view . . . As well might the defendants be charged for the expenses of entering and maintaining a man at the university, in order to render him competent to decide in matters of science. As for these preparatory experiments, if the process was new, the plaintiffs ought to have known the effect of it before they ventured to use it; if they knew the effects, the experiments were superfluous. As for the allowance of time, it is difficult to say why physicians should have such an allowance . . . But, at all events, the exception has extended no further.[29]

Rhetoric aside, the real issue before the court was to decide the legal status of these men of science who dominated the proceedings. In order to be paid for their time in court, these experts had to be recognized as professionals like physicians, who were an upper-class oriented profession, defined by a highly formal and specialized training.[30] The legion of experts produced in the trial could profess none of these traits. Socially, they were faceless, intellectually they had just been proved incoherent, and their religion, while tolerated, was still considered deviant by many. Only a few of them were university educated. Their expertise was not based on any regulated training but rather was self-taught, and many of them earned their living with their hands as chemists or chemical technologists. Thus, speaking for the court, Chief Justice Dallas denied these men of science the legal rights of professional men. Instead, he classified them as "men of skill," a broad legal category that included all other traditional experts—mechanics, navigators, and so forth—who derived their expertise from their own private experience and were not entitled to be compensated for loss of time. The request for compensation for expenses incurred in the process of furnishing the experimental evidence was also denied. "It is true," Dallas stated, "that evidence of persons of skill, is not only admissible but highly desirable, but are they to acquire knowledge by any experiments they may think proper to make, at the costs of the party? I think not."

The court directed its clerk to review the plaintiffs' bills "on the ground that no allowance ought to be made for the expense of experiments, nor for the time of scientific witnesses, unless they were medical men, such as physicians or surgeons." Twenty years earlier, such a scientific gallery would not have gathered in the courtroom. Now, they took central stage and their experiments, prepared especially for the case, became primary irreplaceable evidence. Alas, their conflicting theories and experimental results deeply frustrated the court, and it was not without a happy undertone of score-settling that Chief Justice Dallas refused to recognize them as professionals. It would take chemistry at least another half-century to reverse this decision.[31]

Nuisance Litigation and the Birth of Public Regulation

The entrée of modern men of science into the courtroom, in the Wells Harbor case, was through nuisance litigation. One would have expected that during the first half of the nineteenth century, as the consequences of the Faustian deal England had made with the industrial revolution began to unfold, that nuisance litigation would dominate the business of the courts and men of science would play a major role in protecting the countryside and civilizing the cities. That, however, was decisively not the case. The relative dearth of nuisance litigation, especially that involving industrial pollution, is, in fact, a remarkable feature of nineteenth-century Common Law. Various reasons have been suggested for this: the slowness of nuisance litigation, its high cost, *laissez faire* ideology, the reluctance of royal judges to serve as social engineers, and the legislative protection of industrial interests.[32] Two more reasons will be advanced here: the malleability of scientific evidence and the link that had developed between industrial pollution and prosperity. The terms and significance of this association were clearly articulated in the so-called "Great Copper Trial," held in South Wales in 1833.

The Great Copper Trial

For most of the nineteenth century, the port town of Swansea and a handful of its neighbors along the coast of south Wales smelted virtually all of Britain's, and much of the world's, copper. The smelting industry brought prosperity to the region but at a substantial cost. Copper ores are notoriously impure, and the smelting of the metal produced mountains of slag and clouds of noxious vapors infused with by-products, such as carbonic acid and nitrogen gases from the coal used, and fluoric, arsenious, and sulfurous gases from the ores. Once exposed to the moisture in the air, the gases condensed and descended in destructive clouds of acid upon the once picturesque valley. "On a clear day," according to one eyewitness account, "the smoke of Swansea valley may be seen at a distance of forty or fifty miles and sometimes appears like a dense thunder cloud."[33]

Swansea's copper king was John Henry Vivian, F.R.S., the owner of the giant Hafod works. Heir to a mining family, John Henry was sent in 1803, at the age of 18, to the famous mining institute of the University of Freiburg in the Black Forest. There, he became a pupil of the celebrated German geologist, Abraham Werner, with whom he studied mineralogy, geology, chemistry, and metallurgy. All these subjects were to prove of inestimable value in the decades ahead, during which he turned his family firm into the largest British exporter of finished copper, while still maintaining an active career as a scientific gentleman. Though he was no pure philanthropist, Vivian invested much time and money in efforts to suppress the volume of the smoke from his smelt-works and to eliminate its poisonous components. In 1810, he started to experiment with various mechanical and chemical means of doing this. The poisonous components, he soon learned, could be effectively destroyed in the laboratory, but the cost of the necessary chemical agents rendered any such large-scale application impractical.[34]

In October 1821, a bill of indictment was presented against Vivian at the Court of Quarter Sessions in Swansea, declaring his Hafod works a public nuisance and demanding that they be shut down. The local jury promptly threw the bill out, but it later resurfaced at Cardiff. Although nothing came out of these legal actions, they precipitated decisive action on Vivian's part. He promptly organized a subscription to remunerate any person who might offer a practical way to obviate the injurious effects of the copper smoke. By the beginning of November, the new fund was up and running, offering a £1000 award to the person who could provide a solution to the copper smoke problem. A respected committee was nominated, headed by David Gilbert, vice president of the Royal Society, to adjudicate the contending proposals.[35]

A year passed, and in November 1822, the committee announced that none of the proposals it had reviewed had been found to offer an acceptable solution. The only effort the committee found worthy of commendation was that of Vivian himself, who had carried out in his Hafod works a systematic large-scale program of experimentation. He had built long horizontal flues between the furnaces and the smoke stacks and directed the smoke passing through the flues into various chambers, where it was subjected to experimentation with various agents such as fire, coal, steam, lime, nitrate, and so on. To help him with his experiments and to strengthen their credibility, Vivian hired the services of two of the best experimental minds in the kingdom: Professor Richards Phillips, chairman of the Chemical Society, and Michael Faraday of the Royal Institution. Phillips visited the Hafod works in January 1822 and again in the following summer to advise Vivian in his experiments. Faraday's services seemed to include the chemical analysis of the results of the various experiments.[36]

Vivian presented the prize committee with a detailed account of his experiments, appended with Phillips's and Faraday's report on the results of their experiments. The possibility of further attempts to indict his smelt-works as a public nuisance was clearly on Vivian's mind as he summarized the results of this pioneering collaboration between factory and laboratory:

> If we cannot flatter ourselves that we have absolutely, entirely, and effectually got rid of every particle of matter which has been considered as producing inconvenience, we may at least affirm, that we have abated it to a degree beyond the possibility of its producing, as far as our Works are concerned, future cause for complaint . . . We could not but find our security in the conviction, that a measure so fatal to the best interests of the Town of Swansea and County of Cornwall, as an attempt to stop the Copper Works, would never be really restored to and persisted in; or if it were, in the certainty that of such importance is the manufactory to the country generally, that it would be found necessary to afford it legislative protection.[37]

Vivian had good reason to be worried about legal action against his works. In spite of his attempts to present his experiments as a success, he failed to break new ground. The only solution found applicable on a large scale was passing the smoke through showers of plain water. This eliminated much of the soluble gases, but failed to eliminate the sulfurous acid gas, by far the most abundant and toxic by-product.

The town's corporation seemed to consider the noxious vapors a price well worth paying for their prosperity. In an official resolution following Vivian's report, the Swansea Corporation made known "the high satisfaction they feel at the success that has attended the spirited and scientific exertions of Messrs. Vivian and Sons." Not lagging behind, the *Fund for Obviating the Inconvenience rising from the Smoke produced by Smelting Copper Ores* also made known its satisfaction with "the indefatigable industry" with which Vivian had devoted his time and attention to the subject, and of the full scientific explanation he provided of it. The results of these experiments, David Gilbert, chairman of the fund's prize committee, wrote to Vivian, "must sooner or later moderate public expectation, and induce the nation to tolerate one great source of its wealth, of its power, and of its prosperity; a source from whence tens of thousands of its most industrious members derive their subsistence; at a price of a small and unavoidable inconvenience."[38]

The neighboring farmers took a different view of this "small and unavoidable inconvenience," which was slowly but surely annihilating their properties. In 1832, a group of eleven farmers indicted Vivian's famous Hafod Works for constituting a public nuisance. Knowing that no jury in Glamorgan, Swansea's industrialized county, would indict, let alone convict, the powerful Vivian family and its prosperous copper works, the plaintiffs initiated their action in the neighboring county of Carmarthen. There they succeeded in getting a grand jury finding for an indictment against Vivian and Sons, declaring their works a public nuisance.

The Great Copper Trial was held in the Carmarthen assizes in the spring of 1833. Led by a local barrister named John Evans, one by one the farmers stepped onto the witness stand and recalled better times, "not before the flood, but before the smoke." They described the bad effects that the constant bombardment from the defendant's battery of chimneys had on their deteriorating property; how the land had become barren and the little it produced was destroyed by acid rain; and how the animals, feeding on the poisoned pastures, sickened and died. Their counsel even produced a scientific expert, an experienced metallurgist by the name of Perceval Johnson, who testified to the lethal effect that sulfurous acid gas has on the vegetation.

But even in the supposedly disinterested county of Carmarthen, the Glamorganian farmers and their local barrister stood little chance against the powerful Vivians, who imported from London the illustrious Sir James Scarlet (whom we have already met in the early 1820s insurance series) to represent them in the trial. Sir James built his defense on the economic threat posed by the suit to the prosperity of the region. A verdict of guilty, he warned the jury, would impoverish tens of thousands of souls, and its effects would ripple so far as to jeopardize even the Royal Navy, whose vessels were protected from worms by sheets of copper. And for what? For a bunch of backward farmers who decided to blame the progressive smoke for their poor coalfield soil, instead of improving their crops with better husbandry and saving their animals by a more vigilant herding.[39]

The law of public nuisance required proof that the works were injurious on a large scale. But, as Scarlet reminded the jury, the value of the land and property around Swansea had increased when the copper works came to Swansea. As for the copper smoke itself, not only was it not injurious but, on the contrary, it was a blessing, a shield against cholera and other diseases. Children in copper districts, Scarlet informed the jury, were renowned for their cheerful disposition and "healthful and florid countenance." To this end Scarlet, of course, had his own scientific expert to produce, a Swansea surgeon who swore not only that he had never seen a disease directly attributable to copper smoke, but also that the smoke was an effective antiseptic against ague and malaria.[40]

Scarlet's ingenious rhetoric could not have been effectively disproved, since the effect of the various gases on the lungs was not yet well understood. A decade later, in 1843, testifying before a Parliamentary committee, Michael Faraday was still of the opinion that the discharge of carbon monoxide and dioxide gases was not "so much a question of health as of cleanliness and comfort."[41] Equally important, Scarlet's rhetoric fell on ears eager to be persuaded. After all, choking the copper industry, the county's main source of prosperity, for the sake of cleanliness and comfort, threatened the welfare of many. Thus, when charging the jury, the presiding judge made it clear that the distress of a handful of farmers, who had clearly failed to prove their point, had to be balanced against the well-being of the entire community. To the surprise of no one, the jury gave a verdict for the defense.

Swansea's Great Copper Trial set an early example for the course of nineteenth-century nuisance litigation. It was no defense in the eighteenth century to argue that an interference with someone's property was reasonable within the circumstances, or that it was of public benefit. If it produced interference, the courts treated it as a legal nuisance. Remarkably stable since the Middle Ages, this conservative doctrine mutated during the first half of the nineteenth century into the modern doctrine of reasonable use, which enquired whether those affected by the alleged nuisance ought reasonably to be expected to put up with it. This was a rather flexible standard, which in the rapidly industrialized and urbanized nineteenth-century English society frequently boiled down to the examination of relative values, that is, whether the benefits from the continuation of the nuisance exceeded the costs to the victims of either tolerating it or eliminating its effects.[42]

The heaviest polluter was the rapidly growing chemical industry, especially the so-called alkali industry, which produced sodium carbonate for the soap, glass, textile, and gun-powder industries. The production consisted of a series of connected processes centered around the Le Blanc process. In the first stage of the process, common salt was treated with sulfuric acid to produce sodium sulfate. The chlorine in the salt went up the chimney as hydrogen chloride gas, which interacted with atmospheric moisture and came down as acid mist or rain that destroyed vegetation and corroded metal and stonework. In the second stage of the Le Blanc process, the sodium sulfate was roasted with coal and limestone to produce soda and "alkali waste," which included calcium sulfide, a gelatinous solid that smelled like rotten eggs and yielded, upon exposure to acid in the atmosphere, toxic hydrogen sulfide gas. To these two pollutants, one should add nitrous acid and hydrogen sulfide from the related manufacturing of sulfuric acid, and carbonic acid from the imperfect combustion of the steam boilers.[43]

At mid-century, the alkali industry was already under attack for the pollution of air and water. Thus, when Cox, Score & Co. decided in 1852 to set up their alkali works on the banks of the Severn River just outside Bristol, they turned to Edward Frankland, a professor of chemistry at Owens College in Manchester, for advice on how to minimize their noxious by-products. Frankland provided them with many practical suggestions on to how to minimize pollution. Among these was the advice "to make the surrounding neighborhood believe you have perfect condensation, for without this faith stupid farmers will detect muriatic acid [i.e., hydrogen chloride] in every diseased ear of wheat, or decayed branch of quick thorn, and old fish-women will be quite certain their oysters and mussels are in the last agonies of death from the same cause." Another helpful suggestion was to take the toxic waste in boats and drop it into the Severn.[44]

Frankland's dubious advice testified, among other things, to the growing public awareness of the problems of pollution and to the emerging political pressures on the industry to minimize contamination. Indeed, growing public protests over the problems of smoke, sewage, sanitation, air, and water pollution, and food and drug adulteration had been accumulating since the late 1830s. Edwin Chadwick's 1842 *Report on the Sanitary Conditions of the Labouring Population of Great Britain* furnished the rising protest with focus and detailed documentation. Soon the quality of air, water, and food were all being called into question.[45] Facing growing public pressure, in 1847 the government introduced a smoke-control article in its Town Improvement Clauses Act, which forbade excessive commercial smoke and required the installation of abatement apparatus. Much similar legislation followed in the next twenty-five years, but most of it was permissive in character and thus was adopted only sporadically. The main policy of the legislation was to transfer many of the enforcement duties to the local authorities and to empower them to prosecute violators. Consequently, most of the large industrial towns began to employ smoke inspectors to record violations and report them to the authorities. No reliable monitoring technology was available, however. Instead, in prosecuting offenders, the local authorities had to rely on inexact visual sightings, which, they soon found, had little status in the courts.[46]

Still, local inspection was not totally without fangs. That was demonstrated in 1857 when Spence and Co., the world's largest alum manufacturer, was indicted for constituting a public nuisance and endangering the health of Her Majesty's subjects.

Regina v. Spence

Between 1836 and 1850, Peter Spence, a Scotsman of humble origins, perfected a new process that reduced the notoriously time-consuming process of alum production from a year to a month by treating the calcinated alum shale with sulfuric acid and ammonia instead of the traditional alkaline salts. First, a sulfurous acid gas was produced at the works by burning sulfur in closed chambers together with nitrate of soda. The gas was then conducted into a lead chamber and condensed into sulfuric acid. The condensing process required a certain proportion of nitrous gas, and if there was not enough of it some of the sulfurous acid gas, which smells like rotten eggs, could not condense and escaped. Ammonia was also produced at the works, distilled from "ammoniac liquor," a particularly offensive residual of gas making. The liquor was boiled until decomposed into ammonia and sulfate of ammonia, releasing during the process large quantities of hydrocarbons that carried a variety of disagreeable odors. The ammonia and the sulfuric acid were then combined with the alum shale and kept at boiling temperature for many hours until alum was produced together with two waste products, a small portion of sulfurous acid gas and a larger portion of the highly toxic sulfurated hydrogen gas. These residuals were duly passed along a flue and over the coals of the boiler in order to destroy the sulfurated hydrogen gas. If the fire was properly adjusted and if enough oxygen was present, the whole of the sulfurated hydrogen was supposed to decompose into water and sulfurous acid gas.

In 1846, Spence set up his alum works in Pendleton, near Manchester. By the late 1850s, the works were producing over eighty tons of alum per week and discharging from its high chimney a perpetual white sulfurous smoke that consisted of carbonic acid from the imperfect combustion of the coal boilers, hydrocarbons, and sulfurous acid gas, laced occasionally with other toxic gases such as sulfuric dioxide and hydrogen sulfide. The smoke rolled with the wind and, wherever it struck, injured the vegetation and caused burning eyes and sore throats. Like their Glamorganian predecessors, the residents of Pendleton complained bitterly of the offensive fumes that emanated from the works. Unlike Vivian's works, however, Spence's works were located near an upper-class neighborhood, and the protesters included bankers, lawyers, surgeons, and other middle-class professionals who carried considerable weight with the local politicians.

In 1855, the town council sent its smoke inspector to inspect Spence's works. Although no immediate legal action was taken, Spence began to take the complaints more seriously. In 1856, he erected a new, higher chimney and hired the services of his chemical associates from the Manchester Literary and Philosophical Society, Edward Frankland, John Leigh, Angus Smith, and Frederick Crace Calvert to advise him on possible solutions. However, like Faraday and Phillips before them, these experts offered no practical solutions. The protests continued and finally led to a criminal indictment for public nuisance against Spence's works. Spence was convinced that the alleged nuisance was exaggerated due to the "the sanitary smoke-consuming mania," flared by the anti-smoke sanitarians. His lawyers suggested that the indictment would be better handled by a professional committee chosen by both sides, and requested that the court have the case submitted to such a committee and that Spence be allowed to follow its recommendations at his own expense. The prosecution refused, and the case of *Regina v. Spence* was tried at the South Lancashire assizes in Liverpool in late August 1857.[47]

During the trial, the familiar spectacle of two parties presenting flatly contradictory scientific evidence before the jury unveiled itself once again. The prosecution presented a Mr. Booth, a surgeon, who "had found that the gases evolved at the alum works were injurious to health." The defense, of course, presented its own medical witness, a surgeon named John Southam, who testified to the contrary. However, the effects of the various gases on the lungs was still not well understood, so the main scientific battle was carried on by the chemical experts.

The leading scientific witness for the defense was Dr. Angus Smith of the Manchester Royal Institute, perhaps the best-known sanitary chemist in the kingdom. Smith prepared himself well for the trial. He inspected the premises twice in the two months before the trial and found no trace of sulfurated hydrogen and only small amounts of sulfurous acid, which were far less than what other neighboring works were emitting. He also collected and compared about a hundred samples of rain water from various parts of Manchester and found that the water next to the works contained less than half the amount of sulfurous acid that was present in the other waters. He even tested the smoke from the works with litmus paper and found it to be less acidic than the air in the streets of Manchester. Professor Crace Calvert, John Leigh, a consulting chemist to the Manchester Corporation, and Dr. Burnese, fellow of the Chemical Society and a professor of chemistry in St. Mary's Hospital, London, also appeared for the defense and corroborated Smith's testimony with their own negative results.

The prosecution, however, pulled an ace from its sleeve. It had secured the services of Edward Frankland, Spence's chief consultant, who acknowledged from the witness stand that the works indeed emitted noxious vapors. In Frankland's hands, unlike in Smith's, the litmus paper did turn red, indicating "a large impregnation of sulfurated hydrogen." Frankland told the jury that when performing other tests he had also detected various species of hydrocarbons, carbonic acid, and sulfurated hydrogen as far as half a mile away from the plant. Frankland's testimony was highly controversial. Not only had he performed his tests for the prosecution incognito while still working for Spence but he did not find it necessary to inform the defendant of his findings, assuring Spence instead "that he need not fear any prosecution." His scientific colleagues and his ex-employer perceived his testimony as an act of betrayal. "Dr. Frankland had become a deserter," the defense counsel complained bitterly. "[He] had been induced to join the ranks of the enemy." The position of the defense experts was not beyond criticism either. They were all Spence's employees and his colleagues at the Manchester Literary and Philosophical Society. Smith, in particular, was a personal friend of Spence. Such a complicated web of loyalties was not an uncommon phenomenon in Victorian England and often muddied the legal waters.[48]

The resolution of Spence's trial depended more on legal doctrines than on the adulterated scientific evidence. Indeed, this time it was the prosecutor's turn to remind the jury of the crucial fact that Spence's potential expense for losing his lease and relocating his works "were nothing in comparison with the value of the property of others which had been impaired." Charging the jury, the judge duly reminded them that they had before them "a most important case." The defendant's alum production was "of great importance to the manufacturing interests of the district," but, on the other hand, it was equally important that gentlemen engaged in such production "should not, for their own profit, annoy or injure their neighbors." After deliberating for an hour and three quarters, the jury delivered a guilty verdict regarding nuisance but declared as not proven the charge that the nuisance was injurious to health. Spence's landlord refused to extend his lease, and the Pendleton Works were soon shut down and moved downwind from Manchester.[49]

Regina v. Spence constituted the first successful legal attempt to prosecute the chemical industry. "By bringing the law to your aid," the bitter Spence would argue ten years later, in 1866, before the National Association for the Promotion of Social Sciences, industrial pollution "will not be abated to any great extent till it can be done economically, and you only compel the manufacturer to take up the first patent nostrum that is forced upon his distracted attention, and which as soon as your legal fangs are removed, he throws up in disgust." Spence was right. A few more polluters were prosecuted successfully, but they remained the exception.[50] Judges and legislators remained sympathetic to industrial interests, who were facing increasing foreign competition, and the malleability of the scientific evidence continued to hamper the attempts to prosecute offenders. An unsuccessful and unremarkable nuisance action in 1862, brought in the Stafford assizes by a certain landowner, Mr. Timmins, against the Birmingham and Staffordshire Gas Company for fouling his well with leakage from their gas tank might serve us as an additional example.

The scientific experts recruited by the plaintiffs were Mr. Duncalfe, a surgeon from West Bromwich; Alfred Bird, a practical chemist from Birmingham; and William Mattieu Williams, a professor of chemistry at the Midland Institute of Birmingham. They testified that they had used the most delicate test available to detect the presence of benzol in the well's water. They boiled the water with nitric acid, and the color turned magenta, which was a proof that benzol was in it. The scientific experts for the defense included Dr. Wrightstone, a professor of chemistry from Birmingham, and Dr. Henry Letheby, the chief examiner of gas for London under the Board of Trade. They testified not only that they failed to detect any traces of gas in the plaintiff's water, but that the nitric acid was no test at all for the detection of benzol and that it would not turn magenta or any other color even if there was benzol in the water. The gas company had to explain, however, what sickened the animals drinking the water. To that end, its scientific experts argued that the well was indeed contaminated but with sewage, not gas. They testified that they had tested the water with sulfate of copper and chloride of lime and that these created the characteristic precipitation and discoloration that reveal the presence of phosphates, typical of sewage. The experts for the plaintiff, however, not only testified that they had failed to find any trace of phosphates, but also swore that sulfate of copper and chloride of lime were no test at all for phosphates.[51]

Such cases fed the growing contempt for the performance of science in the courtroom.[52] To illustrate the absurdity of expert testimony, one journal recalled the testimony of an expert, who defended a suspected water supply with the argument that it was not the water that was bad but the things that were in it.[53] By the late 1860s, a general understanding had emerged that it was unrealistic to expect the existing legal machinery to put the brakes on the relentless march of industry. What was needed was a different meeting ground, outside the adversarial courtroom, where a dialogue between science, government, and industry could be developed in order to find better solutions to the problem of pollution. The Adulteration of Food Acts of 1862 and 1870 and the Alkali Acts of 1863 and 1874 articulated the political terms for this new moral economy and established for the first time a regular official mechanism to enforce it. They assigned a leading role for central and local government as regulators of industry and articulated an alternative function for the scientific expert in the regulatory process—a civil servant whose job was to turn the advancements of science into instruments of social policy and to represent its ends in the courtroom.[54]

Terms of Disenchantment: Patent Litigation

The courts may have been unwilling for most of the nineteenth century to recognize men of science as professionals and compensate them accordingly, but there were plenty of others who were ready to pay them for their services in court. This was especially true of the soaring technical industries of the nineteenth century, which were more than happy to pay men of science extravagant amounts to represent them in court in their brawls over patent rights. In 1853, for example, August Hofmann, the German professor of the Royal College of Chemistry, was able to write to his patron, the great Baron Justus Von Liebig, that his court appearances augmented his yearly income by eight to nine thousand pounds. A major figure in the field of organic chemistry and a leading consultant to the chemical industries, Hofmann made this small fortune in patent litigation, which became a most lucrative side-line for many leading men of science. So much so that while writing about the spectrum of Victorian science patronage, the historian W. H. Brock could not help wondering why was it that "on more than one occasion it would seem . . . that all the chemists in London were being shared by aggrieved parties in a patent action."[55]

Patent law required the patentee to make public the invention and to disclose its essence by means of formal specification. In return, the patentee was granted fourteen years of royal protection.[56] The Patent Office rarely assessed the validity of the proposed patents. If the patent were not useful, it was assumed that the market would reject it. If it was not novel or if its specifications were bad, it would be contested in court. The wrangling in court, however, exposed both law and science as poorly equipped to handle the complexities involved. The industrial processes were often ahead of the relevant science and baffled the scientific witnesses. Handsomely paid, these witnesses tried their best to suit their explanations to the needs of their clients. The result was often a divergent and incomprehensible array of scientific evidence.

The principles behind the patent system only served to further highlight the malleability of the scientific evidence. Patent letters were designed to cover as much ground as possible, while disclosing as few details as possible. The specifications of the patents were designed to be accurate enough to satisfy the clerk of the Patent Office but obscure enough to prevent their simple reproduction by the reader. The terms used to describe the patents needed to be vague enough to allow for the coverage of future possibilities, as yet undiscovered, but clear enough to allow for their defense in court and to prove infringement. This environment gave free play to what one historian called "the prodigious capacity of scientific rhetoric to defy common sense and to make simple things utterly unintelligible."[57] The following series of patent cases, which decided the fortunes of the English synthetic dyestuffs industry provides an apt testimony to this capacity.

Simpson and Others v. Wilson and Another

The industrial revolution, convention holds, took off with the mass-production technology developed in the textile industry. A lot of cloth was manufactured in England in the early nineteenth century. Still, its colors were the same boring colors that had been in use for the previous two centuries. Moreover, the main dyes, such as blue from indigo and red from the madder plant, were present in nature in concentrations of only a few percent, and their slow and expensive extraction could not sustain the growing market, which was hungry for cheaper and better colors that could feed the new high-speed printing technology.

Mauve, the first artificial dye, was synthesized from aniline in August 1856 by William Perkin, Hofmann's nineteen-year-old assistant at the Royal College of Chemistry. It was received with much enthusiasm. Distilled from coal tar, an abundant waste product of the coal-gas industry, this majestic bluish-purple artificial dye was inexpensive to make, resisted soaping and light, and adhered strongly to cloth and silk. Coming on the market in 1858, it was an instant huge commercial success, which precipitated a feverish entrepreneurial scramble to realize the vast commercial potential of aniline. In a wild wave of experimentation that knew neither method nor theory, the aniline was subjected to the action of almost every known oxidizing agent in the hope of seeing the magical transformation of that dark-colored sticky base into a majestic color. Quite a few succeeded. Still, to get a look at the pot of gold at the end of the rainbow one had to produce either a new hue or a known one but with a new agent. Then, one had to pray that the chemical reaction would prove commercially feasible.

Hofmann, the godfather of the new coal-tar chemistry, the basis of the synthetic dye industry, took notice of such a new crimson shade while reacting aniline with carbon tetrachloride. Hofmann, however, did not go farther than mentioning it in the proper journals. Not so Émanuel Verguin, a schoolteacher from Lyon, France, who independently happened on the same color around the same time, while reacting aniline with stannic chloride. Recognizing its commercial value,

Verguin secured a patent and sold its rights to the Lyon silk dyers, Renard frères & Franc, who marketed the new dye under the name fuchsin, after the flower fuchsia whose shade it resembled. The vivid crimson color, one of the more stunning new colors that had begun to appear on the market, was an immediate success all over Europe. Introduced into England around the time of the Battle of Magenta, the color, which resembled the shade of blood, came to be known in England as "magenta."[58]

Soon, numerous processes that used other salts and metal oxides to oxidize aniline and produce the popular magenta were also patented. One patent, taken out by a certain Heilman, listed almost two hundred oxidizing chemicals, none of which had made its mark on the market.[59] Edward Nicholson, a former student of Hofmann, discovered yet another way to produce the dye, using arsenic acid. However, when he tried to secure a patent in July 25, 1860, he found that another former student of Hofmann, Henry Medlock, had just patented the same process eight days earlier. "I mix aniline with dry arsenic acid," Medlock specified in his patent letter, and heat it "to or near its boiling point until it assumes a rich purple colour, and then I mix it with boiling water, and allow the mixture to cool; when cold, it is filtered or decanted. The aqueous solution which passes through the filter contains a red coloring matter or dye." Arsenic acid turned out to be the cheapest and most manageable among all patented agents, having the additional advantage of being easily recoverable for reuse. Unable to secure the patent for itself, Nicholson's firm, Simpson, Maule, and Nicholson, the largest manufacturer of dyes in the kingdom, bought the patent from Henry Medlock in October 1860 for the considerable sum of £2000 and vowed to "take proceedings against all persons who employ arsenic acid for the production of 'magenta' dye."[60]

In July 1861, Simpson, Maule, and Nicholson filed a bill in the Chancery Court, requesting it to restrain another large chemical manufacturing company, Wilson and Fletcher, from infringing their patent by using arsenic acid to produce magenta dye.[61] That was only a preliminary legal move. Whether an infringement really existed was considered a question of fact to be decided by a jury in a Common Law court. This practice had long been under dispute. Many thought patent disputes were beyond the comprehension of ordinary tribunals and recommended that a special scientific tribunal should be formed. Noticeable among the critics was James Watt, whose experience in patent disputes in the 1780s had led him to believe that ignorant juries should not decide them. Still, the prevalent opinion among the legal profession was that men of science should remain witnesses rather than judges, and that the lay jury, helped by the judges, could handle patent cases.[62] Following this convention, Lord Chancellor Richard Baron Westbury and Vice-Chancellor Page Wood ordered the magenta dispute to be brought before the court of the Queen's Bench to be resolved as a matter of fact before a jury.

The trial was scheduled to start in June 1862 but the lawyers for Simpson, Maule, and Nicholson argued that the case should not be tried before a common jury and asked for a special jury that would be able to follow the merits of their case. That delayed the trial, which finally took place in December 1862 before Lord Chief-Justice Cockburn and a special jury, and became an immediate chemical cause célèbre.[63]

To appreciate the dynamics of the trial, we need to acquaint ourselves with some of the scientific nomenclature and chemical recipes of the day. Heating together arsenious acid with nitric acid produces arsenic acid, the main chemical agent of the patented process. The action of the heat, it was understood, caused the portion of oxygen to part from the nitric acid and to increase the strength of the arsenious acid until it turned into arsenic acid. A by-product of the process was moisture. Chemists differentiated between three types of arsenic acid. The first was the original moist product. Driving the moisture off by means of heat created a second type of acid, dry to the touch. Still, a certain amount of moisture (11 to 15 percent) remained in the arsenic, and much additional heating was needed before the arsenic would release it, creating in the process a third kind of acid—anhydrous arsenic acid. Lacking a clear theoretical understanding of the chemical processes involved, the tripartite classification of arsenic acid was explained by the empirical distinction between mechanical and chemical combinations with water. The mechanical combination existed in a solution and was easy to break; the chemical combination was much harder to break and generally involved a change of chemical characteristics. Indeed, some chemists held that anhydrous arsenic acid was not a proper acid because of its different type of reactions.[64]

The complex reaction of arsenic acid with aniline was also a long way from being understood in 1862. It was assumed that the acid dehydrogenated the aniline, producing arsenite aniline which, when dissolved in water and boiled down, produced the prized magenta. It was also empirically recognized that arsenite aniline could not be formed without the presence of water, and thus that the arsenic acid could not be used in its anhydrous state to produce the dye. The defense contended that the term "dry arsenic acid" used in the patent specification meant anhydrous arsenic acid and therefore that the specifications were misleading and the patent bad. The plaintiffs argued that the term "dry" was flexible and generally also included the second type of arsenic acid, which was not wet or moist but nevertheless contained 11 to 15 percent of "chemically combined" water, enough to sustain the desired reaction.

The legal battle over the control of the flourishing magenta market in England boiled down, therefore, to the meaning invested in the term "dry" as used in Medlock's patent specifications. The scientific champions for the plaintiffs included August Hofmann, who had close business connections with Simpson, Maule, and Nicholson;[65] Edward Frankland, then a lecturer on chemistry at St. Bartholomew's Hospital in London; and William Odling, F.R.S., secretary of the Chemical Society and later Faraday's successor at the Royal Institution. These three chemical luminaries stepped onto the witness stand and unanimously swore that chemists and chemical technologists applied the term "dry" indifferently to both hydrate and anhydrous bodies. The defense presented an equally respected scientific roster, which included William Allen Miller, F.R.S., the author of the multivolumed, multieditioned *Elements of Chemistry*; Alfred Swaine Taylor, a professor of chemistry at Guy's Hospital; Henry Letheby, the medical officer of health of the City of London; Dugald Campbell, lecturer on chemistry in University College, London; F. A. Manning, a well-known practical chemist; Frederick Crace Calvert of Manchester, an authority on the chemistry of dyes; and Arthur John Phillips whose *Treatise on Chemistry of Metals* was in its third edition by the time of the trial. One by one, these experts stepped onto the witness stand and swore to the contrary. In the practical context of Medlock's process, they told the jury, "dry" meant only anhydrous.[66]

While contemplating the information that dry substances were not necessarily dry, the jurors also learned that wet stuff was not necessarily wet. Not all waters, Frankland explained to the bewildered jury, affect the wetness or dryness of the substance. "It is only the water that is mechanically combined that affects the wetness or dryness." Dry acid as it is sold, Odling reiterated, contains about 12 percent combined water, but is perfectly dry. "A substance may be anhydrous and still not dry," Miller explained to the by now dazed jury, for it could be a liquid free from water. Dry arsenic acid, however, was a solid and was therefore generally understood by chemists to be anhydrous.[67]

The plaintiff's scientific witnesses concurred that dry arsenic acid, which contained around 12 percent of combined water, had been a well-known commercial article for many years. Anhydrous arsenic acid, on the other hand, was a chemical curiosity not readily available commercially. Alas, the scientific witnesses for the defense were equally positive that anhydrous arsenic acid was not only commercially available, but also that every manufacturing chemist could supply it upon receiving an order for dry arsenic acid. Chief Justice Alexander Cockburn, who was very troubled throughout the trial by the fact that dry things could be wet and wet things dry, suggested that this last issue, at least, could be resolved by the simple practical test of ordering dry arsenic from a nearby manufacturing chemist. After all, Cockburn remarked, "it is with this class of person that a patentee is dealing—not with professors of science." He soon learned however, that even this task was far from being simple. Arsenic acid, the scientific luminaries explained, was highly deliquescent and readily absorbed moisture from the atmosphere. Thus, ordering dry arsenic acid and finding some water in it would constitute no proof.[68]

The defense lawyers presented also a second line of defense. Even if the patent was valid, they contended, Wilson and Fletcher still was not guilty of infringement, for it was not using dry arsenic acid in its process but a solution of it. Moreover, the defense was able to show on cross-examination that in practice the patentee too used a solution instead of the prescribed dry arsenic acid. In return, the plaintiffs argued that the usage of a solution instead of dry arsenic acid had no significance. True, everybody used a solution, but that was only because it was easier to mix and because that was how the acid was obtained from the manufacturing chemist. Receiving the solution, it was more efficient to mix it with the aniline and then evaporate the mechanically combined water, than to first evaporate the water, then mix the dry arsenic acid with the aniline and then reheat it. These two processes, Hofmann assured the jury, were chemically equivalent. The dry arsenic acid was the substance that produced the dye, and its chemically combined water was all that was needed for the production of the dye.

The defense had a different story to tell. It contended that the usage of a solution instead of solid arsenic acid constituted a critical difference between their process and the patented one. To that end, Campbell, Miller, Letheby, and Taylor performed experiments whose parameter was the percentage of water in the heated mixture. They all found that a mixture with 10 to15 percent water produced little outcome, while 66 percent water produced the best and most dye. They offered two explanations as to why this was so. According to Campbell and Dr. Miller, the water lowered the mixture's boiling point and thus prevented the destruction of the color. You can get the color without the arsenic acid being in a solution, Campbell told the jury, but unless it was in a solution the process would not be commercially profitable. Too much aniline would be destroyed. Drs. Letheby and Taylor reached the same conclusion, but had a different explanation for it. To get the color, they explained the jury, you need first to get the arsenite of aniline. But with a little water you only get a little of the arsenite, not enough for the conversion of the whole of the aniline into color.

The plaintiffs responded by summoning Dr. Frankland to testify that in his experiments all the aniline was converted to magenta with the use of dry arsenic acid that had but 13 percent water. They also summoned Mr. Hands, a Coventry dyer who had sponsored Medlock's original experiments and was the first to apply it, to testify that using Medlock's process with dry arsenic acid he had successful dyed "upward of 100,000 pounds' weight of silk with colour made under this patent." On cross-examination, he was forced to admit that there was "a certain waste of aniline in the process . . . [but] we made a satisfactory profit, nevertheless."[69]

The trial lasted thirty-four days, during which each side also produced in court various learned chemical treatises that allegedly agreed with their experimental data and theoretical interpretations. When the parties finally rested their cases, Chief Justice Cockburn tried to make sense of the mountains of conflicting evidence for the jurors. First, he instructed, they needed to decide the validity of the patent. That entailed two questions: (1) whether the specifications were accurate or misleading, and (2) if the specifications were proper, whether the process they described was commercially profitable. To answer the first question, Cockburn instructed the jury, a conclusion must be reached on the meaning of the word "dry." Unfortunately, he pointed out, the scientific evidence on this question was so conflicting that laymen were "bewildered, perplexed, and left in despair as to knowing how to decide." To answer the second question, concerning the commercial availability of the patented process, the jurors needed to weigh the practical evidence of Mr. Hands, the manufacturer who said that although there had been considerable waste, yet, on the whole, the process was commercially profitable, against the "theoretical evidence given by gentlemen who had not dealt with the matter commercially," but nevertheless opined that the patented process could not be profitable. If the jurors found the patent valid, then they needed to decide the issue of infringement, that is, whether the patented process and the one used by the defendant were substantially the same. To answer this, they needed again to decide between the conflicting scientific explanations of the processes involved.[70]

Like Chief Justice Dallas before him, Lord Cockburn made clear his frustration with the scientific evidence: "It was very sad, that you [the jury] and I—or, rather, you (for it is you who will have to decide the question), who are laymen, will have to decide such a question without having for our benefit and assistance a unanimity of opinion among the scientific witnesses to whom we naturally resort for information." "He was glad," the *Chemical News* summarized its report of Cockburn's instructions, "the jury, and not he, had to decide the case." The jurors too were not happy to decide the case. Having deliberated for nearly two hours, they declared themselves unable to agree on a verdict and were promptly discharged.[71]

Simpson v. Wilson turned out to be but the opening shot for a long series of legal battles over the fortunes of the British synthetic dyestuffs industry. Throughout the series, which lasted to the end of the 1860s and involved various companies, the confusing nature of the scientific evidence remained unchanged. In one of these trials, for example, the plaintiffs produced Dr. Frankland with the results of a novel and superior scientific procedure, spectroscopic analysis, which proved the identity of the defendants' and the plaintiffs' products. Alas, the defense immediately produced William Crookes, the reputable editor of the *Chemical News*, whose spectra proved the opposite.[72] Such cases, where the court and the jury found themselves again and again in the paradoxical position of being expected to decide on issues about which they knew absolutely nothing except what the clearly partisan scientific experts had told them, appalled the courts. Consequently, from the mid-1860s on, the English courts began to divert patent cases from jury trials to the juryless Chancery Court. All the dyestuffs cases subsequent to *Simpson v. Wilson* were held in Chancery. The legal system learned its lesson. By 1875, Common Law judges were officially granted unfettered discretion in civil actions to order a trial without a jury in any matter requiring scientific evidence that, in their opinion, could not be handled by the credulous jury.[73]

The growing judicial frustration could also be traced in the judicial treatment of the scientific evidence proposed in court. The royal judges were ready to tolerate, to a certain limit, theoretical disagreements among the scientific experts. After all, the courts had always maintained a skeptical view concerning the opinions of experts. Whether a fever that raged in a certain neighborhood was or was not caused by the fumes of a factory in the vicinity was a matter that admitted no demonstration, and it was considered to be speculation admitted as evidence out of necessity. Experimental evidence was a different area, however. It had been considered, at least since Francis Bacon's time, to be among the surest species of evidence. The judges found it therefore exceedingly difficult to accept the fact that similar experiments were constantly producing antithetical results when conducted by opposed experts. Such conflicting experimental results, they believed, reflected the partisanship of the scientific experts who produced them, and since these experts were highly paid for their services, their conduct was perceived as the prostitution of science, of selling its credibility to the higher bidder.

Thus, as the nineteenth century advanced and the legal use of scientific expertise grew exponentially, the court developed a skeptical view not only of the opinion testimony of the scientific experts, but also of their experimental data. Not because nature could lie, but because its torturers could. The legal wars of James Young, the Scottish paraffin magnet, through which he attempted to establish his exclusive rights to the manufacturing of paraffin oils and wax from bituminous coal, could serve us as a measuring stick for these important changes.

The Paraffin Wars

During the first half of the nineteenth century, the growing demand for oil for lubrication and burning from both domestic and new industrial users had pushed up the price of natural oils, particularly of whale oil, and turned inventors to the problem of finding alternative sources of oil. In the early 1830s, the German *Naturphilosoph* and chemist, Karl Von Reichenbach, discovered a curious new mineral oil during his studies of industrial wood-coke.[74] His investigations were designed to yield greater amounts of harder charcoal from the available wood by dry distillation. One of the liquid by-products of the distillation was what came to be known as paraffin oil. Paraffin, it seemed, had the same percentage of carbon and hydrogen as the most luminiferous portion of coal gas. In addition, it was capable of being condensed into a solid-state wax, thereby making a perfectly constructed portable energy source. Finding a way to produce paraffin commercially, Liebig remarked in 1841, would certainly be one of the greatest discoveries of the age.[75]

As with aniline, numerous patents were taken out for the production of paraffin oil. The prevailing opinion was that shales, or schists as they were called, were the best material and produced the largest proportion of paraffin. Other patented processes specified materials such as clay, asphalt, and peat as materials from which paraffin could be extracted. But whatever material was used, none among the many laboring practical and manufacturing chemists had been fortunate enough to find out how to manufacture paraffin in commercial quantities.[76]

In December 1850, James Young, an industrial chemist from Manchester, was the first to patent a practical mode of obtaining paraffin oil by low-temperature pyrolysis. A Scotsman of humble origins, Young came to London in 1837 to serve as Professor Thomas Graham's assistant at University College. However, he soon was attracted to Lancashire's prospering chemical industry. Having worked for local firms there for ten years, Young decided in 1847 to go into business on his own. Following the advice of his good friend, Dr. Lyon Playfair, then a professor of chemistry at the New School of Mines, Young took a lease on a petroleum spring in Derbyshire and began to produce lubrication and lamp oil. Speculating about the origin of the petroleum, Young concluded that it was the result of the natural distillation of bituminous substances by the heat of the earth, condensed through the sandstone formation. Encouraged by his theory, Young commenced a long series of experiments with the slow dry distillation of bituminous substances. This led him, after two years, to the process he patented in 1850.[77]

Young found that cannel coal, the prime raw material of the gas industry, produced the best results. Upon distillation at a high temperature cannel coal yielded highly illuminating gases in considerable quantities. Upon long distillation at a low temperature, it produced paraffin oil. The coal, Young prescribed in his patent specifications, was to be put into a common gas retort that was gradually heated up to "a low red heat," and kept there until all "volatile products" ceased to come off. The vapor was then distilled by a worm pipe that was attached to the retort and was kept at a temperature of about 55°F by a stream of cold water. The distillation process yielded paraffin wax and, following additional filtering, paraffin oil. Both products promised to be of great commercial value. The paraffin oil was ideal for the growing market of cheap and reliable lamps, and the paraffin wax allowed, for the first time, the mass production of candles.[78]

Young found a newly mined product of the Scottish lowlands, known as Boghead coal, to yield particularly high quantities of paraffin oil. He went into partnership with Edward Binney, a lawyer and geologist who served as the president of the Manchester Literary and Philosophical

Society. In February 1851, they began oil production at works in Bathgate, Scotland, where the Boghead coal was mined. Soon though, Young's patent came under attack from an unexpected source. William Gillespie, the owner of Turbane Hill Estate near Bathgate where the Boghead coal was mined, became convinced that he was being exploited. He had given James Russell & Co., a firm of iron masters, the rights to mine coal from his land for the fixed royalty of 13 shillings and 6 pence per ton. Realizing in retrospect the unexpected value of the Boghead coal, Gillespie attempted to exclude it from the lease on the grounds that it was not truly coal but shale. After a set of inconclusive negotiations, Gillespie sued Russell and Co. for mining a mineral not included in their lease. The case, which was heard in 1853 in Edinburgh's Court of Session, constituted a direct attack on Young's patent. If the famous Boghead coal was found not to be a coal, Young's patent would not cover it, and anyone could use it freely, even according to Young's specifications.[79]

The parties assembled a large crowd of scientific men to debate the true identity of the Boghead coal. Mine experts, geologists, chemists, microscopists, and other sorts of men of science referred to the smell of the substance, to its color, streak, structure, luster, fracture, and even sound in an effort to prove its identity. The chemists, who included August Hofmann, Edward Frankland, Thomas Graham, and John Stenhouse, seemed to agree that the great element in coal was carbon. They disagreed, however, on how to measure it. One complication was that the quantity of fixed carbon (that is, what remained after distillation) depended on the quantity of hydrogen that was in the coal—the more hydrogen, the less fixed carbon remained, since the carbon interacted with the hydrogen to produce a permanent gas. But the hydrogen varied from 1 percent in anthracite to 8 percent and more in gas coal. Anthracite had the largest proportions of fixed carbon (90 percent), yet by the strict geological classification it was actually a metamorphic rock and thus excluded from the category of coal. On the other hand, some of the cannels, which were categorized as coal, had as little as 21 percent of fixed carbon, which was less than in some of the shales.[80]

It had been agreed by all scientific factions that coal was of organic origin. The court hoped then that the classification of the Boghead coal could be reduced to the examination of its organic content. This, all agreed, could be ascertained by the microscope. Alas, the microscopic evidence was as contradictory as the rest of the scientific evidence. One set of microscopists swore that there was vegetable tissue in the substance, while the other set was equally sure that there was none.

The Lord President of the Court of Session, Duncan McNeil, was exasperated. "I do not care what you call it," he reprimanded the experts. "I do not care about theories of the formation of coal—I do not care about what chemists choose to call it." Whatever it was, he reasoned, Gillespie agreed to lease it. Thus, the verdict was given for the defendant.[81]

Gillespie moved for a new trial on the grounds that the contradictions and discrepancies in the scientific evidence "could only be accounted for on the hypothesis that the specimens submitted to the examination had not all of them being genuine." Since the date of the trial, his counsels informed the court, the German tax authority had investigated the same problem to see if the duty on coal ought to be paid also on the imported Boghead coal, and the German scientific investigators had decided that it was not coal. Thus, argued the counsel, it was essential to the justice of the case that a new trial should be allowed and that the court should order suitable specimens of the mineral to be provided for a new investigation.[82]

A similar situation had occurred thirty years earlier, in 1820, in the insurance case of *Severn, King and Company v. Imperial Insurance Co.*, where the judges of the Court of Common Pleas agreed to postpone litigation until further experimentation would be able to clarify the science involved.[83] By the 1850s, however, the judiciary no longer believed that new experimental investigation would reduce the discrepancies among the scientific witnesses. "They are all agreed upon the theory," mocked the Lord President, "but they all disagreed on what they looked at with their own eyes. I see no ground too for assuming that they are different specimens because parties differ in the result about them. My opinion is that they would differ in the result to the end of time." "Are we sure," iterated a second judge, Lord Rutherford, "that they will be ever agreed? Are we going to get better microscopes and better eyes? Shall this branch of science, not only new in its name, but in its scientific terms, become new in a much more remarkable feature—in the unanimity of its professors? I cannot expect that. I do not anticipate it." Gillespie's request for a new trial was rejected.[84]

Their patent secured and its precious raw material legally declared coal, Young and Binney proceeded to turn their works into one of the largest chemical industries in the world. Still, tempted by the large profits to be made, infringers of Young' patent multiplied, and Young and Binney had to fight a persistent legal battle in order to establish their monopoly over the manufacturing of paraffin oil and wax from bituminous coal. Three of their most celebrated infringement suits were: *Young and Co. v. Hydrocarbon Gas Co.*, which took place in 1854, in the Court of the Queen's Bench before the Chief Justice of England and a special jury; *Binney and Co. v. Clydesdale Chemical Co.*, conducted in 1860 before the Lord President of the Court of Session in Scotland and a Scottish jury; and *Young v. Fernie*, which took place in 1864 in the jury-less Court of Chancery before Vice-Chancellor Stuart. This last case was also the last of Young's legal battles for the control of the paraffin market and the culmination of them all. Seven active counsels represented *Young and Co.* and six the defendant. The trial spanned thirty-four days, during which about two hundred witnesses were examined, including chemists, geologists, various technologists, mining experts, and other men of science. August Hofmann alone, the German professor of the Royal College of Chemistry, who served as the principal expert witness for Young, was kept for three days on the witness stand.[85]

The defense strategy was archetypal. It argued first that the process described by the letters of patent was not new and therefore the patent was bad. It then argued that even if the patent was good, the process it specified was different from the one used by the defendants and therefore that they were not guilty of infringement. The first line of defense consisted of an attack on the originality of the patent. To that end, the chemical experts for the defense, which included a score of practical and manufacturing chemists led by the towering figure of Dr. Alfred Swaine Taylor, testified that neither Young's process, his materials, nor his products were new. His patent letters included nothing that had not been publicly known or used before the date of the patent. His distillation process had been long used in the gas industry for the distillation of gas from coal. His products, paraffin oil and wax, had been obtained by other persons by a similar process from other substances, such as bituminous shale. Thus, Young's patent was nothing but the application of a well-known process to an analogous substance (coal) to obtain the same product (paraffin). Ample legal precedents existed in patent trials, argued the defense counsels, which declared such patents bad. Voluminous scientific literature, some of it rather curious, was read in the courtroom by the defense in order to prove a lack of novelty. A passage from Glauber's 1658 *Opera Chymica* was quoted as revealing the germs of Young's invention. The method of obtaining "oil of bricks" in the *Pharmacopoeia Londinensis* of 1678 and in the works of De Gensanne and Morand, who allegedly distilled coal already in 1763 and 1781, respectively, were quoted as illustrations of the process. D'Alambert's 1792

Encyclopedie, Ure's 1839 *Dictionary of Arts,* Kane's 1841 *Elements of Chemistry,* Henry's papers in *Annals of Philosophy,* quotes from Dumas and Poggendorf, and, of course, much of Reichenbach's work were also brought forward.[86]

The plaintiffs' counterstrategy was to draw a clear distinction between curious scientific discoveries and viable products suitable for commercial purposes. True, the defense argued, Reichenbach had ascertained the existence of paraffin in coal, but he had not shown how this paraffin could be commercially produced. The defense may have unearthed many sources containing or anticipating parts of Young's discovery. Still, knowing the whole was different from knowing its parts. The fact was that none of the experts working in the field, familiar as they were with previous sources, similar processes, and analogous substances, knew how to put them together so as to create a successful process by commercial standards. As for previous precedents forbidding the patenting of new applications of known principles, these related to mechanical principles and did not apply to chemical patents. "Chemical patents depend on the molecular action of matter, and cannot be predicted a priori," the defense argued. "A brush that will brush a coat will brush a waistcoat; but, except for experiment, it cannot be known that nitric acid will have the same effect upon soda that it has upon potash. The law of application to analogous substances of a similar process does not apply to chemical patents."[87]

The defense's second strategy consisted of the attempt to show that the defendants were not guilty of infringement because they had used a different and better process. The dispute on this point boiled down to the question of Young's specifications of temperatures and materials for his process. The plaintiff's experts, which besides the inevitable August Hofmann included also William Odling, John Stenhouse, Sir Robert Kane, the editor of the *Philosophical Magazine,* and Lyon Playfair, who held the prestigious chair of chemistry at Edinburgh University, had all attempted to persuade the judge that the originality of Young's process had to do with its choice of bituminous coal for its raw material and with its careful manipulation of temperatures. No other material produced oil in commercial quantities like cannel coal, the Boghead type in particular. At too low or high a temperature, the process yielded no or very little results. If the temperature in the worm pipe were kept too low, the product of distillation would congeal and stop the pipe. If the retort got too hot, the products of the process were converted into permanent gases.

The scientific experts for the defense labored hard to construct an impressive line of experiments to show that the defendants worked with lower temperatures than those specified in Young's patent, and that these lower temperatures produced more and better-quality paraffin oil. However, the presiding judge, Vice-Chancellor Stuart, totally dismissed the experimental narrative presented to him by the defense on the ground that "experiments conducted for the express purpose of manufacturing evidence for this case are to be looked at with distrust." Instead, Stuart fully adopted the plaintiffs' double distinction between chemical and mechanical laws and between scientific discoveries and practical inventions. "Inventions in mechanics," he wrote in his verdict, "are as widely different from inventions in economical chemistry as the laws and operations of mechanical forces differ from the laws of chemical affinities and the results of analysis and experiment in the comparatively infant science of chemistry, with its boundless field of undiscovered laws and undiscovered substances." Stuart thus gave a full verdict for the plaintiffs, laying down a clear principle upon which such cases should be decided: "Something therefore remained to be ascertained in order to [make] the useful application of this article for economical and commercial purposes. This illustrates the important distinction between discoveries of a merely scientific chemist and of the practical manufacturer who invents the means of producing in abundance, suitable for economical and commercial purposes, that which previously existed [only] as a beautiful item in the cabinets of men of science."[88]

We learn two things from Vice Chancellor Stuart's verdict. First, that in 1863 chemistry was still considered more of an art than a science and its laws more of empiric recipes than universal *a priori* laws. Second, that by 1863, the court had reversed its position concerning the experimental evidence produced in court. Earlier in the century, experiments conducted by men of science "for the express purpose of manufacturing evidence" were considered by the courts to be new and powerful instruments to appraise the disputed facts of the cases before them. In 1824, as we saw, the Court of King's Bench was willing to postpone further litigation until "the results of certain proposed experiments affecting the point in dispute be made known." Thirty years later, in 1854, the Scottish Court of Session saw no ground for assuming such an affect and refused to allow for a new round of experimentation. A decade later, in 1863, even the close-at-hand results of such investigations were "looked at with distrust" and dismissed. This judicial change of mind was the result of the growing frustration of the courts with the constant spectacle of leading scientists contradicting each other from the witness stand. A process that by mid-century brought the legal system to the view that "hardly any weight is to be given to the evidence of what are called scientific witnesses; they come with bias on their minds to support the cause in which they are embarked."[89]

Ironically, the process of disillusionment that the legal system was going through in its relations with science was not the result of some sort of philosophical relativism or skepticism. On the contrary, it was the result of hard-dying positivism that saw in the scientific method the best passage to truth. "All [legal] inquiries into assertions as to matters of fact rest upon the same foundations as assertions about physical science," Sir James Fitzjames Stephen, the leading mid-century authority on evidence, wrote in 1863. "At bottom they rest upon the same great assumptions—the general uniformity of nature and the general trustworthiness of the senses. The logic on which each proceeds is the same." Still, scientific knowledge, according to Justice Stephen, was much closer to certainty. This is because "in physical inquiries the relevant facts are usually established by testimony open to no doubts, because they relate to simple facts which do not affect the passions, which are observed by trained observers who are exposed to detection if they make mistakes, and who could not tell the effect of misrepresentation if they were disposed to be fraudulent." No wonder then that Stephen also interpreted the constant spectacle of leading scientists contradicting each other from the witness stand, not as legitimate debates but as a sign of moral decadence: "As to want of will to speak the truth," he wrote, "the case of experts is as strong a one as can be mentioned. No one expects an expert, except in the rarest possible cases, to be quite candid. Most of them—are all but avowedly advocates, and speak for the side which calls them."[90]

The Public Scandals: Criminal Litigation

The scientific community and the increasingly irritated judges both remained relatively oblivious to the uncertainties of scientific evidence as long as these were limited to civil litigation. But by mid-century, the problems of expert testimony began to surface also in the criminal system, where not just property but life and liberty were at stake. This was especially true for the highly publicized murder trials that introduced the wider public, who had not closely followed the technical details of civil litigation, to the curious spectacle of eminent scientists zealously skirmishing on the witness stand.

The leading theater was by far that of the "mad doctors," that is, the growing crowds of eminent medical experts called into almost every case of atrocious and brutal crime to persuade the jury that certain facts of physical or mental disturbance in the life of the defendant should or should not raise reasonable doubt as to the defendant's sanity. I will not discuss this category of expert testimony here. It has been well investigated. Moreover, many, then and now, have considered it to be only partially scientific and therefore less of a challenge to the credibility or authority of science. Nevertheless, it should be mentioned that the public display of insanity trials frequently verged on legal and scientific chaos and that the protection that science seemed to offer violent criminals contributed significantly to the deterioration of the status of scientific testimony in the eyes of both the general public and the legal profession.[91]

After "mad doctors," the category of scientific evidence in criminal litigation that attracted most attention with its intense confrontation was forensic chemistry. Two notorious poison trials in particular—the 1856 trial of Dr. William Palmer and the 1859 trial of Dr. Smethurst—presented to the stunned public all the ailments of scientific expert testimony. Alarmed by the damaging effects of the scientific partisanship displayed in these trials, the scientific community criticized for the first time in public the false scientific statements made in court by leading scientists; a criticism to which the law eventually reacted with a pardon.

Palmer's Trial

In 1856, Dr. Palmer, a medical practitioner and a racing man, was charged with poisoning with strychnine his intimate acquaintance and gambling buddy, Mr. Cook. He was also indicted for the murder of his wife and brother and was suspected of murdering many other people in the same way and for the same reason—to take over their property and money. Palmer was respectably brought up and was a model of physical health and strength. The fact that such atrocious wickedness was consistent with good breeding and education attracted unprecedented public attention to his trial. The trial was considered by the legal historian, Sir James Stephen, to be one of the greatest trials in the history of English law. Dr. Palmer was to be tried by an Assize Court in Staffordshire but public prejudice was so great that it was felt that Palmer would not have a fair trial there. Thus, a special Act of Parliament was passed to allow the trial to take place at the Old Bailey, the central criminal court in London. Three judges, Lord Chief Justice Campbell and Justices Cresswell and Alderson, were assigned to the trial, and the bar on both sides consisted of some of the greatest advocates of the day.[92]

There had been no trial for poisoning by strychnine before Palmer's, and there were only three known cases of fatal poisoning by strychnine. Therefore, little was known about the symptoms and effects of the poison. That, however, did not stop thirty-nine witnesses of the highest rank in the medical and chemical professions from arguing enthusiastically about whether the symptoms displayed by Cook were typical of strychnine poisoning and whether traces of the strychnine should have been found in Cook's blood. Many animals died in pain during the weeks before the trial as the scientific witnesses armed themselves with an instant last-minute first-hand knowledge on the action of the poison and its symptoms.

The first scientific issue at the trial was on the medical side. One set of doctors for the prosecution testified that the symptoms displayed by the late Mr. Cook were precisely those that showed themselves in the reported cases of poisoning by strychnine. A second set of doctors swore for the defense that Cook's symptoms differed significantly from those reported as being caused by strychnine. Palmer had been seen giving Cook two pills that were supposed to contain the strychnine, but no symptoms were observed until almost two hours later. This was unusual, and quite a few experts swore for the defense that this interval rendered it impossible that the symptoms could have been caused by strychnine. Still, witnesses for the prosecution testified that the coating of the pills may have delayed the action of the poison and that cases were known where it took the poison longer than two hours to make its presence known. Many of the scientific witnesses for the prosecution also testified that they knew of no other cause that could explain Mr. Cook's symptoms except strychnine. The witnesses for the defense, however, testified that these symptoms were consistent with a variety of diagnoses from epilepsy to tetanus to angina pectoris.

The second scientific issue created even more confusion, if that was possible. Strychnine, a free alkaloid, is probably the easiest poison to detect on account of its stability. Putrefaction does not change it, and it had been detected in decomposed or exhumed bodies after eight weeks, and in one reported case, after eleven years.[93] By 1856, various reliable color tests were available for the discovery of as little as one half-millionth of a grain of strychnine in its pure state.[94] Still, the two experts who operated on Cook's body—Dr. Alfred Swaine Taylor, F.R.S., professor of chemistry at Guy's Hospital and the highest authority of his age on medical jurisprudence, and Dr. Rees, professor of *materia medica* at Guy's Hospital—failed to find any strychnine in it. They accounted for the absence of the poison with the theory that the strychnine had been fully absorbed into the blood and diffused in the tissues, where it underwent chemical changes that made it impossible to detect. To support their theory, they performed a few experiments on animals killed by strychnine and satisfied themselves that if given in a small enough quantity (less than half a grain) the poison would not leave a trace. Their theory was backed in court by experts such as Dr. Christison, professor of *materia medica* at Edinburgh, medical adviser to the Scottish Crown, and the highest authority on toxicology in the kingdom, who testified that under the particular circumstances of the case it was reasonable that the strychnine could not be detected.

The defense found itself facing a serious dilemma. It had to attack Taylor's creditability in order to rebut his authoritative conclusion that Cook's symptoms were consistent with strychnine poisoning. Yet, it also had to uphold his creditability as an analytical chemist so that they could take advantage of the fact that he did not find strychnine in Cook's body. The strategy the defense chose was to admit Taylor's skill but deny his good faith. Thus, a second set of experts, led by William Herapeth, professor of chemistry and toxicology at the Bristol Medical School, rejected Taylor's theory that strychnine totally decomposes in the blood as utterly unworthy of credence. If even the minutest quantities were administered, they all testified, the strychnine should be detected, and no degree of putrefaction or fermentation in the human body should so decompose the poison that it would not respond to the unerring tests used for its detection. The fact that Taylor did not find any strychnine, they argued, constituted a proof that there was none. Unwilling to accept that, Taylor had contrived his theory especially for the trial.

The scientific evidence formed the largest part of the evidence of the trial, and its inconsistency and contradictory nature left the public frustrated, since the circumstances of the case were such that all were morally convinced that Palmer was guilty. To make things worse, Herapeth was made to confess on cross-examination that he had again and again bragged among friends that he too thought that Cook was poisoned by strychnine but that Taylor did not know how to find it. The attorney general, Sir Alexander Cockburn, took advantage of the public frustration and constantly inveighed against the "partiality and partisanship" of the scientific witnesses for the defense in general and their leading chemical expert, Herapeth, in particular.

> I have seen that gentleman not merely contenting himself with coming forward, when called upon for the purpose of justice, to state that which he knew as a matter of science or experiment, but I have seen him mixing himself up as a thorough going partisan in this case, advising my learned friend, suggesting question upon question, and that in behalf of a man he believed to be a poisoner by Strychnia. I do not say that alters the fact; but I do say that it induces one to look at the credit of those witnesses with a very great amount of suspicion . . . I abhor the traffic in testimony to which I regret to say men of science sometimes permit themselves to condescend.[95]

Chief Justice Lord Campbell also castigated the defense witnesses "whose object was to procure an acquittal of the prisoner." Instructing the jurors, he reminded them that "it is material, in the due administration of justice, that a witness should not turn into an advocate, any more than an advocate should be turned into a witness. It is for you to say whether some of those who were called on the part of the prisoner did not belong to the category which I described as witnesses becoming advocates." The newspapers that eagerly followed every detail of the trial quoted this unprecedented rhetoric of the attorney general and chief justice widely. Palmer was convicted and with him much of the public image of the scientific community.

Smethurst's Trial

In 1859, Dr. Smethurst, a practicing surgeon, was charged with illegally marrying a second wife and slowly poisoning her for her money with small doses of arsenic and antimony. His trial involved half the doctors of Palmer's trial (only twenty) but twice the scandal. The scenarios of the two trials were similar. First, the doctors debated the meaning of the medical evidence, that is, the symptoms displayed by the deceased, and then they quarreled about the chemical evidence. Nine doctors on the Crown's side attributed the death to the slow administration of some irritant poison and swore that they knew of no natural disease that would account for the symptoms of the deceased. Seven doctors testified for the defense that some of the well-known symptoms of slow poisoning by arsenic or antinomy, or by both, were absent and that the symptoms that were there—obstinate and irrepressible vomiting with violent purging—were not only inconsistent with slow poisoning but similar to those exhibited in cases of acute dysentery made worse by an early stage of pregnancy.[96]

It was, however, the chemical evidence again that ignited the scandal. During the previous coroner's investigation, Drs. Taylor and Odling had testified that they had detected antinomy in the intestine and arsenic in two places—in the evacuation of the deceased and in a bottle that contained a seemingly harmless saline mixture of chlorate of potash intended to help the action of deceased's kidneys. To detect the arsenic, Taylor and Odling had used the so-called Reinsch's test, which consisted of boiling a solution of the suspected material with a fragment of pure copper and hydrochloric acid. If the solution contained arsenic, the copper turned black or grey from the formation of an alloy of copper and arsenicum, which could later be oxidized into crystals of arsenic by drying the copper and heating it in a glass tube. Using a standard copper gauze, they found that the chlorate of potash dissolved the gauze as soon as they put it in. Determined to exhaust the chlorate, they continued to throw in copper gauze until the solution no longer possessed the power to dissolve the copper. Then, they put in another piece of copper and at once received the incriminating results. Dr. Smethurst, Taylor explained in his testimony before the investigating magistrates, had mixed the arsenic with the chlorate on purpose so that the latter would help flush the incriminating ingredient from the body as quickly as possible.

Later, Taylor discovered that he and Odling had been mistaken and that the bottle contained no arsenic. The arsenic, he found, came from the impure copper that he and Odling kept dissolving in the chlorate mixture.[97] Taylor reported the mistake before the trial and also notified the prosecution team. Embarrassing as it was, the mistake was not critical, for Taylor and Odling had also detected arsenic in the evacuation and antinomy in the blood and tissues of the deceased. The real danger was that the defense would capitalize on the mistake and undermine Taylor and Odling's credibility in an attempt to discredit the rest of their findings.

The defense indeed laid much stress on the mistake and asked the jury to dismiss Taylor's and Odling's evidence from their consideration. Anticipating this move, the prosecution called good old William

Brande, the former professor of chemistry at the Royal Institution, who forty years after the insurance case of *Severn, King and Company v. Imperial Insurance Company* was still practicing chemistry, to testify that the mistake was not a sign of incompetence or prejudice. "If I had applied Reinsch's test," the fatherly Brande told the jury, "and the result appeared as it did to Drs. Taylor and Odling, I should have come to the same conclusion, that there was arsenic in the substance. The matter that has happened appeared to a certain extent new to the chemical world. We have always been aware of the presence of very minute quantities of arsenic in copper, but we have never considered it as interfering in any way until this particular case." The presiding Judge, Lord Baron, was persuaded. "The failure of Dr. Taylor's analysis in one instance," he charged the jury, "arose from a new and hitherto unknown fact in science, and did not in any way invalidate his testimony." Thus charged, it took the jury only twenty minutes to find the defendant guilty.[98]

The truth, however, was that the attempt of the defense to undermine the scientific evidence had much weight to it. To start with, Taylor's and Odling's procedure, which included dissolving piece after piece of untested copper, was, to the say the least, highly questionable. Moreover, Taylor's testimony before the investigating magistrates that the copper had been previously tested and found pure was untrue and misleading. Worse still, Brande's cover-up testimony seemed to be intentionally misleading. The knowledge that chlorate of potash rendered Reinsch's test unreliable was far from being "new to the chemical world." It was well known that chlorates, nitrates, arsenates, and other oxidizing agents interfered with Reinsch's process. Thus, when Taylor found the copper dissolved and realized that he had dealt with a chlorate, he should have used another test instead of Reinsch's or prepared the solution by sulfurous acid first. Many in the scientific community were well aware of all this, and as soon as the verdict was given, the protests began to pour in.

Julian Edward Rodgers, the former professor of chemistry at St. George's Hospital in London, who testified for the defense in both Palmer's and Smethurst's trials, was the first to protest in public against the "errors of the greatest importance, and more or less apparent" that had occurred in the trial. "The most prominent of these errors," he wrote less than a week after the verdict was given in a letter to the *London Times*, "is the statement given in evidence by Professor Brande that it is a new fact in chymistry that chlorate of potash dissolves copper."

This statement is untrue and required correction, the more so as the learned judge urged this as a strong point on the consideration of the jury; and again, I repeat not true, for this mixture has long been known as one of the most powerful solvents, actually used to dissolve and separate copper from its ores . . . Again, Reinsch's process was presented as the most efficient known, but it is ill-adapted where blood organs and tissues form the subject of analysis, and is totally inapplicable in all cases unless copper, perfectly free from arsenic, be employed.[99]

Two days latter, in the same newspaper, William Herapeth joined the protest. The mischief done is not limited to the Smethurst trial, he wrote, for Taylor admitted that he had used the same copper for many years. "If the same impure copper has been used for 20 years and evidence given upon it," he asked, "what shall be said of the justice of the convictions and executions which have taken place during those years upon Dr. Taylor's evidence?" Taylor, according to the vengeful Herapeth, was not just careless but also an incompetent experimentalist and, most probably, a falsifier. The biggest discrepancy in Taylor's evidence, Herapeth pointed, was his strange calculations:

Could the copper wire-gauze dissolved by 7 grains of chlorate of potash and its associated hydrochloric acid deposit one grain of arsenic? In the face of all England, I say it could not. The hundredth part of a grain of arsenic in that quantity of copper would render it so brittle that it could not be drawn into wire at all, much less into fine wire fit for gauze. The fact is the whole set of operations were a bungle. Reinsch's process is inapplicable where nitrates or chlorates are present.[100]

The next day it was the turn of the well-respected Dr. Henry Letheby, professor of chemistry and toxicology at London Hospital and the medical officer of health of the City of London, to join the protest. Nothing can be more dangerous to the community, he trumpeted, than such scientific fallacies as were displayed in Smethurst's trial, for they inspired the criminals with confidence and the public with dread. "It is manifest," wrote Letheby, who had also testified for the defense in the Palmer case, "that if there had been but a reasonable share of scientific skill in the management of the [Smethurst's] case, the learned judge would never have been led into the commission of these errors."[101]

The protests continued to pour from both the legal and the scientific communities, many of them directed to Lord Chief Justice Baron and to the Home Office Secretary, Sir Lewis Cornewall. Under the mounting public pressure, Sir Lewis referred the documents of the case to Sir Benjamin Brodie, who had been for many years senior surgeon to St. George's Hospital. Upon Sir Benjamin's acknowledgment "that there is not absolute and complete evidence of Smethurst's guilt," Sir Lewis pardoned the prisoner.[102]

Who Shall Decide Where Experts Disagree?

By the early 1860s, the continuous spectacle of eminent doctors, chemists, geologists, engineers, microscopists, and other men of science contradicting each other on the witness stand and exchanging accusations of moral decadence cast serious doubts on their integrity and on their science in the eyes of the public. An anonymous writer to the *Chemical News* captured this frame of mind in 1860:

> The Palmer case, the Turbane Hill mineral case, the Smethurst case, are instances in which scientific men have been led to exhibit science to the world as utterly unworthy of reliance in such cases. The public had been taught to believe that in judicial investigations the chemists and the microscopists would be able to place the truth before the court in such a manner as to secure justice, and it was a terrible blow to find that the professors were at variance among themselves as to the truth . . . At present, it must be confessed, neither the judge, the jury, nor the public, have any confidence in the scientific evidence in cases of poisoning.[103]

The growing mistrust of scientific evidence that has been traced in this chapter over a wide range of scientific and legal fields resulted in widespread condemnation. "It is a fact," wrote the conservative *Saturday Review* in January 1862, that

> in all matters which require to be investigated through the evidence of expert witnesses, the same remarkable discrepancies show themselves. Hardly a single patent case is ever tried in which men of the highest scientific eminence do not appear to contradict one another flatly on the newness of the invention, or of some of its parts or stages, and the commonest disputes concerning architects' and engineering bills are constantly calling forth similar

conflicts of skilled testimony. Even in criminal cases, where the point to be decided is whether a particular poison was administrated, or whether a death was caused in a particular way, the evidence of the experts is generally more contradictory than would be supposed from the nature of the inquiry; and, in short, judges and lawyers are rapidly coming to the conclusion that skilled testimony, which ought to be the most decisive and convincing of them all, is of all the most suspicious and unsatisfactory."[104]

To suppose that courts can do without such evidence, the *Review* admitted, would be "a stolid and ignorant prejudice, for expert witnesses can supply materials for judgments not to be obtained from any other source." Still, "there is no doubt that a system has been growing up of late years under which men of special knowledge are consulted under such circumstances as to render their opinions almost worthless."[105] Twenty years later, the London *Times* could not agree more:[106] "One of the most unsatisfactory parts of our law is that which relates to the admission of the testimony of 'experts.' It is impossible to shut out such evidence altogether; but there is nothing which brings more discredit upon the administration of justice. There is one consequence of its admission which is common to all cases in which it occurs. It is, that no difficulty has ever been found in obtaining any amount of evidence of this description on either side of any point in issue."

Not all commentators placed the blame on the shoulders of the scientific man. "When one is aware," a well-known columnist sarcastically wrote in the Scottish *Blackwood's Magazine*, that

the "learned Serjeant" knows as much about chemistry as a washer woman does of "wave theory," the display of impromptu learning he makes is positively astounding. Armed with an hour's reading of Beck and Orfila, the great man comes down to court to puzzle, bewilder, and very often to confute men of real ability and acquirement; to hold them up to the world as hopelessly ignorant of all that they had devoted their life to master; and in some cases to exhibit the very science they profess as a mass of crude disjointed facts, from which no inference could be drawn, or safe conclusion derived . . . A pitiable specimen is that poor man of science, pilloried up in the witness box, and pelted by the flippant ignorance of his examiner! What a contrast between the different caution of the true knowledge, and the bold assurance, the chuckling confidence, the vain-glorious self-satisfaction, and mock triumphant delight of his questioner![107]

Still, nobody expected the barrister to be without a shade of partiality. But everybody did expect that from the scientific man, including the scientific man himself. Indeed, impartiality and objectivity were key features in the rhetoric of Victorian science, which promoted the scientific method as the yardstick of truth, or, at least, of certainty, and portrayed the scientific man as the impartial and disinterested keeper of this truth. Thus, the growing public mistrust of scientific knowledge, and even more so of the integrity of the men of science, in such an important cultural domain as the legal system deeply troubled the scientific community. The damage done by the scandals in court, leaders of the scientific community emphasized again and again, was great not only to the administration of justice but also to the public image of the Victorian scientific community, which was toiling hard to achieve professional status and seeking to expand its influence into the public domains of education, industry, health, administration, and culture in general. Consequently, heated debates evolved concerning the growing problems of scientific witnesses for hire; debates that called to the witness stand, for a thorough examination, the epistemological, ethical, and social conventions of the Victorian scientific community. We turn to these debates in the third chapter.

3

Who Shall Decide Where Experts Disagree? The Nineteenth-Century Debates

Though every branch of science has been advancing with sure and rapid strides, it is perhaps not too much to say that from the time of Lord Mansfield, and *Folkes v. Chadd*, to the present; there has been a steady decrease in the credit awarded to the testimony of scientific witnesses.

~Anonymous, "Expert Testimony," *American Law Review* (1870)

L ORD CHIEF JUSTICE DALLAS, who presided over the 1820 insurance case of *Severn and Co. v. Imperial Co.*, considered the conflicting opinions and experimental results of the scientific witnesses to be a "humiliation to science." However, Arthur Aikin, secretary of the Society for the Encouragement of the Arts, thought the situation to be fully representative of "the uncertainties of the scientific knowledge of the day." "I had attended to all which was said by the scientific gentlemen on both sides," Aikin told the jury, "and for the talents and abilities of all of them I had the highest opinion." The same was true of Dr. John Bostock, F.R.S., who also "was not surprised by the [contradictory] results, considering the general uncertainty of all experiments with oil." This difference in interpretation was most telling. What looked to the legal mind like the deplorable results of men of science "drawn up, not on one side, and for the maintenance of the same truths, but, as it were, in martial and hostile array against each other," looked to the scientific mind like a fruitful meeting of minds that constituted "the first step of an investigation which will probably lead to many curious results."[1]

Both sides, to a certain extent, were right. The complex chemistry

of oil decomposition was far from being understood in 1820, and
the series of insurance cases excited a surge of experimentation and
publication that led to an improved understanding of the chemistry
involved.[2] The series, on the other hand, served also as an early example
of the problems that arose when scientific gentlemen moved across
professional and institutional boundaries from their exclusive laboratories and lecture rooms to the public courtroom. All the problems we
saw in the previous chapter became increasingly common as the nineteenth century progressed and the field of scientific testimony was
extended: the embarrassing display of disarrayed definitions, contradictory opinions, and inconsistent experimental results; the double standard of the legal community, which prided itself on its adversarial
ideology, but interpreted scientific disagreements as signs of moral
corruption; the equally ironic partisanship of the scientific experts, who
prided themselves on their collaborative ideology and unprejudiced
search for truth, but performed in court as zealous advocates; the
manipulations by the legal counsels who controlled the proceedings
and conducted the questioning of the experts; the inability of the lay
jury to appraise the conflicting scientific claims.

By the early 1860s, the scientific community was no longer able or
willing to explain the frequent in-court controversies among its members as part of "the uncertainties of the scientific knowledge of the day"
and to brush them off as legitimate disagreements. Three decades earlier, English men of science had sought no government support, saw
very limited employment opportunities, and their leading organization,
the Royal Society, was little more than a fashionable club controlled by
amateur gentlemen of independent means who conceived of science
mainly as a personal calling. They felt, therefore, little need for public
apologetics. By the early 1860s, however, a self-conscious scientific
community had already been forged that had successfully challenged
the intellectual authority of religion and metaphysics. Its speakers were
toiling hard to persuade the public that the progress of science was to
be equated with the progress of civilization and that its method of
critical analysis and empirical verification was a new and superior norm
of truth.[3]

The physical sciences, according to Prince Albert, the great benefac-
tor of Victorian science, are fundamentally different from the moral and
political sciences. The latter involve opinions and feelings, and their
discussion frequently rouses passion. The physical sciences, in contrast,
deal only with facts. "There may for a time exist differences of opinion

on these also," he explained in his famous 1859 presidential address to the British Association for the Advancement of Science,

> but the process of removing them and resolving them into agreement is a different one from that in the moral and political sciences. These [the moral and political sciences] are generally approached by the *deductive* process; but if the reasoning be ever so acute and logically correct, and the point of departure, which may be arbitrarily selected, is disputed, no agreement is possible; whilst we proceed here by the *inductive* process, taking nothing on trust, nothing for granted, but reasoning upwards from the meanest fact established, and making every step sure before going one beyond it, like the engineer in his approaches to a fortress. We thus gain ultimately a roadway, a ladder by which even a child may, almost without knowing it, ascend to the summit of truth and obtain that immensely wide and extensive view which is spread below the feet of the astonished beholder.[4]

That this scientific ladder to the promised summit of truth collapsed time and time again in the courtroom under the weight of the adversarial proceedings, was very troubling to the scientific community and forced it to review its basic epistemological and ethical assumptions. The first part of this chapter follows the fierce debates within the Victorian scientific community, and those between it and the legal profession, about the meaning and the solutions to the growing problem of expert testimony. The debates are situated within the extensive ongoing discussion about the character of the Victorian scientific community and the values by which it judged the work of its members.

Meanwhile, a similar problem had developed across the Atlantic. The sale of expert advice did not become widespread in America until the middle decades of the nineteenth century. Once it did, though, the deployment of expert testimony in American courts of law grew quickly, and with it all its familiar woes. As in England, the growing deployment of men of science in divergent areas of litigation turned the American courts into a important arena for scientific activity, and, as in England, it put on public view the curious spectacle of leading scientists zealously contradicting each other from the witness stand. In contrast to the judges in England, however, the American judges lacked the authority and some of the tools that allowed their royal colleagues to control the usage of expert testimony in their court. Consequently, as the second part of this chapter describes, late nineteenth-century American courts saw the crisis of scientific expert testimony escalating with no resolution in sight.

The Campaign for a Reform of Expert Testimony

The public scandals of scientific expert testimony in the 1856 trial of Dr. William Palmer and the 1859 trial of Dr. Smethurst drove Dr. Henry Letheby to campaign for the reform of the legal system. "The apparent contradiction of science [in the courts]," he wrote in 1859 to the London *Times*, "the seeming uncertainty of its results, and the conflicting testimony of its alumni, are such as to deprive it of that value which it ought to have in the estimation of the public . . . it is clearly time that something should be done to better this condition of things."[5] Letheby, a lecturer on chemistry in London Hospital, was one of the more prolific and technology-minded chemists of his age. His career included serving as the medical officer of health and analyst of food for the City of London and as chief examiner of gas for London under the Board of Trade. These positions had earned him intimate familiarity with the mills of justice. If science is ever going to become a true ally of the law, he concluded, it must be taken out of the hands of the interested parties and employed methodically as a state engine. In 1859, Dr. Letheby brought the matter before the National Association for the Promotion of Social Science [NAPSS], a coalition of social reformers that intended to do to the moral and political sciences what the British Association for the Advancement of Science had done to the mathematical and physical sciences. There, he met Robert Angus Smith, F.R.S., who was also campaigning for the same cause.[6]

We too have met with Smith—a student of Liebig, a coworker of August Hoffman and Lyon Playfair, and one of the leading sanitary chemists in the kingdom—he served as the leading witness for the defense in the 1857 pollution case of *Regina v. Spence*.[7] Smith must have found his experience in that case very disturbing for he never again appeared in the courts of law. Instead, he began to lobby for reform in the legal procedures of expert testimony. He initially brought the subject up before the NAPSS in its first meeting in Birmingham in 1857. Two years later, in 1859, he brought the matter again before the NAPSS at its second meeting at Bradford, where he met Dr. Letheby. The same year, he also brought the subject before the Chemical Section of the British Association for the Advancement of Science at Aberdeen and before the Society for Promoting the Amendment of the Law. Smith's campaign culminated on January 20, 1860, when he was invited to present his views before the Royal Society of Arts, which included among its members many leading members of both the legal and the scientific communities. His address attracted much attention and the Society's journal reported on it and on the tempestuous discussion it triggered in details. The report offers us a unique entry into the mid-Victorian discourse on the sore subject of scientific expert testimony. We shall follow it closely.[8]

The Meeting of the Royal Society of Arts

"All professions and all classes have differences of opinion," Smith opened his address before the members of the Royal Society of Arts, "and none more so than the legal body. Why then do we object to the opposition of physicians, chemists, and engineers? Because it is repugnant to our feelings to see great questions from nature played with, distorted and hidden for selfish purposes."[9] The root of all evils, Smith told his audience, lies in the partisan position men of science have come to occupy in the adversarial courtroom:

> The scientific man in that case simply becomes a barrister who knows science. But this is far removed from the idea of a man of science. He ought to be a student of the exact sciences, who loves whatever nature says, in a most disinterested manner. If we allow him or encourage him to become an advocate, we remove him from his sphere; we destroy the very ideal of his character; we give him duties which he never was intended to perform . . . we lead men who were educated to one mode of thought to act under the auspices of another. We teach them to study impartially and then tell them to practice with partiality. Such a division of the moral nature of man is extremely hurtful, both to the individual and to society.[10]

Worse still, even if the scientific witness attained the moral power to resist the circumstances and temptations that impelled him to turn advocate, he would still find himself unable to communicate his knowledge to the judge and the jury. No matter how carefully he arranged his testimony or how well he rehearsed his words at home with scrupulous attention to exact phraseology—in court, under the pressure of cross-

examination, he would still find "that the great and the small are con-
founded together by the extreme skill of the practiced examiner."
Hence, the first goal of any attempt to reform expert testimony must be
to make the position of men of science in court independent of the
parties to the suit. In addition, the scientific expert should also be
allowed to express his own opinion independently of the advocate.
Only then, the expert could serve the court, "not as a mere tool of a
barrister or his client, but as a student of the exact sciences, as one
whose business it is to weigh and to measure to small fractures, and to
deal with laws which are not controlled by his emotions or influenced
by his promises."

Correcting the form by which the scientific expert gave his evidence,
Smith continued, was but one element of the reform, which, alone,
would not suffice to guarantee the courts clear scientific ground upon
which to base their decisions. What was accepted in court as scientific
evidence and the way it was evaluated in courts also needed reform. As
things stood, the courts often rejected the best class of the evidence men
of science have to offer because it could not be substantiated by direct
observations. However, science, like every other walk of life, possesses
a common stock of intuitive knowledge that is shared by the cultivators
of that specific science; knowledge that has arisen out of the praxis of
their science and often constitutes "the best known part of that
science—that which 'all believe.'" Instead of rejecting such evidence,
Smith suggested, "merely because not proved by the individual observation of the witness, it is to be sought as the best of all."

Alas, Smith moved on, even if such intuitive scientific knowledge
were accepted, neither the judge nor the jury was qualified to evaluate
it. The jury represented the common sense of mankind, that "region of
instinct and of unexplained and unconscious reasoning," in court. But
who, he posed, represented the common sense of science? The only
way that the judge could cope with this task would be "by being made
into a scientific man, or by having a scientific adviser." Smith, however,
had no desire to make the judge into a scientific man, nor the scientific
man into a judge. He was an enthusiastic proponent of science but not
a blind follower. He considered physical science to be "the ultimate
referee" of hard facts but he also recognized that "the use of science in
the State must, like everything else, be regulated by law." Thus, Smith's
final analysis consisted of a three-part reform: giving the scientific men
who represented the parties an independent status in relation to the
barristers; allowing the scientific men to give their evidence in writing,
with examination and cross-examination by the barrister to follow; and

having a scientific assessor on the bench besides the judge, to examine the witnesses and advise the judge. "My belief," Smith concluded, "is that no class of men will so fully agree with each other as the scientific, if not kept separate by the present corrupting system, and no class will spread a more beneficial influence over society if not contemptuously treated by counsel, as they often are, in a witness box."[11]

Smith's close friend and ally Edwin Chadwick, the apostle of governmental science and its application to public health, spoke next. He agreed with Smith's conclusion but not with his premises. What kept the class of scientists from fully agreeing with each other, according to Chadwick, was not the corruption of the legal system, but that of his fellow men of science, who let out their testimony for hire to any side that could pay for it. Chadwick's rich experience with the politics of industrial pollution had left him with harsh things to say about the role of scientific expertise. "A trade in scientific evidence had arisen, in consequence of the increase in the cases affected by it," he told the members of the Royal Society of Arts, "and this nefarious trade required legislative and administrative arrangements for the protection of the public, as well as for the character and position of science." In a maneuver that was probably coordinated with Smith, Chadwick read a long passage from an official report that condemned the habit of scientific witnesses giving contradictory evidence before different parliamentary committees, sometimes during the same session of Parliament. The practice of trading evidence, Chadwick noted, had become so prevalent in Parliamentary committees and in other areas such as the patents system that "true scientists went with extreme pain and reluctance to give evidence, under the apprehension of receiving a taint as participators in it."

Unlike Smith, Chadwick advocated the constitution of scientific tribunals before which "the true scientist might befittingly and confidently appear." As a supplement to such tribunals and, in their absence, for ordinary tribunals, he recommended the French practice of referring cases to officially approved scientists, who would then submit their reports in writing. The proper experts, he suggested, could be found among "those whose qualifications were made manifest by the success in the particular branch or subdivision to which they applied themselves." Such a reform, he stated accusingly, would work well but is blocked by the lawyers, "who made profit of the existing system, and who well knew that the trade . . . could not continue as it was, before a tribunal conversant with the special subject," and thus blocked all reform plans.[12]

Mr. Thomas Webster, F.R.S., a barrister with a considerable Parliamentary practice and a leading authority on patent law who played a central role in many attempts to reform patent law, took it upon himself to respond to Smith's and Chadwick's harsh accusations. The great majority of disagreements among scientific experts, he reminded the audience, was not about evidence of fact but of opinion. Experience had shown that "there were classes of minds, some of which drew their inferences dealing with similarities, and others drew their inferences dealing with dissimilarities. Therefore, upon scientific questions, men of science had the same conflict of testimony resulting from honest differences of opinion as upon any other subject." Thus, although "half-a-dozen scientific men might be found who, if they were paid money enough, would swear to anything," he did not think it proper to blame men of science, as a body, of corruption. He also would not blame the legal system for its corruptive influence, for he knew of no better system to investigate opinion than that of examination, cross-examination and reexamination.

Webster approved of Smith's and Chadwick's suggestion of scientific assessors, but he warned that their function was incompatible with that of the jury. They had assessors in Admiralty cases and in similar cases without juries, and he certainly wanted to see them also in patent cases because "nothing could be more absurd or more unsatisfactory than the present system of submitting questions of Patent Right to the arbitrement of juries." He also agreed with Chadwick that the assessors should not be permanent officers, but experts in a particular vocation. But he resented Smith's and Chadwick's repeated attempts to monopolize expertise by drawing a line between scientific and nonscientific men, and by referring to the students of the exact sciences as the only true representatives of the laws of nature. "A man who had experience in shipping and judging cotton," Webster argued, "a man who acquired a particular kind of knowledge by long training—was just as much a scientific man in his particular art as the man who contributed to those wonderful discoveries of science at which we all so much rejoice." Thus, it was wrong to speak of scientific men as a class to the exclusion of such skilled witnesses.[13]

Sir Thomas Phillips and James Anderson Rose, two well-known lawyers and scientific gentlemen, whose principal practices lay in Parliamentary committees, also stood up for their profession. They both assured the audience that the current system was the best possible and that "it was rare that any miscarriage of justice happened" under it.

Both also emphasized Webster's observation that if they had assessors they must dispense with the jury, because the two would be wholly inconsistent. The unaired implication was clear, at least in criminal trials where the jury was indispensable, assessors were out of the question.

Sir Henry Holland, F.R.S., Queen Victoria's physician and long-time president of the Royal Institution, was clearly annoyed by the lawyers' response. Experience showed, he asserted, that many experts distorted the truth when giving their evidence. "Would any one," he asked, "who was acquainted with scientific evidence in courts of law and especially in parliamentary committees, deny that there was this tendency on the part of the witness? It was a fact that, year by year, men were increasing in numbers, who gained large incomes by being witnesses for whoever would employ them." The great question was not whether the problem existed, but how to guard against it. Webster told them that "there was no real difference between scientific evidence and ordinary evidence; but they all knew that practically there was very little difficulty in distinguishing what was called scientific from ordinary knowledge." Perhaps they could not prevent the selection of quacks as witnesses and the distortion of truth, but at least they should try to check these practices as far as possible. Smith's plan, Sir Henry concurred with Chadwick, "would prevent the growth of witnesses who were paid merely to advocate particular cases," but the lawyers would not approve of it because "it would probably make a difference in the amount of their fees."[14]

Professor Alfred Swaine Taylor asked to speak next. As the most sought after scientific witness in Britain, and one whose performance was frequently at the eye of the storm, he could have easily interpreted much of the discussion as a personal attack on himself. Still, perhaps as a man of both the scientific and the legal worlds, he preferred to try to temper what was clearly turning into an open skirmish between the two fraternities. "There could be no doubt," he agreed with the one side, "that scientific men, in courts of law in this country, occupied a disagreeable position to themselves, and an unsatisfactory position to the public." But, he agreed with the other side, "the differences amongst scientific men were rather those of opinion, than of facts." An important reason for these differences, Taylor added, was that the facts were often laid before the experts in such a manner "that they had not a half even—if they had a quarter—of the truth of the case." It was his experience, "in many trials, both in cases of patents rights and of murder, involving questions of the greatest importance to society, that for the first time, he heard in the court facts which would have materially altered his opinion, so that men of science were entirely at the mercy of those who instructed them." If only for this reason, Taylor agreed with Smith and Chadwick, that having scientific assessors independent of the parties involved in a case would elevate the character of scientific expertise. The scientific witnesses also would probably give their evidence with more caution, and the public would be more satisfied, since they would feel that they had a judge in science as well as in law.[15]

Mr. William Hawes, a veteran member of the society and a well-known philanthropist who was engaged in schemes for the management of hospitals, would have none of these niceties. The main difficulty, he was convinced, was not in the courts but in the scientific witnesses themselves. Men of science, he disagreed with Taylor, were not entirely at the mercy of those who instructed them, and if they would simply investigate every point brought before them "purely as scientific men and not as witnesses in special cases, they would overcome many of the difficulties which now existed." Also, it appeared to him quite impossible that the opinion of assessors would be preferred to that which was given under oath and that the experts had been cross-examined on. They had assessors in the Admiralty Courts, Hawes reminded the audience, but their presence did not prevent expert witnesses from swearing as strongly and contradictorily as in other courts. No other mode of giving evidence was likely to reveal the truth better than that presently in use. True, there were problems, especially since science advanced so rapidly that the best man for one case could be a very bad witness in another, but "they had better put up with such inconveniences as existed rather than run the risk of those great changes which had been suggested that evening."[16]

Finally, the chairman Vice-Chancellor Sir Page Wood, vice-president of the Society, took the stage. Sir Page Wood was one of the more scientifically-minded royal judges, and, as a young man, had translated the *Novum Organum* for Basil Montague's sixteen-volume edition of Bacon. He felt it incumbent on him to protest against the accusations that the legal profession distorted scientific evidence. "It was a great mistake to suppose that a witness had difficulty in making a clear connected statement upon scientific matters in the witness box. A witness, after having been examined by counsel, was always allowed by the judge to do so." If a witness gave his testimony in writing, Wood argued, it would be as if he came "with his opinion cut and dried when he was acquainted with half the facts." As the adversarial system worked, the witness, having heard all the points of the case on the other side, was in a better position to give his final opinion. As for the question of assessors, Wood was ready to concede that "it might be beneficial, in strictly scientific cases, to appoint assessors to sit with the judge," but only as long as the final decision always remained with the judge.

Critical of much of Smith's plan, Sir Page was very careful not to excite the contention anew. Where the scientific gentlemen before him saw incompetence, he saw a situation in which the skills of a man of science and an expert witness were different, and "the legal gentlemen who conducted the case would naturally select the person who would make a good witness." Where the scientific gentlemen saw corruption, he saw a scientific community that like everybody else "might honestly entertain different opinions upon the same facts." These polite formulations, however, did not enter his final analysis. "As a judge," Wood told the members of the society, "he had come to the conclusion that he could set very little value upon evidence of opinion given by witnesses brought forward either for a plaintiff or a defendant, and in many cases he had availed himself to the privilege which was accorded to the judges of the Chancery Court, of calling in disinterested witnesses in matters of opinion." Having said that, he passed a unanimous vote of thanks to Smith for his courageous paper and concluded the meeting.[17]

Sir Page Wood's remarks constituted a proper conclusion to the tempestuous meeting, which brought to light many of the differences and tensions that existed not only between the legal and the scientific communities but also within the scientific community itself. Representing the majority of the scientific community, Smith, Chadwick, and Holland argued that the opposing views of scientific witnesses reflected not real scientific disagreement but the corruption and incompetence of both the legal system and the witnesses themselves. Representing the majority of the legal profession, Wood, Webster, Phillips, and Rose rejected the allegations. They did not think that men of science possessed a special faculty that allowed them to agree with each other better than other mortals. A great proportion of experts' disagreements, they argued, was not about fact but of opinion, on which men of science, like all mortals, had legitimate differences. But whatever the causes alluded to and regardless of whether the experts, the legal mode of their deployment, or simply human nature should be blamed—all of those involved in the discussion concurred that the frequent disagreement among the scientific witnesses was detrimental to both justice and science. The first group, that of men of science, represented it as pathological and demanded a quick cure, while the second group, legal professionals, represented it as a chronic disease that society should learn to live with.

Odling's Critique

The report of the meeting published in the Society's *Journal* precipitated a lively exchange among its readers. Almost all of the reactions shared the widely accepted convention that expert disagreement was detrimental to both justice and science.[18] Only one commentator challenged this view. William Odling, Taylor's deputy both at Smethurst's trial and at Guy's Hospital, was prevented by illness from being present at Smith's talk. As a popular scientific expert in his own right, Odling felt personally attacked and wrote a strong letter to protest against "the calumnies where with the characters of scientific witnesses were then so freely assailed." It was wrong to say, he protested, that scientific witnesses supported any cause for which they could get paid. On the contrary, scientific men were constantly refusing to give evidence in support of opinions they did not entertain. Alas, the public were only aware of their appearances in cases they accepted and knew nothing about the cases they rejected. Odling pointed out that Drs. Graham, Hoffman and Frankland, the authoritative triumvirate of British chemistry, and Mr. Brande and Dr. William Miller who had served as vice presidents of the Royal Society and presidents of the Chemical Society, were some of the busiest scientific witnesses. "These, I suppose," he wrote, "are the men who were branded the other evening as mere trading witnesses engaged in a nefarious traffic; these are the men who receive thousands of pounds to support an opinion by their name and the cunning which others of less name know to be false."[19]

But Odling had much more to offer the debate than mere indignation. The legal lesson acquired by the experience of many centuries, he argued, was not only that "opposition, either as to fact or opinion, does not necessarily imply falsehood, and may be credible to all parties," but also that the truth of a disputed matter "can only be arrived at by the conflict of testimony." These lessons, Odling maintained, were true also of science. Webster and Vice-Chancellor Wood had already pointed out that scientific men could honestly oppose one another on questions of opinion. Odling not only concurred with them but went a step further and argued that scientific witnesses are also "not infrequently opposed to one another on questions of fact, and very proper it is that they should be." He gave an example of a recent patent case where three eminent men of science swore for the patentee and supported with works of authority their claim that a certain chemical reaction could not take place. Nevertheless, desperate to prove otherwise, the scientific witnesses for the defense succeeded after various attempts to devise a method by which the reaction was accomplished. "This illustration," Odling argued, "shows the importance of having a subject investigated by men desirous of establishing different conclusions. Had this case been referred to an independent commission, they would probably have decided it by the mere knowledge of the day and this new reaction with its important consequences would have been altogether overlooked."

Men of science forget, Odling added, that "a court of law is not the arena in which scientific distinctions are to be achieved." It was not the job of the man of science who undertook to give evidence to decide upon the justice of the claims maintained by the side by which he is retained. His duty was merely to testify to the scientific truths and deductions that bore upon his side of the question, and to admit or refute in his cross-examination the position that bore upon the opposite side of the question. Being a scientific assessor was also not the right job for the scientific man, for it was the job of the jury, not the court, to decide which set of scientific men was telling the truth. The job of the court was to balance for the jury the truths brought to bear on one side of the question, by one set of witnesses, against the truths brought to bear on the other side of the question by another set of witnesses. And the judge, Odling claimed, was educated to do that job "to an extent that no scientific man possibly could be." Not only was he skilled in evolving the points of a complex case, no matter how technical, but he was also accustomed to abandoning remorselessly his first impressions upon the establishment of fresh facts, and he was trained to make decisions on the basis of reason, not prejudice.[20]

Eloquent as they were, Odling's arguments failed to persuade many. The notion that the judge is educated to the duties of an assessor to an extent that no scientific man was and that it was not for the scientific witness to decide on the justice of the claims made in court carried little weight with the Victorian scientific community, which claimed a superior capacity to sift evidence and to derive rational conclusions from it. They championed the expertise of the scientifically educated against the incompetent amateur in its attempts to penetrate and dominate the industrial and public domains. The suggestion that adversarial proceedings might result in the production of new scientific knowledge also went against the ideological grain that portrayed the scientific discourse as a neutral domain of knowledge where all could meet and work together for good social ends. The early Victorian scientific community was not interested in adversarial procedures. Science, in their view, was a voluntary enterprise, shared by the blue- and red-blooded, Tories and Whigs, Anglicans and Unitarians, landed aristocrats and industrial manufacturers, university professors and working savants. Often at odds with each other, men of science felt that the study of God's work should transcend their private goals and open the way for common fellowship.

Most radical, perhaps, was Odling's notion that the truth of a disputed matter "can only be arrived at by the conflict of testimony." This extreme position went against the grain of moderate positivism, which won its battles against its metaphysical and theological contenders and provided the best intellectual solvent to clean science of the metaphysical and theological. Although facts belonged to everybody, Prince Albert graciously allowed before the members of the British Association for the Advancement of Science, they still remain "the same facts at all times and under all circumstances." While there may exist for a time differences of opinion about them, eventually "they can be proved, they have to be proved, and when proved, are finally settled."[21]

Thus, almost all scientific commentators seemed to share the view that the disagreements among the scientific witnesses did not reflect uncertainties within the body of scientific knowledge itself. Most of them believed that the disagreements were largely created by the improper procedures by which the legal system processed scientific knowledge. Others found the source of the problem to be an unscrupulous minority of men of science who were ready to sell their testimony to the highest bidder. Both interpretations rejected Odling's high opinion of the adversarial legal system. Even those who were ready to concede that scientific opinions may legitimately differ did not believe that the judge, let alone the jury, could reliably assess and screen such differences. While each commentator had his own idiosyncratic reform proposal, they all seemed to agree on at least one of two central elements—that the central civil and criminal courts should be allowed to call their own independent scientific witnesses and that they should be allowed to appoint, at least in civil cases, scientific assessors who would sit next to the judge and advise him on technical matters as was regularly done in Scotland and, on various occasions, also by the English Court of Chancery and by Parliamentary committees.[22]

The Conflicting Agendas of the Scientific and the Legal Professions

The reform campaigns and the debate among the members of the Royal Society of Arts concerning the proper deployment of science in the courts were triggered by the general feeling by the early 1860s that the crisis had reached intolerable proportions. "The evidence of Experts," William Crookes, the editor of the *Chemical News*, complained in 1862, "is just now the object of general derision:"

> Smart newspaper writers, wishing to indite a telling sarcastic article, select the discrepancies in scientific evidence for a theme; noble Lords, anxious to enliven the dull debates of our hereditary legislators, find nothing so provocative of laughter as a story about the differences of "mad doctors"; and barristers, ready to advocate any opinion, and anxious, perhaps, for a monopoly of the "any-sidedness," when addressing a jury, dilate with a well-simulated indignation on the fact that eminent scientific men are to be found in the witness box on opposite sides. But there is nothing in these differences to excite astonishment. Scientific men are constantly called upon to express opinions on matters which do not admit of demonstration, and about which men may conscientiously come to different conclusions . . . On all such points, men do, will, and may reasonably differ.[23]

The problem with expert testimony, Crookes maintained, rests not with the scientific witnesses but with the legal system that allows the "profoundly ignorant" judge and jury to decide on the scientific evidence. "The fact is," he wrote, "the machinery of ordinary judge and jury for the trial of cases which depend on chemical evidence is simply 'a mockery, a delusion and a snare.' A chemist might just as reasonably sit at Westminster to decide points of law."[24]

Crookes' opinion reflected that of most men of science. The efforts by the scientific community to solve the crisis of expert testimony were directed, therefore, toward amending the legal procedures relating to scientific evidence. In 1860, the British Association for the Advancement of Science, which had established itself as the major voice for the scientific community, nominated a committee headed by Rev. William Vernon Harcourt and the Parliamentary barrister Thomas Webster to consider the subject and produce a comprehensive report that would point out the defects of the system and the manner in which they could best be corrected. The report, the Association hoped, would serve as a basis for a Parliamentary bill for the amendment of the rules of evidence concerning expert testimony. Two years later, at the 1862 meeting of the Association in Cambridge, the committee presented its report before the section on Economics, Science and Statistics, the most political section of the Association, which had always attracted crowds and provoked controversy. The committee castigated the legal system for the very great evil done "both, as regards the reputation of public professions and the proper administration of the law," and recommended getting rid of the jury in civil cases of technical character. It also proposed to create by legislative act a system similar to the Admiralty Court, where the bench would consist only of a judge and up to three skilled assessors, who should give their opinions on the statements of the witnesses. This court, the committee further recommended, would also be allowed to call on witnesses independently of the parties.[25]

The discourse that followed the committee's report should sound familiar by now. The scientific gentlemen, led by Chadwick who chaired the session, and Reverend Vernon Harcourt who presented the report, castigated the unscrupulous experts who tailored their testimony to their employers' needs, and criticized the legal mechanism that brought scientific matters "before judges, who, however conversant with the law, knew nothing about the particular branch of science involved, and counsel, witnesses and jury who were all equally ignorant of the matter." The legal gentlemen, led this time by a former attorney general, Mr. Whiteside, saw various difficulties with the plan suggested and cautioned that "the remedy should not be worse than the disease."[26]

Indeed, even those in the legal profession who empathized with the frustrated scientific community were well aware of the operation of fundamental postulates of the legal system that rendered the reforms proposed by the scientific community unworkable. Getting rid of the jury ran against the fundamental political right to a trial by a jury of one's peers, and the suggestion of allowing the court to call in assessors or witnesses independently of the parties was contrary to two other equally fundamental postulates—the right of the parties and their lawyers to present all evidence and the neutrality of the court. Such suggestions were considered therefore by the legal gentlemen to be "remedies far worse than the disease."

With no resolution in sight, the two camps remained belligerent. The legal profession remained disturbed by the scientific partisanship displayed in the courtroom, while the scientific community remained frustrated by the awkward position it occupied in the court. Still, the deadlock did not stop the increasing deployment of expert testimony in the courts. The growing scope and accuracy of scientific knowledge; the constantly increasing application of its principles in the arts, trades, and business of everyday life; and the ever-increasing subdivision of human pursuits, had inevitably expanded the uses of experts and tended to make the courts more and more dependent on their advice. Still, the increasing tendency of lawyers to fortify their cases with the testimony of experts did not reflect appreciation for their excellence but the requirements of the rising culture of Victorian professionalism. The result was an ironic schism that was clearly emerging during the second half of the nineteenth century—the same increasingly indispensable expert opinions that were treated in everyday life as safe and reliable under the mere good faith of social and business reputation, were considered unsafe and unreliable when given in court under oath.[27]

The Enemies from Within: Scientific Patricians and Technological Plebeians

While almost all scientific commentators agreed that the problems of scientific testimony did not reflect problems within the scientific body of knowledge itself, many of them shared the view that it reflected the deterioration of the moral standards of the scientific community. For these commentators, the problem of expert testimony was but a reflection of the two ominous problems that faced the Victorian scientific community as a whole—the overpowering of the idea of science as a search for truth by the utilitarian passions of the age, and the eroding of the scientific ethical code of gentlemanly voluntarism by the rising new culture of professionalism. These anxieties were not new. The correlation of science and utility had been the target of criticism earlier in the nineteenth century by thinkers such as Samuel Taylor Coleridge and Thomas Carlyle, who condemned the moral and cultural effects of industrialization. Such criticism conceived of the popularity of science as symptomatic of the mechanistic and materialistic values of the middle class, who were interested only in the practical benefits that followed from the control of nature.

This Romantic critique was shared throughout the nineteenth century by other commentators, who referred to the innate tensions between the advancement of science and its promotion. Earlier in the century, these were the gentlemen of science who conceived of science as a personal calling whose end was spiritual betterment. They were afraid that the attempt to popularize science would lead to its plebifi-cation, and that its capacity to enlighten would be overwhelmed by the demand for immediately useful knowledge.[28] The dignity of science, William Herschel warned in 1831, would be degraded if it were only considered as a "mere appendage to and caterer for our pampered appetites." A similar warning was sounded by another leading astrono-mer, Reverend Thomas Romney Robinson, at the 1849 meeting of the British Association for the Advancement of Science:

> To know is not the sole nor even the highest office of the intellect; and [science] loses all its glory unless it acts in furtherance of the great end of man's life. That end is, as both reason and revelation unite in telling us, to acquire the feelings and habits that will lead us to love and seek what is good in all its forms . . . But if it be perverted to minister to any wicked or ignoble purpose—if it even be permitted to take too absolute a hold of the mind, or over-shadow that which should be paramount over all, the perception of Right, the sense of Duty—if it does not increase in us the con-sciousness of an Almighty and All-Beneficent presence—it lowers instead of raising us in the great scale of existence.[29]

As the century advanced, the rapidly growing applications of science served to foster these anxieties. Many were afraid that the progress of science was threatened by what they conceived of as the gradual offset in the balance between research and application, and the easily acquired technical information that seemed to take precedent over careful study and scientific reasoning. "Original research," complained the gentle-men journal *Athenaeum* in 1848, "except in a few rare cases, has been entirely neglected; and in many cases experimental inquiry has been carried on under the prevailing influences of the day." A generation later, Maxwell repeated the warning that "while the number of profes-sors and their emoluments are increasing, while the number of students is increasing, while practical instruction is being introduced and text-books multiplied, while the number and caliber of popular lecturers and popular writers are increasing, original research, the fountain head of a nation's wealth, is decreasing."[30]

The quickly growing species of professional men, who earned their living by directing the perpetual flow of new scientific facts and pro-cesses to fulfill the various wants of society, frustrated the attempts to maintain the traditional scientific code of ethics forged in an era when science was not used as a means of livelihood. The scientific gentleman was supposed to labor for the love of knowledge not for money, and his heart was supposed to be in his researches, oriented toward communal interests rather than toward individual self-interest. No matter how useful the professionals were, if their object in life was to obtain money, they were morally tainted. And among these professionals, none was more repugnant than the so-called scientific expert who had made his living from his appearances in court, tailoring his opinions to the wants of his clients. These accumulating tensions were finally ignited in 1885, and it was our acquaintance, William Odling, who again stood in the midst of the controversy.[31]

Odling's Lecture

On June 13, 1885, the Institute of Chemistry, which had existed and represented the professional needs of its community since 1878, was finally recognized by a royal charter as a professional body ready to be charged with certain public duties and responsibilities and entrusted with the corresponding rights and privileges. A few months later, on the afternoon of November 6, 1885, the members of the newly incorporated institute gathered at the rooms of the Chemical Society in London to celebrate what they considered to be the long-awaited recognition of chemistry as a profession and not just a branch of knowledge. In an address delivered for the occasion, their elected president, William Odling, congratulated the society's members on the formal recognition of their "public utility and importance." He then took the opportunity to disparage those "very superior persons, whose happy mission it is to put the rest of the world to rights," and who held that "there was something derogatory to the man of science in making his science subservient in any way to the requirements of his fellows, and thereby contributory to his own means for the support of himself and those depending on him."[32]

All of us, Odling told his festive audience, are qualified for higher things in life, such as enlightening humanity with our knowledge. Unfortunately, most of us do not have the luxury of jogging leisurely along in life without responsibilities and anxieties, and have a higher duty than enlightening humanity—to feed our families. It is therefore important to realize not only that the need to earn our daily bread is rarely an impediment to scientific achievement but that experience has shown to the contrary, that the fear of poverty and the stern necessity to succeed provide constant stimuli for the cultivation of the moral qualities, the judgment, the determination to succeed, and the self-denial necessary to achieve the highest things.

Odling complained bitterly of the hypocrisy of the scientific community. This community considered it praiseworthy for the man of science to contribute to his means of livelihood by the dreary work of conducting examinations in elementary science for all sorts of examination boards, by teaching elementary science at schools and colleges, by giving popular expositions of science at public institutions, by compiling manuals of elementary science, and by writing popular works for the more general public. But it considered it derogatory, "if not indeed a downright prostitution of his science, that he should contribute to his means of livelihood by making his knowledge subservient to the wants of the departments, corporations, and individuals alike of great and small distinction, standing seriously in need for the special scientific services that he is able to render them."

Odling's message to his professional audience was clear. "While the investigation of nature and the interpretation of natural law are admittedly among the highest as they are among the most delightful of human occupations, the right application of natural law to effect desirable objects is in itself a scarcely less worthy occupation." His related message to the scientific community at large was that the time had come for "the self-engrossing science . . . to be humanized by its association with the cares and wants, and the disappointments of an outside world."[33]

Nature's Response

Reporting on the celebration at the Chemical Society in the scientific journal Nature, Norman Lockyer, the founder and editor of the journal and one of the more vocal champions of Victorian science, felt it his "bounden duty to make a protest" against Odling's address, which "all men of science should read with pain." Lockyer was worried about its effect on the image and progress of science. From Odling's address, he wrote,[34] "It would appear that the life of a chemist should be divided into two periods—seed time and harvest. Research may be the seed, the harvest must be gold. The continued pursuit of truth, the continued love of science for its own sake, may be left to the unwise. The ideal chemist is one who uses research only as an investment . . . We wish it to be known, therefore, that the spirit it breathes is an alien spirit, repugnant to students of pure science in this country." The true English spirit, according to Lockyer, was breathed by the students of pure science, who dedicated their life to research, teaching, and writing books for others. These researchers and teachers were "the highest benefactors of our race and the founders of our modern civilization. The nation remembers them because they forgot themselves." Following these "guardians of the sacred fire," who kept the engine of civilization turning, were the professional men of science, who treated their knowledge as "merely a stock-in-trade to which they look for their livelihood." These were the "Youngs" who made money out of the paraffin discovered by the "Reichenbachs" by turning it into candles. They were semi-parasites who were kept alive, fostered, and made useful for mankind only through the knowledge produced and disseminated by the former class.

In Lockyer's eyes, the appearance of men of science as witnesses in a court of law was the worst example yet of the professionalization of science. Odling's own career was a case in point. Holding the Waynflete professorship of Chemistry at Oxford, Lockyer wrote, Odling neglected his research and teaching duties for his court appearances. The lust for money caused him to forsake "the fair fields of knowledge," and "by thinking only of self and pelf," he was "dragging her reputation through the mud" by becoming "the friend of the manufacturers and a *persona grata* to limited liability companies." Like Smith before him, Lockyer found the role men of science played in court totally incompatible with his idea of science. Turning into mercenaries, their "devotion" to their employers caused men of science to contradict the statements of yet other men of science on the other side, doubtless equally "devoted." Consequently, they developed a highly cultivated faculty of throwing dust in the eyes of the jury by emphasizing certain statements and evading others, "the effect of which was actually worse than lying:"

> It is consoling to think that the qualities most valuable in an expert, since experts there must be, are not those for which men of science are best known. Coolness under cross-examination, verbal dexterity, a ready wit, not too much knowledge or conscience, the fidelity of a partisan or rather "professional devotion," and a dash of impudence, are quite as frequently the passport to the "professional eminence" of an expert as scientific ability. Surely it is not

necessary for us to point out the sophistry and fallacy of the argument that "the right application of natural law to affect desirable objects is in itself a scarcely less worthy occupation" than "the interpretation of natural law," when such applications are made at the instigation of an individual—a client—who pays for such application of natural law at the rate of so many guineas a folio; and who, if it suits him, may then proceed incontinently to suppress the right application of "natural law."[35]

Frankland's and Crookes' Responses

Lockyer's spiteful attack on the professional men of science in general and on Odling in particular precipitated immediate protests from certain corners of the scientific community. In a letter to the *Chemical News*, Edward Frankland, the first president of the Institute of Chemistry and the leading force behind the professionalization of English chemistry, called Lockyer's view of science "flabby sentimentalism;" complained that it was shared by many in the scientific community, especially in the academy; and argued that it was "high time that its hollowness should be exposed." Lockyer, Frankland wrote "waxed grandiloquent over the high dignity of the investigator" whose whole heart is in his researches. "As a matter of fact, however," Frankland reminded his readers, "researchers have been and still are endowed from the national Exchequer to enable them to live whilst making their researches." As for Lockyer's bitter criticism of expert witnesses, Frankland repeated the standard view that the fault "is to be attributed much more to the disgraceful state of the law relating to their evidence," than to the experts themselves.[36]

William Crookes, the editor of the *Chemical News*, joined the complaints about both Lockyer's unfair criticism and the status of the scientific expert in the courts. Lockyer, he argued, mistook the problem's tail for its head. True, the expert witness had to possess coolness under cross-examination, verbal dexterity, and ready wit, but that was only because the legal system treated him as a partisan who was prepared to lie for a fee. The real problem, he echoed the late Angus Smith, was that "the expert occupies a totally anomalous position in court. . . . Technically he is a mere witness; practically he is something between a witness and an advocate, sharing the responsibilities of both, but without the privilege of the latter. He has to instruct counsel before the trial and to prompt during its course. But in cross-examination he is more open to insult because the court does not see clearly how he arrives at his conclusions, and suspects whatever it does not understand."

As things stood, Crookes warned, the spectacle of men of scientific standing contradicting or seemingly contradicting each other in the interests of their respective clients is a grave scandal to science. "Men of the world are tempted to say that 'Science can lay but little certainty, and is rather a mass of doubtful speculations than a body of demonstrable truth.'" The legal system suffers too, because it loses access to many of the most eminent men in every department of science, who promptly refuse to be mixed up in any affair that may expose them to the humiliation of cross-examination. Thus, at the end of the day, everybody was losing. "The outside public is scandalized; experts are indignant; the bench and the bar share this feeling, but unfortunately are disposed to blame the individual rather than condemn the system." The only way out, Crookes laid down in his reform program, which did not stray afar from Smith's a generation earlier, is that "the expert should be the adviser of the court, no longer acting in the interest of either party. Above all things he must be exempt from cross-examination. His evidence, or rather his conclusions, should be given in writing and accepted just as are the decisions of the bench on points of law."[37]

The Journal of Gas Lighting, Water Supply, & Sanitary Improvement

Oddly enough, William Odling, the person who had started the commotion and who was always happy to enter a debate, seemed reluctant to enter this one. His contribution consisted only of a short letter to *Nature* protesting Lockyer's yellow journalistic style, which included supplementing "criticism of an author's performance with flippant insinuations as to his personal conduct and career," and suggesting that Lockyer's malevolence had to do with the contempt with which Lockyer's own contribution to chemistry had been met.[38]

Odling's views on scientific expert testimony were nevertheless represented by a strong editorial that appeared the following week in the professional publication *Journal of Gas Lighting, Water Supply, & Sanitary Improvement*. The unsigned editorial took the opportunity to vent the long-accumulated frustration of the rising new species of English professional and technological men of science:

If any man wishes to know the exact high-water mark of contemporary progress in science, manufacture, art, or trade, he should
not go to the meetings of societies, or listen to lectures and papers.
He must attend an action in the Law Courts. This building, or
rather mass of buildings, in the Strand, is more of an educational
institution than the Royal Society, the British Association [for the
Advancement of Science], the Institution of Civil Engineering,
and all other learned societies of the Kingdom, put together. This
is a truth which has been in a measure appreciated for very many
years; but it is only in recent times, when specialism in science has
become so marked, that the full sense of the educational value of
Law Reports has been brought home to an interested public."[39]

The author brought forward a few reasons for his radical contention.
First among them was that social equality prevailed in the courts, where
"there are no professors, tutors, or classes, and where everybody is
admitted to learn as much as he likes or can retain of the instruction
provided at other people's expense." A second reason was the different
reward systems. The anonymous author decried the bankruptcy of the
scientific system of nonmonetary rewards that served as ends in themselves and practically ensured that no man who worked for a living
could receive the honor and recognition they conferred. The inefficiency of such a system, he argued, is clear to anyone who dips into the
transactions of the learned societies, which were full of "scrappy papers
and memoirs upon infinitesimal points of details . . . [and] arid disqui-
sitions scarcely distinguishable in character and aim from those to be
found in the same publication for ten, twenty, or thirty years back."[40]

It was only in the courts that one could find "progress in science so
clearly epitomized and contemporary knowledge of things in general so
simply expressed." The reason for that, according to the *Journal*, was
"simply the magnitude of the [pecuniary] prizes which are contended
for in this arena." These forced the litigants into an all-out battle that
served best to expose all deficiencies and contradictions of the evidence
proposed. In the scientific establishment, where hypocrisy and double
standards prevailed, "the great man of science may ordinarily, in his
orations and papers, take advantage of his position and reputation, and
circulate rubbish that would ruin an unknown speaker." In the courts,
however, the cross-examining counsel did not care "that the witness
whose statement he is turning inside out is a president of this or that
learned society, and has a long tail of initial letters after his name.

Nothing of this was sacred to an acute lawyer who means to win. Hence, the author concluded:

> It is that the Law Courts, after all, are the scenes where the greatest triumphs of science are placed within the mental reach of the student. Here, sooner or later, he is sure to see all the authorities in natural philosophy and applied mechanics—not to mention the masters of other orders of knowledge, not as they are in the habit of doling it out in instruction to their pupils, but struck out in the course of a fierce conflict of wits which leaves no mercy for the vanquished.[41]

Round II

The wholesale attack on the scientific establishment and its mores prompted a strong response from William Crookes. In developing his astounding claims, Crookes wrote, the anonymous author of the editorial in the *Journal of Gas Lighting* displayed "an utter misconception of the functions and the results of our learned societies and our scientific journals, as well as the character and the habits of scientific men." Crookes dismissed altogether the argument concerning the educational superiority of the courts. The scientific societies, he wrote, are not educational institutes and their function is not to diffuse knowledge but to create new knowledge. On the other hand, how could the courts be educational if the highest generalizations and epoch-making research do not come within the purview of the courts at all?[42] "Who ever heard of the doctrine of organic evolution, the atomic theory, the cosmogony of Laplace, the law of continuity, the conservation of energy, or the like, being bequibbled in a court for the edification of a bewildered jury? What occupy the attention of the courts are incidental matters, such as telephones, torpedoes, and gas engines . . . things which, however grave in a pecuniary sense, are but as the crumbs that fall from the table of science." The savant who reads his paper before a scientific society speaks to an audience who is familiar with the same kind of research and who can judge the value of the evidence adduced and how it fits with what already exists. In the law courts all this is reversed and everything has to be explained to men who have little or no previous knowledge of the subject. "The main difficulty which the expert witness has to

grapple with," Crookes repeated another of Smith's arguments from a
generation earlier, "is that frequently a large part of the evidence by
which he reaches his conclusions cannot be expressed in words, though
it could easily be pointed out to another expert in consultation." The
hostile instrument of cross-examination is hardly the tool to facilitate
better communication in the courtroom. In fact, Crookes argued, the
deficiencies and contradictions of the scientific evidence referred to by
the *Journal of Gas Lighting* were not exposed by the adversarial machinery of examination and cross-examination but created by it. "Let this
system be done away with," he concluded, "let experts be no longer the
partisan of the litigants, but the impartial advisers of the court and those
unseeingly exhibitions which throw our friend into rapt excitement will
be at end."[43]

The feisty writer of the *Journal of Gas Lighting* did not back down. He
came out swinging again even before Crookes' critique was published.
What had provoked him this time were Frankland's and Crookes'
responses to Lockyer's attack on professional science, which were published while the first editorial of the *Journal of Gas Lighting* was still in
press. From his point of view, Frankland's and Crookes' positions were
not much better than Lockyer's. They all were demolishing bogus
cases—Lockyer, that the experts were morally corrupted; Frankland
and Crookes, that the adversarial proceedings were counterproductive
to scientific evidence. Their responses were characteristic of the scientific establishment, which looked down on expert testimony from all its
corners. Once a respectable man of science agrees to take a side in an
argument in court, complained the anonymous writer, his colleagues
interpret it as a "lust of lucre" and feel compelled "to regard all his
statements respecting the matters at issue as subject to a plus or minus
sign, according as they make against or for the side upon which he is
retained." However, the contradiction between the values of the
respectable man of science and the expert was only an illusion. "Truth,"
after all, "is not less truth because it is made to serve the immediate
profit of interested persons." It is only by applying the obsolete code of
scientific voluntarism to a free market situation, that the scientific community persuades itself otherwise.[44]

Whether the scientific establishment likes it or not, the anonymous
writer developed his position, the market of scientific evidence is
an open market where clients shop around among the various experts
who are offering their services. "Natural law, like the law of the land, is
susceptible of more than one interpretation," he argued, "and men of

science, like lawyers, contradict one another daily upon every known issue. It is indeed the natural way and most notorious habit of men of science. Why, then, should there be any peculiar loss of dignity accruing to the fact of contradicting and being contradicted in court?" Once we accept that the market for scientific evidence is an open market we find that the question of the proper conduct in such a situation is neither new nor unique to science. It was asked by many professions and had found its resolution in the well-tested liberal code of the free market:

> How far an advocate may go in the interest of his client . . . How far may an auctioneer go in the way of forcing up his sales by commendation of the wares? Nay, the same question applies to all sellers of marketable commodities, down to the costermonger in the street. There is no reason, therefore, for condemning the system of expert witnesses in lawsuits because to their case, as soon as ever they accept a retainer, the maxim "caveat emptor" may be applied as aptly as to any other commercial transaction.[45]

The writer forwarded a similar argument against the frequently suggested remedy of appointing scientific assessors. The free market, he argued, offered not only the best mechanism for proof testing, but also the best protection from the abuse of power. "Natural law, of course, is immutable; but it cannot be said that its interpretation by short-sighted mortals is the subject of such universal agreement that the *dicta* of this or that man, however prominent, would be accepted upon all points of difficulty . . . under the present system, if an expert misleads the bench or the jury, whether by accident or willful intent, he can be corrected on the spot; but how is an ignorant assessor to be reached?"

Neither a wise assessor nor an impartial one could be taken for granted. "There are," after all, "in scientific matters, prejudices and fancies which lead men to think and act very strangely, and just as mischievously as though they were paid to take partial views." In other words, "learned professors may occasionally swear that black is white, like mere vulgar perjurers." But this is where the function of the legal system is best exercised, for it is so constructed that "the wrong is generally detected and turned back upon the perpetrators." Thus, the writer concluded by repeating his attack on the scientific oligarchy: "We still maintain that the high-water mark of the arts and sciences is indicated in the courts; although these may usually be the haunts of the bagmen of science, rather that of the highest rank of philosophers. The bagman, however, generally knows the extent and quality of the scientific stock-rooms, and under proper checks, he is likely to prove a more instructive talker than the more retiring master."[46]

The Problem of Expert Testimony as a Stage for the Victorian Science Wars

There was yet a third round of editorial altercation between the *Journal of Gas Lighting* and the *Chemical News*, but we will not follow it. Each side just dug in further in their positions, of which we have learned enough. To a large extent, the whole of the 1885 debate was but a replication of the debates that had occurred a generation earlier. Lockyer took Chadwick's position, arguing that the source of the problem with expert testimony was the unscrupulousness of some men of science who were ready to sell their testimony to the highest bidder. Crookes reprised Smith's position that the problem was largely created by the improper procedures by which the legal system handled scientific knowledge. And the editorials at the *Journal of Lighting* editorials replicated Odling's position, which denied the existence of a problem, to such an extent that one could not help wondering if their anonymous writer was not indeed Odling himself.[47]

The fault lines dividing the 1885 controversy also repeated those of a generation earlier. While there was some difference of opinion about whether expert disagreement in the courts was caused by the corruption of the experts or by the legal mode of their deployment, almost all commentators acknowledged that this disagreement was detrimental to the cause of science. Lockyer blamed it on the moral degeneration caused by the professionalization of science, while Crookes blamed it on the adversarial procedures of the courts. Both, however, agreed that the partisanship displayed in court transgressed the scientific ethical norms and prevented the appropriate resolution of the scientific issues presented in court, and thus should be avoided one way or another. The minority position was represented this time by the mysterious writer for the *Journal of Lighting*, who, like Odling a generation earlier, refused to acknowledge the frequent scientific disagreements in court as a transgression of either the ethical or the epistemological norms of science. These controversies, he maintained, reflected real scientific disagreements that had not been acknowledged previously by the scientific community and were forced into the light by the superior ability of the adversarial legal procedures to disclose areas of uncertainty, untested presumptions, and interpretive conflicts within the seemingly robust scientific claims.

It was no coincidence that Lockyer was a leading figure in astronomy, the supreme field of Victorian science, while Odling was a leading figure in chemistry, one of the more utilitarian and, therefore, inferior links of the Victorian great chain of science. Indeed, underlying the epistemological and ethical debate about the issue of expert testimony was an enduring power struggle over the direction and shape of the thriving Victorian scientific enterprise. Thus, Chadwick and Smith, the apostles of public health, inveighed in 1862 against the system of expert testimony and lobbied for scientific assessors as part of their campaign for state-sponsored science, whose authoritative representatives were to be the official guardians of the public against the perils of industrialism.[48] Lockyer's wholesale denunciation of the expert witness in 1885 was part of the power struggle between researchers and practitioners over what it meant to be a man of science in an increasingly professional world. Such were also the proclamations of Odling and the anonymous writer in the *Journal of Gas Lighting*. Both celebrated the normative commitment of the legal system to develop two sides to every story as a means to disparage the hypocritical scientific establishment, which, they felt, had turned into a stagnant bastion of social privilege that suppressed rather than expedited scientific progress. The problem of expert testimony, they argued, reflected neither on the expert witnesses nor on the legal system but on the outdated ethical and epistemological codes of the scientific community, which led it to misinterpret plurality as disharmony and honest competition as moral deterioration.

The Problems of Expert Testimony in the United States

In 1866, six years after Dr. Angus Smith had brought the problem of expert testimony to the attention of the Royal Society of Arts, Judge Emory Washburn, who also taught law at Harvard, brought the same subject before the members of the American Society of Arts and Sciences. An American judge or lawyer "may consider himself fortunate," he told his audience, "who has not, when trying a case involving inquiries to be answered by experts, again and again felt that the opinions advanced by the witness had more to do with promoting that cause of the party in whose favor he appeared than the furtherance of justice."

Like Smith, who found it "repugnant to see great questions from nature played with, distorted and hidden for selfish purposes," Washburn found that "it has often disturbed and distressed sensitive minds, when seeking to ascertain what the truth was, to be obliged to resort to the opinions of men who have seemed to regard their line of duty as lying in the direction of the success of the one who employed them, rather than that of the discovery and establishment of truth." And like Smith, Washburn also decried the anomalous position of the scientific witness in the courtroom, complained about the rules of evidence that subjected the witness to the manipulations of the counsels, and advocated that the scientific expert be "like an arbitrator or a commissioner, who, standing indifferent between the parties, is to hear the facts which they had to offer; and then, by applying his knowledge as a man of science, to form an opinion, and declare it, as a guide for the court and jury in forming a final judgment in the case."[49]

The close resemblance between Smith's 1860 plight before the Royal Society of Arts and Washburn's 1866 address before the American Society of Arts and Sciences was not accidental. The American courts observed the same adversarial procedures of Common Law, while the American scientific community advertised the same high expectations from the scientific method, as did their English counterparts. These two features ensured that, in spite of the significant differences in the institutional and social dynamics of the legal and scientific communities between the two countries, the problem of expert testimony would develop in nineteenth-century America in the same basic pattern as that displayed in England. Thus, as in England, the growing deployment of men of science in divergent areas of litigation turned the American courts into a major, certainly lucrative, arena for scientific activity. And as in England, this arena soon put on view the curious spectacle of leading scientists disagreeing with each other from the witness stand, a view that served to cast doubts on the integrity of the experts and their science.[50]

By 1870, a study on "expert testimony" that took first prize at the Harvard Law School was already able to report in detail on an "unmistakable tendency on the part of eminent judges and jurists to attach less and less importance to testimony of this nature," and to explain it by "the surprising facility with which scientific gentlemen will swear to the most opposite opinions upon matters falling within their domain."[51] Many shared this bleak view. "Whoever has read the reports of trials or been present at them," Morrison Remick Waite, Chief Justice of the U.S. Supreme Court wrote in 1874, "in which experts are arrayed against each other, prostituting at times the science which they professed to represent needs not be told, that the subject of expert testimony as now understood, is one of no ordinary importance."[52]

Like their English colleagues, American men of science were much concerned with the damage that the scandals in court were inflicting on the public image and credibility of their emerging community, and like their English colleagues they were bitter about the adversarial legal machinery that placed them in the awkward position of partisan witnesses.[53] "No class connected with the administration of justice is more frequently misunderstood, or abused," complained Charles Himes, a professor of physics and chemistry at Dickinson University. The improper position of science in court seemed to have turned the scientific witness into a legal annoyance, "a sort of intractable, incompatible, unharmonious factor, disturbing the otherwise smooth current of legal procedure; too important or necessary to be ruled out, too intelligent and disciplined mentally to yield without reason to ordinary rules and regulations of the court . . . and at the same time possessing an undoubted influence with a jury that it is difficult to restrict by the established rules and maxims of legal procedures."[54]

The typical English class-awareness and the wars it precipitated within the English scientific community were absent from the American scene. Earning money was never frowned upon in America, and, lacking a tradition of self-supporting gentlemen of science, American men of science saw the application of science to the arts, trades, and business of everyday life as natural and praiseworthy. Indeed, if anything, in America it was the "pure men of science" who were often derided by the "practitioners." As William P. Mason, vice president of the American Association for the Advancement of Science, professor of chemistry at the Rensselaer Polytechnic Institute and a noted sanitary expert, complained in 1897:[55] "Scientific witnesses run up against all sorts of popular superstitions and are inveighed against as 'professors' by those who consider themselves the 'practical' workers of the time, and let it be noted, the burden of proof is uniformly laid upon these 'professors' shoulders, while most astounding and occult statements made by the 'practical' men may be received without verification."

Both the English and the American legal systems were well aware of the need to protect the credulous jury from charlatans. Still, neither system was able to lay down a precise rule for determining who was and who was not a competent expert. The only legal criterion was eight hundred years old: those persons are qualified to speak as experts who possess special training and experience in the subject in question.

Everything beyond this point remained purely a matter of discretion with the presiding judge, who had to decide afresh in each case whether the particular person offered as an expert witness would be admitted or not. In most cases, it was very hard for the judge to satisfy himself as to the qualifications of the persons offered as experts. Scientific titles and diplomas, and professional reputation, carried little judicial meaning during the nineteenth century, and preliminary examinations were impossible to make. The judges, therefore, were continually forced to decide on the spur of the moment, and often in relation to the most difficult subjects, whether the persons offered as experts were creditable. Unable to discriminate with any reasonable degree of accuracy between experts and charlatans, the actual practice of the courts came to be to admit everybody presented as experts, leaving it for cross-examination to expose quackery, and for the jury to be the judge of the ensuing battles between expert witnesses and lawyers.[56]

No one, of course, trusted the jury to be able to do this job properly. The jurors "are expected to exercise judgment and discrimination of facts that require training in the most favorable surroundings," pointed out a doctor who studied the psychology of the jury. However, "in reality the jury is selected from active working men unused to confinement, and unable to think and reason continuously on any topic outside their everyday life. They are untrained to discern the probable facts in contested cases, and to understand the real from the apparent in the arguments of the counsel."[57] Here, however, came into play two critical differences between the English and the American situations. While the English legal system recognized the jury as the final adjudicator on the facts of the case, it nevertheless granted its judges the freedom to take part in the questioning of the witnesses, advise the counsels in the framing of their questions, and comment fully on the weight of the evidence and the credibility of the witnesses in their charge to the jury. The English judges did not hesitate to use these legal instruments to control the usage of expert testimony in their court and to guide the jury in its assessment of the scientific witnesses and their evidence. In addition, by the 1870s the English judges were granted unfettered discretion in civil actions to order a trial without a jury in any matter requiring scientific investigation that, in their opinion, could not conveniently be made with a jury.[58]

The American courts lacked access to these instruments. The eighteenth-century English notion of the institution of the jury as a mainstay of liberty was adopted with added zeal in the American colonies, and the fact that many judges were laymen with no special claim to legal competence only added to the prominence of the jury. The Jacksonian faith in the ability of the common man and the enduring political philosophy that supported maximizing citizen participation in government, kept this enthusiasm alive throughout most of the nineteenth century to an extent that was unknown in England.[59] Consequently, early nineteenth-century American juries did pretty much as they pleased. The second half of the nineteenth century saw a growing pressure by the bar and the business community for more predictability and rationality in the operation of the jury, but the pressure also bred popular fears of undue influence on the jury. The result was a practical compromise, which was attained by the sharpening of the law-fact dichotomy and the corresponding spheres of the judge and jury. On the one hand, the power of the jury to determine the law, especially in civil cases, was eroded. On the other hand, fears of undue influence on the jury were eased by legislative and constitutional restrictions on the power of the American courts in charging juries. By 1889, in twenty-one out of the forty-nine states and territories composing the United States, judges were expressly forbidden by constitutional provisions to charge the jury on questions of facts. And in about half of the remaining twenty-eight states and territories, the courts had voluntarily adopted the same restriction. Only in federal courts and a minority of state courts were judges allowed to comment on the weight of the evidence in their charge to the jury.[60]

American men of science decried the absence of a judicial hand that would guide the jury in its difficult task of assessing the scientific evidence of the case. "If it be necessary to give juries authoritative instruction on points of law," wondered *Scientific American* in 1872, "how can it be less necessary that they should be similarly instructed in matters involving scientific knowledge?"

> To bring before them A, who swears to one thing, and swears to the truth, and then bring B, the charlatan, who looks and talks twice as wisely as A, and denies under oath all that A has asserted, is not to instruct but to mystify them . . . The jury must decide, or rather make a guess, as to what is right or wrong; and the average juryman is rather more likely to guess wrong than right in matters of science . . . Is it any wonder that the public is beginning to mistrust the value of this kind of evidence? Such mistrust is based upon good grounds enough. As now presented to juries, the

testimony of both competent and incompetent witnesses only
serves to muddle their intellects, and to complicate rather than
make plain the facts.[61]

The nineteenth-century American scientific community lacked the
organization, status, and political resources needed to challenge the
legal system and its procedures.[62] Most attempts to reform the legal
procedures of expert testimony were initiated by members of the medical and the legal professions.[63] The reform of expert testimony became
one of the hottest topics in the meetings of the various bar associations
that mushroomed in late-nineteenth-century America, and many bills
were drafted to remedy the evils of expert testimony. It was suggested
that experts be chosen by the court, either reserving or denying the
right of the parties in the case to call additional witnesses; that the
selection of the courts be unassisted or made from an official list that
would either be permanent or drawn up specially for each case. In
regard to the examination of witnesses, it was recommended that the
examination be done by the court, with or without the right of the
parties to cross-examine, or that there be no examination at all and that
the expert submit a report. In regard to making a decision when experts
disagree, it was recommended that a jury of experts be selected, or that
an expert sit with the judge during the trial and advise him.[64] Alas, the
American legislature and judiciary seemed even more reluctant than
their English counterparts to dissent from the legal axioms of the adversary system. Most reforms bills did not pass the legislative stage, and the
few that did were promptly held to be unconstitutional. Unable either
to control the selection of the experts or to guide the jury's assessment
of their evidence, nineteenth-century American courts concentrated
their hopes on the law of evidence, in an attempt to check the growing
problem by regulating the processes through which the experts communicated their information in court.[65]

The Failure of the Law of Evidence

The judicial process of expert testimony could be likened to a communication system. The expert could be compared to the source or the
transmitter of the data, the testimony to the communication channel,
and the jury to the receiver. Both English and American legal systems,
as we have seen, were unable to control the transmitter (the expert
witness) during the nineteenth century. The dominant English royal

judges were able, to a certain extent, to compensate for the problem by controlling the operation of the receiver (the jury). The American judges, on the other hand, were not allowed to do so. Thus, the only outlet left for them in their effort to control the problem in their courts was to attempt to control the communication channel, that is, the testimonial forms by which the expert witnesses were allowed to communicate their opinions to the jury. This, indeed, was where the main effort of the American judiciary was directed during the nineteenth century. American law of evidence was expected to provide the tools needed to contain the problem of expert testimony.[66]

One major legal doctrine sought to protect the credulous jury from being uncritically influenced by the expert's view by preventing the expert from giving his opinion upon the "ultimate issue," that is, the precise factual issue before the jury. To permit that, it was held, would put the expert in place of the jury and invade their province. Rational as it may sound, the application of this doctrine created great confusion and led to absurd consequences. In theory, it made irrelevancy a ground for admission, and relevancy one for exclusion. In practice, the "ultimate issue" was often exactly what the expert testimony was all about. In an insanity defense, the ultimate issue was the mental state of the defendant; in a malpractice suit the ultimate issue was whether the patient was treated properly or not; in forgery cases, the ultimate question was often the genuineness of a certain document, and so forth. It these cases and others, the doctrine seemed to exclude expert evidence exactly where it was most needed. Consequently, the courts developed various ways to bypass the rule and allow the witnesses to give their opinion on the ultimate issue.[67]

The most popular procedure was to allow an expert to state whether a certain cause may have produced the result under consideration and leave it to the jury to decide whether it did produce it or not. To enable this, a second evidentiary doctrine came into play. Under the "hypothetical question" doctrine, the expert's testimony was given in the form of answers to hypothetically framed questions. These questions specified a set of factual premises, already submitted in evidence, and the expert was asked to draw his conclusion from them, assuming that they were true. This cumbersome technique was justified on triple grounds: as a means of enabling the expert to apply his general knowledge to facts that were not within his personal knowledge; as a means of allowing the jury to recognize the factual premises on which the expert opinion was based; as a means of allowing the expert to give his opinion on the ultimate issue without "invading" the province of the jury. The jury was then instructed to credit the opinion given only if it believed these premises.[68]

Sound in theory, the technique broke down in practice. If counsel was required to recite all the relevant facts, the question became intolerably lengthy. If counsel was allowed to select the facts, as was the case in most courts, it prompted a one-sided hypotheses. Designed and controlled by the interested parties, the hypothetical question was used more as a means to manipulate the facts of the case than to clarify them for the jury. Forced to assume as true any cleverly defined transcript of the facts of the case, the expert was frequently manipulated to give an answer against his true conviction. "It was a strange irony," Professor Wigmore, Dean of Northwestern Law School noted in 1904, "that the hypothetical question, which is one of the few truly scientific features of the rules of evidence, should become that feature which does most to disgust men of science with the law of evidence."[69]

Even the old and powerful hearsay doctrine, which attempted to limit the testimony of ordinary witnesses to information that was based solely on their first-hand observations, turned out to be problematic in the context of expert testimony. The caution of the courts in admitting opinions not based on observation of the particular facts of the case, combined with the traditional requirement that testimony given to the jury should be under the sanction of an oath and subject to the test of cross-examination, and the fear of misleading the jurors by reading to them scientific statements they were hardly competent to assess, had led early in the nineteenth-century to a more or less general refusal by the courts to permit what many considered the most natural resort to scientific information—standards textbooks, reports, and so forth. The use of these written documents in court was excluded by the hearsay doctrine on the premise that they were statements not made under oath, or that their author was not available for cross-examination. As with other doctrines, the courts had slowly devised ways to work around this one too. Some courts permitted the use of scientific treatises, but only to discredit an expert. Others allowed the expert to "refresh his memory" by reading from standard works. Others even allowed publications of exact science, assuming their statements to be of ascertained facts rather than of opinion, and excluded other treatises, especially medical works. Confusion and inconsistency, again, were rampant.[70]

By the end of the nineteenth century, it was clear that the American law of evidence had failed to control the problem of expert testimony.

Designed to be the crown of modern American jurisprudence, a corpus of legal procedures as analytical and as rational as Euclidean geometry, the law of evidence had turned instead into a highly complicated and technical domain, sagging to the point of collapse under the burden of its own distinctions, exceptions, and exclusionary duties. In 1898, Professor Thayer of Harvard Law School called it a "piece of illogical, but by no means irrational, patchwork; not at all to be admired, or easily to be found intelligible," and by 1904, his famous pupil, John Henry Wigmore, needed four thick volumes to cover it. "There is a full realization now," concluded the *Chicago Legal News* in 1909, after a long historical review of expert testimony, "that in the present practice we have carried a branch of procedure out to the utter defeat of its object, to an absurdity; and that the result has been a wide-spread disgust with methods of legal administration."[71]

Thus, although it first raised its head in the English courts, it was in America that the problem of expert testimony reached its fullest expression. The last decades of the nineteenth century saw the bitter English debates concerning the problems of expert testimony beginning to subside, because of the diversion of technical litigation away from jury trials and the efforts of the authoritative royal judges to keep the thriving business of expert testimony under relative check. Across the Atlantic, however, things went from bad to worse. Unable to check either the selection of the experts or their evidence in court, or the jury's assessment of this evidence, late-nineteenth-century America saw the problem of scientific expert testimony reaching a crisis.[72]

In the next chapter, we shall stay in nineteenth-century America, but shift our focus from the courts of law and public opinion to the scientific domain. There, we will closely follow the efforts of one of the most successful American scientific communities, the microscopists, as they tried to balance their aspiration of becoming a true *amici curia* with the opposing fear of tarnishing their public image by appearing in court. These efforts were not mere rhetoric. As we shall see, they constitute an important chapter in the history of both law and microscopy.

4

Blood Will Out: Distinguishing Humans from Animals and Scientists from Charlatans

'Faith' is a fine invention
When Gentlemen can see
But Microscopes are prudent
In an Emergency.

∽Emily Dickinson

O<small>N</small> J<small>ULY</small> 3, 1879, D<small>R</small>. S<small>AMUEL</small> L<small>AWS</small>, president of the State University of Missouri, received in his evening mail a letter from the Governor of Missouri, John Phelps. It read:

> A criminal case is pending in the circuit court of Clark County, for murder, against Wm. Young. The authorities have possession of a shirt and pair of pants on which there is blood, and the question is whether that blood came from a human being or from an animal. The defendant says from an animal. If the theory of the state is correct, that blood was shed in July or August 1877. Can any one of the professors of the State University analyze this blood, and will he be willing to do so, and then deliver his opinion as a witness in the case? Persons may have reluctance to testify to the results developed by a chemical analysis where the testimony may tend to the conviction of a person for the crime of murder. Young is charged with the murder of the Spencer family (5 persons), and I make these inquiries at the instance of the prosecuting attorney, Ben E. Turner, of Clark County.[1]

A personal request by the governor on behalf of one of his prosecutors was by no means an everyday occurrence. The five members of the Spencer family had been murdered in their sleep, and although there were good reasons to believe the murderer to be William Young, there was no direct evidence against him. After two years of faithfully pursuing every available lead, the officers of the law finally reached their wits' end. Their last hope was tiny dark blood spots caught on Young's clothes in what may have been a deadly struggle with his victims. If science could prove that the blood came from a human being, conviction would be likely. If science could not supply such a proof, a heinous murderer might walk away free. Suspecting that the professors would be "reluctant to testify," Governor Phelps exerted political pressure on their president in order to get the blood analyzed and the professors on the witness stand.

Samuel Laws at once placed the Governor's letter in the hands of John Duncan, a professor of physiology in the medical school. It took Duncan less than a day to answer his president's request:

After examining the authorities on the microscopical examination of blood, especially in the case of murder, I am unwilling to undertake the investigation, and from that to testify before court of justice. In the present state of our knowledge, it is easy to distinguish blood stains from all others; but it is almost impossible to decide between the red blood corpuscles of man and those of the lower animals, especially the mammalia; much more difficult is it to differentiate between them after the blood has become dry. Says Dr. Lionel S. Beale in *The Microscope in Medicine*, published in 1878, p. 266: 'I can hardly think that in any given case the scientific evidence in favor of a peculiar blood stain caused by human blood, will be of a kind considered sufficiently conclusive to be adduced, for example against a prisoner upon his trial.' The same author on page 207 says, in speaking of human blood and that of lower animals, 'That is a serious difficulty, and up to this time I fear we must admit that we are unable to decide with sufficient certainty to justify us giving our evidence in a court of law.' On Page 307 of *Taylor's Medical Jurisprudence*, by Reese, 1872, are these words, 'There are no certain methods of distinguishing microscopically or chemically, the blood of human beings from that of an animal, when it has once been dried on an article of clothing.' From these facts and others which are not necessary to relate, I do not undertake the examination . . . the matter is too grave to admit speculation.[2]

On the very next morning, June 5, 1879, Laws sent Duncan's answer to Governor Phelps together with some additions of his own:

> I am aware that there is a popular impression that scientific men are in possession of the means of determining with precision whether such stains as those on the garments in the possession of the State in the pending prosecution were caused by human blood, and I fully appreciate the grave consequences of the negative conclusion submitted above, but the evidence is overwhelming that we have no means of exact determination even in fresh blood. . . . This incompetency is unquestionably a weakness in the administration of government, but it arises from the nature of things, as man is not omniscient, and it has its counterpart of strength in this, that the secret things belong unto the Lord our God; but those things which are revealed belong unto us and our children forever. (Deut. XXIX 29.) Such overt acts as fall within the cognizance of human law imply overt evidence of their performance; and it is because this wholesome criterion does and must so often fail to detect the actual state of facts, that an intelligent recognition of the moral government of God becomes the undergirding of our social fabric.[3]

God's intentions are difficult to read, however. By 1879, the question whether human blood could be distinguished reliably from animal blood had been agitated among scientists for over 50 years. Taking this persisting controversy as a specific issue exemplifying the debates over scientific expert testimony, this chapter concerns itself with the mechanisms of scientific disagreement in court. Was the controversy largely created, as most nineteenth-century men of science thought, by the improper procedures by which the legal system collected, represented, and processed scientific knowledge? Was it influenced by the unscrupulousness of men of science who were ready to sell their testimony to the higher bidder? Did the controversy reflect, as Professor William Odling maintained, problems within the scientific body of knowledge itself? Situating the controversy within the emerging American community of microscopists, who were struggling to negotiate their norms of objectivity and impartiality with the adversarial realities of their patron, the legal system, this chapter demonstrates how what might seem to have been a simple measurement problem, technical in nature, turned into a complex dilemma. Its theoretical and technical aspects became infused with the financial temptations of expert testimony, with the moral discomfort of giving evidence that might cost a person's life, and with economic and political considerations concerning the public image and credibility of the rising community of microscopists.

First Attempts: 1829–1869

By 1879, science had devised a variety of tests for detecting the presence of small traces of blood. These included changing the color of the blood to deep blue by treating it with a tincture of guaiacum and an ether reagent, forming haemin crystals by adding of salt and acid, and identifying the hemoglobin spectrum with its dark spectral bands.[4] Once blood was detected, however, the accused might try to account for it by saying that it came from an animal. None of these tests could prove him or her wrong. The task of differentiating human from animal blood was therefore of crucial importance in many nineteenth-century criminal trials.

Many scientists had tried to devise a definitive discriminator. The earliest attempts took place in Europe, especially in France, where medical jurisprudence was unrivaled. Thus, in 1829, the French chemist Jean-Pierre Barruel suggested that the human nose could be the key to solving the problem. The blood of each animal, Barruel maintained, contained a peculiar odorous factor similar in smell to the sweat of the animal. According to him, the application of heat to a mixture of three parts of sulfuric acid and two parts of blood would dissolve the connections between this volatile factor and the blood and release the characteristic smell. After much experimentation, he found the factor's strength to be correlated with the color of the hair and more marked in the male than in the female.[5]

Others confirmed Barruel's results. Matthieu Joseph Bonaventure Orfila, a renowned authority on forensic chemistry, also approved of it. The test was adopted all over Europe and decided multiple cases.[6] Barruel's new scientific apparatus, the human nose, had its limitations, however, not least, the need for large quantities of fresh blood and the sense of smell of a sommelier. Soon criticism began to mount.[7] In 1848, Karl Schmidt, a German hematologist from the Russian University of Dorpat, asked six people to identify the animals from which various blood samples had been taken. He found little consistency in their opinions.[8] In 1853, a scandal erupted when Barruel and two other respected experts, Ambroise Tardieu and Jean-Baptiste Chevalier, tried to decide whether blood found in the house of a woman accused of murder had come from a human being or, as the woman said, from a sheep. On the witness stand, the experts contradicted each other. The resulting public embarrassment, coupled with continuing unfavorable reports on the method, diminished the credibility of the sulfuric acid test.[9] Other chemical methods fared no better.[10]

By mid-century it was becoming clear that microscopy, not chemistry, offered the best chance for differentiating human from animal blood.[11] The achromatic lens of the newly available compound microscope revealed clear physical differences among the red corpuscles of different animals. The most obvious difference was their shape. Mammals, except for camels, have red corpuscles in the shape of *circular* discs. Reptiles, birds, fish, and camels have oval corpuscles. In reptiles, birds, and fish the oval corpuscles possess a central nucleus that appears raised. In mammals, including camels, the corpuscles have no such nucleus, and they appear depressed in their center. These differences allowed microscopists to register a few impressive courtroom appearances when defendants argued that the blood they were accused of shedding belonged not to human beings but to reptiles, birds, or fish.[12] However, when the defense argued that the blood found on the accused belonged to domestic animals such as dogs, cats, and pigs, the shape of the red corpuscles could not be used to decide the issue. In these cases, researchers hoped that by establishing the average sizes for the corpuscles of each species, measurements of size could determine the kind of animal from which a blood sample came.[13]

George Gulliver, fellow of the Royal Society and professor of comparative anatomy and physiology at the Royal College of Surgeons, obtained the first systematic knowledge of the exact size of the corpuscles in different animals. Hoping to prove that the corpuscle was the most reliable means for the classification of species, Gulliver measured the corpuscles of nearly eight hundred animals over a period of almost thirty years.[14] His biggest impact, however, proved to be in the legal sphere, where his measurements became the leading forensic standard for differentiating human from animal blood by micrometric measurements. Table 1 displays a sample of Gulliver's measurements, as well as those of investigators who followed him.

Gulliver never gave a detailed account of his measurement procedures. That was unfortunate. Since red corpuscles vary in size within the same body, measurement involved an arbitrary choice of samples. Their preparation introduced many other problems, especially when the blood was dry and had to be moistened. A fluid had to be applied to the dried blood to dissolve the fibrin that glued the corpuscles together

Table 1. The Average Size of Red Blood Corpuscles of Mammals Expressed in Millimeter Units

Mammal	Gulliver 1845	Schmidt 1848	French Medico Legal Society 1873	Woodward 1875	Wormley 1885
Elephant	0.0080			0.0093	
Human	0.0079	0.0077	0.0078	0.0082	0.0078
Seal	0.0078				
Porcupine	0.0075				
Monkey	0.0074			0.0075	
Guinea Pig	0.0071		0.0077	0.0079	0.0079
Dog	0.0071	0.0070	0.0071	0.0078	0.0071
Rabbit	0.0070	0.0064	0.0070		0.0070
Rat	0.0068	0.0064			0.0068
Ox	0.0060	0.0058	0.0060		0.0060
Pig	0.0062	0.0062		0.0060	
Cat	0.0058	0.0056	0.0065		0.0058
Horse	0.0055	0.0057	0.0056		0.0059
Sheep	0.0048	0.0045	0.0050		0.0051
Goat	0.0040	0.0040	0.0046		0.0040

Compiled from Formad, *Comparative Studies of Mammalian Blood;* Wormley, *Blood Stains;* Schmidt, *Die Diagnostik Verdächtiger Flecke;* Bell, "Blood and Blood Stains."

and to restore the corpuscles' shape. However, red corpuscles possess strong osmotic characteristics that render them susceptible to rapid and significant changes in size and shape when introduced into fluids with different specific gravities. The age of the blood stain and the refractive index of the fluid introduced further variations. Water, for example, turned out to be best for old stains but caused a violent endosmosis that destroyed the corpuscles from fresh specimens. Albumen and glycerin, though close to serum in specific gravity, were highly refractive and disturbed the microscopic vision. Human and other mammalian sera were another popular choice but had the obvious drawback of contamination by their own corpuscles. Every microscopist had his own preference: salt solutions, ether, alcohol, chloroform, soda, ammonia, and bone oil, to name but a few.[15]

The nature of the materials on which the blood had dried further complicated analysis. If the material was hard and smooth like glass or porcelain, the fibrin, albumen, and serum of the blood hardened around the dry corpuscles, preserving them in only slightly altered form. If however, the blood dried on soft and porous material such as linen, wool, or other cloth, the fibrin, albumen, and serum were absorbed by the cavities of these materials. Deprived of their preservative environment, the corpuscles disintegrated or permanently shriveled so that their diameters could not be measured. The conditions under which desiccation took place had similar consequences. If desiccation was quick and homogenous, the corpuscles preserved their size and shape much better than if humidity was high and evaporation slow.

In most legal circumstances, the sample to be analyzed was old, dry, and abused, and the lack of clear protocols for restoring corpuscles to their original shape and size made the results problematic. Nevertheless, despite the many uncertainties, some researchers thought they could measure the minute differences in size of the red corpuscles of the various mammals. Professor Karl Schmidt of Dorpat, who in 1848 published the first comprehensive work on the micrometry of blood stains in criminal cases, had such confidence. Working with an amplification of five hundred diameters and drawing on forty measured corpuscles per animal, Schmidt presented his own set of average sizes. He found that only 2 percent of the corpuscles varied significantly from the measured averages (hence, that these averages were reliable) and that the desiccation of red corpuscles followed a uniform rule of evaporation, with a constant relation to their volume (that is, the drying action did not alter the relative differences between the various corpuscles). "This allows us," Schmidt concluded, "to overcome the most difficult obstacle in our inquiry, the diagnosis of specific kinds of dried animal blood from one another and from human blood."[16]

Schmidt's work soon came under heavy attack. Hermann Welcker of Halle challenged Schmidt's first claim, that of rigid averages. "I have always, both in animals and in man, found the transverse diameter of the blood corpuscles of one and the same individual varies from one-fourth to one-half of the mean measurement; and it appears that all the sizes lying between the two extremes are present in tolerably equal numbers." Rudolf Virchow criticized Schmidt's second conclusion. He pointed out that variations in the rate of desiccation can cause significant variations in size, and he warned that a diagnosis based solely on the relative size of the remoistened corpuscles could lead the micros-copist to fatal errors. In England, Dr. Arthur Hassall noted that while Schmidt's measurements might be valid for fresh blood carefully dried in thin layers on glass plates in the laboratory, it would not hold for all blood dried on other materials. Finally, in America, Andrew Fleming pointed out that the blood to be analyzed in legal cases usually was so denatured that the analyst could not restore its corpuscles to their original shape and size. Thus, Fleming concluded, "I am forced to admit that great difficulty arises in attempting to fix its [the dried blood's] origin by the comparative size of the corpuscles."[17]

By the late 1850s, the scientific consensus on both sides of the Atlan-tic held that one could not reliably discriminate between the average sizes of the red corpuscles from human beings and domestic animals and that even were such a discrimination possible, the deviations from the average size resulting from natural variation, diseases, conditions of desiccation, and so forth, would make inference from averages unreli-able. As the German physiologist Hermann Friedberg wrote in 1852: "The solution to the most difficult problem, the diagnosis of dried blood of man and certain mammalia, which Schmidt has claimed as positive in all cases—is still a *pium desiderium;* the means of examination now known do not suffice to distinguish them, as I can say from the many systematic experiments I have made." Hence, as Virchow put it in 1857, "No microscopist will hold himself justified in putting into ques-tion a man's life on the uncertain calculation of a blood corpuscle's ratio of contraction by drying."[18]

The American Debate: 1869–1880

In this skeptical climate, microscopists called as expert witnesses usually testified no more than whether a blood stain came from a mammal and therefore could be the blood of a human being. In 1855, for example, Alfred Taylor, the prominent English authority on medical jurispru-dence, "declined to say absolutely that the stains were caused by human blood, although the corpuscles coincided in measurement with them."[19] Still, in the Anglo-American adversarial system, where scien-tific expertise was sold in an unregulated fashion on an open legal market, experts could be obtained who professed under oath to be able to distinguish between human and animal blood. Although such claims ran contrary to the scientific consensus, they were not necessarily cor-rupt. Indeed, many microscopists believed that under favorable condi-tions they could distinguish human blood from that of other mammals

with significantly smaller corpuscles.[20] In the early 1870s, these microscopists received support from a well respected authority, Dr. Joseph G.
Richardson, professor of pathological anatomy at the University of
Pennsylvania, chief microscopist at the Pennsylvania Hospital, secretary of the Biological and Microscopical Section of the Academy of
Natural Sciences of Philadelphia and one of the leading hematologists
in America. Richardson was the first American to advocate the deployment of the new high-power immersion optics in the identification of
blood corpuscles. He wrote in 1869:

> It is true that the older microscopists who rarely obtained first rate
> definition with their lenses magnifying much beyond 500 diam-
> eters, were probably wise in recommending that none but the
> most expert should attempt a decision between the blood of the
> various mammalia, even when fresh, for the difference between
> the apparent magnitude of $\frac{1}{10}$ and $\frac{1}{12}$ of an inch may well be
> counted too minute to lightly determine a question often so
> momentous; but, as during the last three or four years, opticians
> have furnished immersion lenses of $\frac{1}{25}$ and $\frac{1}{50}$ inch focal length,
> which, with the highest eye piece, give an amplification of about
> 2500 and 5000 diameters respectively . . . it seems as if any careful
> observer might now, with the aid of such objectives, be qualified to
> pronounce a positive opinion.[21]

Having the technology to measure individual corpuscles was one
thing, producing the corpuscles to be measured quite another. The
possibility of separating the corpuscles from a blood clot and reconstructing them accurately rested on the assumption that red corpuscles
have an exterior wall, a capsule, that would survive the turmoil of
desiccation and could be reconstructed. But after two hundred years of
constant attention by microscopists, no one knew with certainty
whether or not the corpuscles had such an exterior wall. Following
Theodor Schwann's influential cellular theory, many older microscopists believed that the corpuscles had walls. But this belief grew increasingly difficult to substantiate as microscopes improved, because "even
through the best microscope of the day, each corpuscle appeared as a
homogenous mass of organic substance of the same color, consistency,
and composition." Thus, the younger researchers generally supposed
the corpuscle to have a jelly-like constitution with an outer hardened
layer of protoplasm.[22]

Richardson acknowledged that if red blood globules were lumps of jelly-like matter, the chance of discovering individual corpuscles in a blood clot, however moistened, seemed "almost as hopeless as the search after individual rain drops in a cake of melting ice." However, looking through his high-power lens at the residuum of a clot, after the exosmotic action of water had washed away the corpuscles' content, he distinctly perceived that what was "so long mistaken under lower powers for a mass of fibrin," was actually an aggregation of empty capsules. His careful observations indicated that these empty capsules had "an inelastic character," which meant that when an individual corpuscle dried, "however much of the cavity is decreased, its limiting membrane suffers no actual diminution in superficial area." Thus, he concluded, by proper moistening and tinting, dried red corpuscles could regain their round shape and be magnified enough for accurate measurement.[23]

Richardson's paper failed to impress the profession. Gulliver and Fleming had already noticed the capsules, the "membranous bases or frames of the corpuscles . . . quite insoluble in water, and so faint as not to be easily seen." However, unlike Richardson they considered these membranes to be "very elastic, exceedingly delicate in structure, and quickly affected by the action of agents."[24] Richardson knew that his microscopical observations might be insufficient "for controverting the opinions of those experienced histologists who deny to the red blood corpuscles a proper cell wall." He attempted therefore to substantiate his hypothesis about the cellular structure of the corpuscles with additional experimentation using the gigantic blood disks of the *Menobranchus*.[25] Microscopists remained unimpressed. Finally, when the seventh American edition of Taylor's authoritative *Medical Jurisprudence* was published in 1873, in Richardson's own city, Philadelphia, without any reference to his papers, he decided to change strategy.[26]

Richardson decided to aim at the wider audience of physicians and lawyers who did not fully understand the technical details of the debate but had an important say in its practical outcome. In a third paper, published in 1874, he declared that, with the powerful $1/50$ immersion lens, the corpuscles of different mammals "can hardly be mistaken for each other, any more than a 12-inch shell could be mistaken for a 6-inch shell." He dismissed the standard objection, that the variations from the averages are too irregular to render their application reliable, as empty rhetoric. The smallest human red corpuscles, he noted, are much larger than the largest corpuscles of most animals. Moreover, since corpuscles shrink when drying, any error of identification would favor the accused, since the bigger human corpuscles could only come to resemble those of animals, never the other way around. As for the moral qualms of queasy experts, Richardson reminded those who demur at the idea of allowing a man's life to hang upon such seemingly insignificant circumstances as a difference in size of blood corpuscles, "how often the reactions of arsenic, afforded by a quantity of the metal too excessively trivial to be accurately estimated by the most delicate balance, have sufficed to bring out the crime of murder, and to aid in securing that just punishment for violation of law in which we all have so deep an interest, because on it all our enjoyment of life and property depends." Richardson's claim, on the basis of investigations "made upon many different kinds of blood, and under great variety of conditions," set a clear challenge to the young community: "we are now able by the aid of high powers of the microscope, and under favorable circumstances, to positively distinguish stains produced by human blood from those caused by the blood of any of the animals just enumerated, and this even after the lapse of five years from the date of their primary production."[27]

The Biological and Microscopical Section of the Academy of Natural Sciences, Richardson's home base in Philadelphia, took up the challenge. In discussing Richardson's research, some members expressed discomfort with his wholesale endorsement of the high-power microscope. Experience, they insisted, proved that the gain in amplification went hand in hand with a serious loss of definition. Moreover, they argued, even if high-power microscopy was indeed superior, it would not ensure reliability. As the founder of the section, Dr. John G. Hunt, put it: "two things are essential in real microscopical work: a good lens at one end of the tube, and a good brain at the other; and wide difference in definition and aperture exists in both."[28] The Lancet was more favorable. "The generally received opinion that the microscope is of little or no service in discriminating between the blood corpuscles of man and the common mammalians animals would seem to be refuted by some recent investigations of Dr. Joseph G. Richardson."[29]

Others, however, were at pains to see the revival of the debate concerning the reliability of micrometric blood test. What had seemed to emerge before the Civil War as a vital civil function of science had become by the 1870s a source of discontent. Instead of bringing the legal and the scientific professions closer, forensic science was drawing them further apart. The courts were growing increasingly weary and wary of the spectacle of men of science contradicting each other on the witness stand; the medical profession was growing increasingly impatient with the courts' practice of forcing doctors, under the pretense of public service, to testify as to the results of their tests without paying them properly; and the lawyers took advantage of both sides, particularly by setting asylum superintendents and neurologists against each other in highly publicized insanity trials.[30]

Colonel Joseph Janvier Woodward, member of the National Academy of Sciences, vice president of the American Medical Association, and one of the most distinguished microscopists of his age, was worried. The notorious Wharton murder trial of 1872 with its intense confrontation over forensic chemistry, which "has been the grave of reputations of medical experts whose opinions previously were esteemed to be as good as other men's fact," was still fresh in everyone's mind.[31] Woodward was determined not to allow the debate over the micrometric blood test to do the same to the reputation of the young field of forensic microscopy. Fearing that Richardson "would be understood as teaching . . . that it can be determined by the microscope with certainty whether a given stain is composed of human blood or not," Woodward raised the alarm:

> Justice, no less than scientific accuracy, demands that the microscopist, when employed as an expert, shall not pretend to a certainty which he does not possess. I suppose no experienced microscopist, who has thoroughly investigated this subject, will be misled by Dr. Richardson's paper, but there are many physicians who possess microscopes, and work with them more or less, to whom a partial statement of the facts on such a subject as this is peculiarly dangerous, and the object of the present paper is to point out to this class of readers that Dr. Richardson's statement of the case, even if all he claims be granted as truth, is, after all, not the whole truth; that there are certain mammals, among them the dog, the constant companion of man, whose red blood corpuscles are so nearly identical in size with those of human blood, that they cannot be distinguished with any power of the microscope, even in fresh blood, much less in dried stains.[32]

The truth, according to Woodward, was that the micrometric test of the red corpuscles could not be trusted. The whole truth was that Woodward thought that the expert microscopist witness could not be trusted either and that the combination of the test and the testimony was a professional disaster waiting to happen. The best way to secure the reputation of the profession from that "late spirit of exaggeration that seems to have possessed certain experts" was to discredit the micrometric blood test once and for all.[33]

Woodward started by attacking the accuracy of the standards used in the test. He noted that Gulliver had determined the average size of a human blood corpuscle to be around ⅟₃₂₀₀ of an inch with the help of an eyepiece micrometer that had divisions of ¼₀₀₀ of an inch. Thus, Woodward argued, Gulliver must have estimated a fraction of his micrometer division. Furthermore, Gulliver's measurements on the same sort of animal had varied frequently during his career. The reason, Woodward suggested, was Gulliver's method of calculating averages as the arithmetical mean of several measurements. No wonder, Woodward wrote, that Gulliver himself had claimed only "relative exactness" for his results and had conceded that "in the absolute accuracy of any micrometer applied to objects so extremely minute, it is difficult to place implicit confidence."[34]

Woodward took on other authorities as well. He pointed out that Hermann Welcker derived his "so often quoted" mean of human red corpuscles from four sets of measurements of his own blood only, two from dry preparations and two from moist blood. He showed (see Table 2) that "the variations between the mean diameter assigned to human blood by different observers are quite as great as the variations recorded by any of them between the blood of a man and a dog."[35]

Table 2. Red Blood Corpuscles of Human Beings and Dogs Expressed in Micromillimeters

Source	Average Human/Dog	Minimum Human/Dog	Maximum Human/Dog
Schmidt	7.70/7.00	7.40/6.60	8.00/7.40
Gulliver	7.94/7.16	/5.60	/8.80
Köniker	7.51/7.09		
Friedberg		5.80/5.40	7.00/8.00
Welcker	7.74/7.73	6.40/6.50	8.60/8.20

Compiled from Woodward, "On the Similarity Between the Red Blood Corpuscles," 53–156.

Woodward concluded his lesson with a warning:

If the microscopist summoned as a scientific expert to examine a suspected blood stain should succeed in soaking out the corpuscles in such a way as to enable him to recognize them to be circular disks, and to measure them, and should he then find their diameter comes within the limits possible for human blood, his duty, in the present state of our knowledge, is clear. He must, of course, in his evidence, present the facts as actually observed, but it is not justifiable for him to stop here. He has no right to conclude his testimony without making it clearly understood, by both judge and jury, that blood from the dog and several other animals would give stains possessing the same properties, and that neither by the microscope, nor by any other means yet known to science, can the expert determine that a given stain is composed of human blood, and could not have been derived from any other source. This course is imperatively demanded of him by common honesty, without which scientific experts may become more dangerous to society than the very criminals they are called upon to convict.[36]

Richardson quickly fired back. His refusal to elaborate on the weaknesses of the micrometric blood tests, he explained, was not the result of dishonesty or even carelessness toward science or justice, but an attempt to protect both:

I felt (honestly if mistakenly), whilst writing both my first paper and its continuation, that should I more than indicate the animals which render our conclusions doubtful, my work might be utterly condemned as prejudicial to the interests of society, and myself perhaps compared (should I emphasize and reiterate the fact that science alone could not detect the falsehood of a criminal's story if he cunningly asserted that suspicious stains were made by the blood of a dog) to a toxicologist publishing a treatise, setting forth most faithfully the method by which poisoners may best destroy their victim with the least danger of detection in their crimes.[37]

The scientific controversy thus rose to competing calculi of trusts and fears. Richardson wanted his fellow scientists, who could be trusted not to abuse the information, to keep the weaknesses of the micrometric blood test secret from those who could not be trusted, that is, "the shrewd-witted lawyer, to be found not only in every city but in every country town," who would use this knowledge to construct alibis for his guilty clients. Thus, according to Richardson's calculus, it was Woodward's hasty criticism that eventually would prove more dangerous to society:

Now that all responsibility for harm has been removed . . . I am glad for the sake of the few who might draw erroneous inferences from my former papers, to say most emphatically, that I believe we cannot at present distinguish positively, in dried stains, between the blood corpuscles of man and those of any mammal in which the disks measure on average over 1/4000 of an inch. Hence, therefore, until further discoveries are made, a microscopist's best efforts at revealing crime can only serve the cause of right and justice in those cases where the criminal's attorneys, in spite of being forewarned and consequently forearmed, fail to prepare or suborn testimony skillfully enough to convince the jury that some dog, rabbit, elephant, monkey, etc., has been killed. . . . I venture, however, to predict that from this explanatory note and the essay [Woodward's] which made it necessary will spring a host of bloody dog tales to account for suspicious stains on clothing, etc. of murderers.[38]

Unlike Richardson, Woodward did not trust his fellow scientists not to abuse their privileged knowledge:

I cannot forget that on more than one occasion in the past, witnesses summoned as scientific experts have been so misguided as to go into courts of justice, and swear positively, on the strength of microscopical examination, that particular stains were human blood; and I think the danger that others may do so in the future, to the prejudice of innocent men, is more to be feared than the possibility that the acquaintance with the true limits of our knowledge on this subject may sometimes be made use of in the unscrupulous defense of real criminals.[39]

Two months later, the father of blood micrometrics joined the acrimonious debate. "However truly a careful observer (Dr. Joseph G. Richardson)," Gulliver wrote, "may have distinguished by comparative measurements of the corpuscles, stains of human blood from those of the sheep or ox, this kind of diagnosis, as Dr. J. J. Woodward observes, would be ineffectual in some probable and more possible cases."[40] Still, despite Richardson's retraction and Gulliver's sanction, the thirty-year scientific debate concerning the micrometric blood test refused to subside. As one participant wisely put it, "all agreed to the scientific facts . . . [but] these were sufficient to satisfy some minds, and not others."[41]

Minds might be satisfied when fortified by proper prejudice. "Science certainly is achieving wonders, and in a few provinces greater than in micro-legal studies of blood," reported the *Medical Times* in 1878, following a Louisiana case in which Joseph Jones, a professor of chemistry and clinical medicine at Tulane University, determined that the blood on two separate items was probably from the same individual, who was suffering from fever: "At the rate experts are traveling, we expect soon that a man will be considered an ignoramus who cannot tell from a blood stain the race of the man it came from, the ailments he had suffered, the street corner he was born on, the date of his christening, and the names of his forefathers, back even unto the fourth generation."[42] The jury's finding of guilty against four African-Americans accused of first degree murder, explicitly referred to Jones's testimony as its main ground of decision.

Losing Control: The 1880s

In 1879, the American Society of Microscopists was formed. It made a fine forum for the big scandal that many had fearfully anticipated. The debate over blood micrometrics immediately jumped from the obscurity of scientific colloquiums and journals to the front pages of daily newspapers nationwide that reported on the celebrated case known as the Hayden Trial.

It all started when the body of a young woman named Mary Stannard was found in Rockland, Connecticut, with contusions on the head and a gash in the throat. After the body had been buried, a rumor spread that the deceased had been pregnant and that the father was Reverend Herbert H. Hayden, a young married Methodist minister and father of two. Hayden was arrested. The main evidence against him were the testimony of Mary's half-sister, who said that Mary had told her about the improper relations, and Hayden's own stained knife. In the ensuing investigation, the state expert, Moses White, a professor of pathology from Yale College, testified that microscopic examination of a small drop of blood on the knife showed that it was human. The defense countered White's testimony with that of its own expert, Leonard Sanford, a professor of anatomy and physiology also from Yale College, who testified that "he would not have expressed so confident an opinion as did Dr. White . . . with so few corpuscles to found it on. He would need three times as many to examine before he would express an opinion."[43]

After two and a half weeks of examination, Hayden was discharged. In his closing speech, the presiding judge mentioned that "he himself had at that moment a knife in his pocket which must have been full of blood resembling human, for he had recently killed a dog with it, and a dog's blood resembles that of a human being. . . . If I were as sure of Heaven and eternal felicity as I am that this man is guiltless of the crime of killing Mary Stannard, I should rest content."[44] The prosecution refused to rest content. Accounting for his movements during the day of the murder, Hayden had told of purchasing arsenic in the neighboring town with the thought of killing rats. The investigators exhumed the body, and White found fifty grains of arsenic in the stomach, enough to kill twenty people. Hayden was rearrested. The arsenic, strangely enough, lay undigested. This led the defense to argue that the arsenic had been put there after Hayden's admission. But White testified that he had found traces of the poison in the brain, where it could not have been carried if put in after death. This time the grand jury indicted Hayden.

The trial that started on October 7, 1879, in New Haven was bound to "take its rank among the causes célèbres of criminal history." The prosecution produced one hundred and six witnesses and the defense seventy. Among these were twelve distinguished professors, eight of them from Yale College, and fifteen doctors of all grades and shapes.[45] The scientific crowd arrayed itself spiritedly on both sides of the three scientific matters as issue: the significance of an ovarian tumor found in the autopsy; the distinguishing of the arsenic Hayden had bought from that found in the body, and the identification of the blood stains. As for the tumor, White swore that he had found one on the left ovary, another doctor was positive it was on the right, a third saw no outgrowth at all, and a fourth seemed to doubt the existence of either a growth or an ovary. All the doctors for the defense swore that such a growth, whether it existed or not, could not have led Mary Stannard to believe that she was pregnant; the prosecution doctors were equally positive that it would and did. The arsenic question, which generated by far the most attention, followed the same pattern. Some doctors seemed to think that arsenic taken into a stomach, dead or alive, must produce inflammation. Others thought not. Others yet opined that it might do so if taken into a live stomach, but not if taken into a dead one; the rest thought the opposite. In short, "the profession maintained its old-time reputation—none of the doctors agreed."[46]

Then there was the battle of experts over the nature of the blood found on the knife. The three champions for the state were Professor White of Yale; Professor Theodore Wormley from the University of

Pennsylvania, an authority on forensic chemistry and a close collaborator of Richardson, and Dr. Joshua Treadwell of Boston, another noted microscopist. Returning for the defense was Professor Sanford of Yale, joined by the formidable Dr. Joseph J. Woodward, who gladly accepted an invitation from the defense to come to New Haven and point to the many defects in the methods of the state's experts.

On the witness stand, Treadwell claimed that he could restore the dried corpuscles and determine whether they were the corpuscles of human or animal blood. Woodward, who was described by the reporter of the *Sun* as talking "as though fed with words from a steam engine," was equally positive he could not. Wormley and Sanford tried to explain to the jurors their conflicting versions of how everything depended on the circumstances and on the animals in question. In short, the debate earlier confined to scientific journals was reproduced in the courtroom. But, unlike the previous rounds, the lawyers ran this one. A veteran in court, Wormley avoided all traps laid by the defense, and came "out of the fog of cross examination with a clear throat and sound lungs." Not so, however, the young and inexperienced White, who was so conscientious, the *Sun* reported, that if asked if he had seen a certain object, he would qualify his answer by saying that his eyes saw it.

"Will you swear that your eyes saw it?" his tormentor would shout.

"I will swear that they saw it to the best of my knowledge and belief," the professor would cautiously respond.

"Will you swear that there is no shadow of possibility that your eyes may have been mistaken?"

"I will swear, to the best of my knowledge and belief, that I do not think they were mistaken," the professor would reply.

"Will you swear, to the best of your knowledge and belief, that there is no shadow of possibility that they may have been mistaken? Come, Sir; yes or no," persisted his interlocutor.

"I cannot give a positive answer," the professor would say.[47]

Worst of all, during the previous grand jury investigation, under time pressure from his own counsel, White had made only a superficial examination before claiming to find human blood on a stone found next to Mary Stannard's body. Later, after the grand jury verdict, he discovered that the "stain" was actually moss. Although he acknowledged his mistake before the trial, the defense made him repeat it again and again from the witness stand. Six times White was called by the prosecution to testify as a scientific expert (on the arsenic, the condition of the stomach, the gash in the throat, the blood on the knife, and to give evidence on maternity and ovarian outgrowths), and each time the moss was set in motion, "to the confusion of the prosecution and the demoralization of the rigidly conscientious professor."

Treadwell, the third expert witness for the prosecution, did not do well either. The defense counsel "annoyed him as a yellow jacket would annoy a mettlesome colt. The Boston expert would snort and cavort until his foot was caught in his breaching, then he would kick himself loose and start off at full speed, with all the lawyers after him, the defense trying to keep him going, and the prosecution to rope him in." In one of his tantrums, he kicked Dr. Joseph J. Woodward, who pointed to Treadwell's unskilled handling of the instruments, and to his photographs "that represented optical delusions which do not exist in the corpuscles themselves." "To thoroughly appreciate the merits of the dispute," the *Sun* explained to the mesmerized public, "the reader should take into account the amount of blood discovered. Take all the corpuscles said to have been found on the Hayden knife and shirt, and they would make a drop of blood 4,999,999 times [*sic*] smaller than this degree mark °." The defense counselor advanced only a slightly different interpretation. "Scientific evidence," he summarized in his closing speech, "is the most dangerous of all evidence. If any scamp wants to sell a silver mine, he will dig at Yale College to get expert testimony."[48]

Whether the jury believed the learned advocate or not, it "did not waste any time over the testimony of experts." The cabinet containing the scientific evidence was "not even opened" during the deliberation.[49] After three and a half months of expensive litigation that exhausted the resources of the legal and the medical professions, and cost the state of Connecticut $30,000, the jury failed to reach a verdict. The disagreement, nevertheless, was treated as tantamount to acquittal, since eleven jurors agreed on a not-guilty verdict. "Mr. Hayden will be accepted by his fellow citizens all the more readily," wrote the *New York Times*, "because few men are ever condemned to pass through a more agonizing ordeal than those long three months of legal investigation."[50]

In the first presidential address of the fledgling American Society of Microscopists, its president-elect, Dr. Richard Ward, lamented the dire state of the forensic sciences, "where persons without character, who are liable to creep into any profession without becoming assimilated to it, find it easiest, for selfish purposes and with impunity, to make themselves the enemies instead of the friends and protectors of society." The editor of the *American Monthly Microscopical Journal* also could not hide his disgust:

> After the mass of evidence which has been given by the scientific gentlemen on both sides it will be strange if any jury of ordinary intelligence can decide what is right or wrong. Really, there is much confusion where there should be none at all; whether the microscopist is or is not able to positively identify human blood, or any other kind of blood, is no matter of opinion, but of fact and experience; and there should be no dilly-dally about it. In our next issue, we will review the evidence, and show how easily a man can make assertions under oath, which, after more thorough study, he would discover to be unfounded.[51]

In its next issue, the *Journal* left no doubt concerning whose evidence it considered to be unfounded. White was criticized for embarrassing the profession with the mistake he had made during the grand jury investigation. The bulk of the criticism, though, was directed toward the presumptuous testimony of Treadwell, who was quoted as testifying, after measuring only four corpuscles (having accidentally lost the others) ranging from 0.0075 mm to 0.011 mm, that he was "quite certain that these were human blood corpuscles, and that they did not belong to the blood of the pig, sheep, goat, horse or cat."

The stage thus being cleared, Woodward, the self-appointed defender of the profession's probity, gave his account of the situation:

> Our problem is not that of measurement. The science of micrometrics has advanced considerably since the days of Gulliver, but it is still producing contradictory results. . . . [The problem is] that the difference in size between the largest and the smallest human blood corpuscles was about as great, relatively, as that between the shortest and the tallest adult man; and as in both cases all intermediate sizes occur. . . . We cannot expect, by measuring ten, or fifty, or a hundred, or even a much larger number of corpuscles, to obtain an average size that will agree with the next set of similar measurements. . . . I did not think the microscopist is warranted in attempting to distinguish on oath, between human and canine blood . . . he will no doubt often come out right, but he will also occasionally come out entirely wrong.[52]

While the microscopists exchanged accusations, it was *The New York Times* that articulated the most reasonable view of the scientific side of the Hayden affair:

It is not necessary to impute dishonesty or mercenary motives to the eminent experts in medicine or other branches of science, whose disagreements put the minds of jurymen in a maze, instead of leading them into the light. But there are experts and experts, there are theories and theories, and there are even facts and facts, in every department of science or special knowledge, and lawyers can ingeniously make their selection, giving prominence to some and keeping others out of sight, and twisting and turning until the inexpert mind in the jury box is in danger of losing all faith in science as a witness.[53]

It was this danger—that of the inexpert mind "losing all faith in science as a witness"—that concerned leaders of the scientific community such as Colonel Woodward and Samuel Laws, the president of the University of Missouri. Startled by the events of the Hayden trial, Laws wrote the Missouri Judge H. S. Kelly, a respected authority on criminal law, about his correspondence with Governor Phelps on the subject of the diagnosis of blood stains.[54] Laws asked Kelly for help in placing the problem "before the public, especially before the bar and the courts, as not only true ends of justice may be served, but a great saving of the public money be effected by the diffusion of reliable information." In addition to Governor Phelps' letter of inquiry regarding whether science can distinguish human from animal blood, and Professor Duncan's and his own negative answers, Laws also enclosed a letter from his chemistry professor, Dr. Schweitzer, who, like his colleagues, pronounced "the decision of the question with the knowledge at present, impossible."[55]

Justice Kelly forwarded the letters to the influential *Central Law Journal*, which published the package under the title, "Can Human and Animal Blood be Distinguished in Case of Blood Stains?"[56] This was but one of several efforts to put an end to the frustrating controversy over the microscopic blood tests. Worried about the detrimental effect the debate might have on the image of the young profession, the elite of the newly organized microscopical community kept pressing for conformity. Leading the campaign was the community's magazine, the *American Monthly Microscopical Journal*. There, in October 1880, Dr.

Charles Curtman of the St. Louis Medico-Chirurgical Society reminded the readers that "not long ago it was imagined by the public and even by members of the profession . . . that human blood might be positively and easily identified by the microscope . . . But by degrees the illusion has vanished . . . especially since the thorough investigation of the subject by Dr. J. J. Woodward, no expert would be found bold enough to assert, in the face of contrary evidence, that he can positively identify human blood. Criticizing "a gentleman, well-known to the readers of this journal," who had testified in a recent trial that certain spots found on a coat were produced by human blood, the editor stressed again that "until experience has shown such evidence to be sure and infallible, no scientific man is warranted in stating that a stain upon cloth is made by human blood from microscopical examination alone."[57]

Unlike other professional communities, the infant community of microscopists was made up of persons engaged in many different lines of work, and not all of them agreed to follow the line laid down by their leaders. Many were annoyed by what they saw as manipulative rhetoric and the misrepresentation of an issue, which had significant areas of black and white as well as a problematic gray middle. Thus, after another, particularly nasty, attack on Richardson, Professor S. Gage of Cornell protested to the *Journal* "that Dr. Richardson's claims are very modest" and that "every one who has made a study of the blood corpuscles with high powers, would think is justifiable for him to do so."[58]

The loudest protest came from those who made money in court as expert witnesses. Mocking the Missouri professors, Dr. Richard Piper, a Chicago microscopist, dismissed their opinions as "an odd jumble of false statements and misapplied facts" gathered from out-of-date authorities." Piper called attention to the fact that neither Treadwell nor Richardson ever claimed to be able to distinguish human blood from that of all animals under all circumstances. "Every case should be tried upon its own merits alone," he argued.

Against Woodward's statistical argument, Piper recruited higher authority. "[Charles] Babbage says there is nothing more uncertain than the duration of individual life, while nothing is more certain than the duration of life as applied to a multitude of individuals. It would seem that this might apply with equal force as to the average size of individuals or their blood corpuscles." In any case, Piper, an experienced expert witness, did not think that precision mattered much. "However important we deem this question of microscopic measurement, in a scientific point of view, still its settlement to the utmost of nicety is of but little consequence as it regards legal cases. The variation of a few millionths of an inch would hardly be taken into account as of any value in a given case."[59]

Piper's aggressive rebuttal was part of his efforts to promote his own system of measurement. Piper's system differed from earlier ones because he measured an aggregate instead of separate corpuscles. He would carefully copy the corpuscles' microscopic images onto a paper, one by one, until he had a pile of a hundred, ordered tightly in ten rows. Then he would measure both the width and length of the resulting aggregate. Piper claimed to be able to reconstruct the dry corpuscles, "precisely in size and form as they exist on the substance on which they were first received." Furthermore, having the aggregate of these corpuscles measured in two different directions allowed him to compensate for any remaining irregularities in the shape of the reconstructed corpuscles. He believed that he could deal with dried blood with the same ease he dealt with fresh.[60]

Piper chose to advocate his system in a "journal of the legal profession, to which profession they seem practically to belong, and leave it to them and the scientific world generally to pronounce judgment."[61] Given the many legal appearances mentioned in his papers, Piper must have enjoyed the confidence of the legal profession. The microscopists, however, dismissed him. No reference to his articles or his method has been found in any of the serious scientific discussions of the time.[62] Piper's rhetoric may have frightened the community, which still carried fresh in its collective memory the controversy between Richardson and Woodward, and the Hayden scandal. Thirty years after the British authority Alfred Taylor "declined to say absolutely that the stains were caused by human blood, although the corpuscles coincided in measurement with them," Professor Wormley, equipped with a microscope ten times more powerful, adopted the same position: "The microscope may enable us to determine with great certainty that a blood is not that of a certain animal and is consistent with the blood of man; but in no instance does it in itself enable us to say that the blood is really human, or indicate from what particular species of animal it was derived."[63]

The New Generation: Late 1880s–1900

Woodward died in August 1884, exactly a year before his peculiar statistical argument concerning the instability of corpuscles' mean measurements was refuted by Marshall D. Ewell, a professor of Common

Law at Union College of Law in Chicago. Ewell's interest in forensic science had turned him into one of the leading microscopists of his generation. He found that the average measured size of a group of red corpuscles converged as the group's size grew. The difference between the smallest and the largest averages of 25 corpuscles was 0.0007 mm; for 50, 0.00038 mm; for 75, 0.0003 mm; and for 100, 0.0002 mm. Thus, he concluded, "there appears to be an average size which varies within very narrow limits." It followed, for him, that "it is reckless to the last degree, if not criminal, to express an opinion upon measurements of less than 100 corpuscles." Ewell also warned that his results demonstrated that "by selecting the corpuscles it would be possible for a dishonest observer to make the average much larger or smaller, without the possibility of detection; a fact, the bearing of which upon the value of expert testimony is so obvious as to need no comment."[64]

Ewell represented the newer generation of professional microscopists who were more at ease than their predecessors with the technicalities of the field. From the mid-1880s, research reports regularly included the manufacturers, the specifications, and other details of the instruments used (the microscope, the screw that moved the microscope stage, the eyepiece micrometer and the stage micrometer that calibrated it, the illuminator, and so on), a discussion of the methods used in the preparation of the samples (age of the blood, the drying process, fluid used, and so forth) and in their measurement (operation of the instruments, position of the illuminator, etc.), and an account of possible sources of error, number of samples needed, the quality of various instrument makers, and the unavoidable rules of thumb of technical expertise.

The march of technology, the accumulation of expertise, and a clearer understanding of the problems involved, kept the hope alive among microscopists that, within certain limits, the different species of mammalian blood could be positively identified by the size of their red corpuscles. Thus, in 1889, the president of the American Society of Microscopists, Dr. W. J. Lewis, provided assurance that "where two or more persons, experts in the use of the microscope, are called upon to testify, there should be no disagreement as to the results of any examination they may make." Lewis's presidential address expressed the growing confidence of the thriving field: "there will be revelations which shall change the whole theory of a plea in civil actions, while in criminal cases, they become a terror to the guilty or a joy to the innocent . . . the time has fully come when counsel and client, courts and juries, must and will give heed to its disclosures."[65]

Photographic techniques first suggested in the 1870s by Woodward and Dr. Carl Seiler supported this optimism with photos of single corpuscles, enlarged to 10,000 diameters. In this format, the diameter of human corpuscles came to 3⅛ inches; guinea pigs, 3 inches; dogs, 2⅘; ox, 2⅓ inches; sheep, 2 inches; and so on. With this technology, claimed in 1888 Dr. Henry Formad, a former student of the late Richardson and lecturer on experimental pathology and morbid anatomy at the University of Pennsylvania, "even the rabbit's and dog's blood can be easily distinguished from human blood, by the quite appreciably smaller diameter plainly seen under high amplification." Moreover, even in cases where the dried blood was badly abused, "the diagnosis should not be declared impossible, as long as there are some perfect bi-concave corpuscles present, even if the bulk of the corpuscles are distorted, for we have seen that even altered corpuscles can be measured."[66]

While some parts of the growing bodies of theoretical and technical knowledge bolstered optimism, others fed skepticism. Physiological research revealed a number of factors that affected the size distribution of the red corpuscles. Already in 1878, a French researcher, Georges Hayem, had concluded that the variation in the diameter of red corpuscles related to gender, health, and age was far greater than previously suspected. The average size of the corpuscles of women undergo large fluctuations caused by slight disturbances; previous conditions of health, such as leukemia and anemia (but also simpler conditions such as fever) has a significant influence on the size distribution, and the number of smaller corpuscles increases with age. Other researchers reported additional factors, such as drugs and nutrition that influence the corpuscles' size.[67]

On the technical side, increasing proficiency brought awareness of a need for standardized equipment.[68] That threw all published standards into doubt. Any microscopist who wanted to apply the micrometric blood test had to make all pertinent measurements by himself—that is, to measure the size of the corpuscles of the blood in question, those of the animal the accused said he or she killed, and, for good measure, those of the victim. Alas, even measurements confined to a single instrument might not have been reliable. Two micrometers were involved in the measurement process: the eyepiece micrometer and a stage micrometer. The eyepiece micrometer was a slip of glass that fit the eyepiece of the microscope and was ruled with fine equidistant lines. The stage micrometer, which came in many shapes, was graduated in a standard fashion. Once the image was in focus, the stage micrometer was used to establish the value of the lines on the eyepiece micrometer. For example, if a single $\frac{1}{1000}$-inch division (the English standard of the day) of the stage micrometer covered 20 lines in the eyepiece micrometer, then each division in the later was taken as being equal to $\frac{1}{20,000}$ inch (approx. 0.00127 mm). The uniform ruling, especially of the eyepiece micrometer, was therefore crucial even for a single measurement. And although much improvement had been made during the half a century since 1840, the corresponding increase in the microscope's power kept the error from disappearing.

Optical progress carried its own uncertainties. With a depressed center and thick borders, the red corpuscle consists optically of two lenses: a bi-concave center, dissipating the light, and a thick band around it, a double convex, concentrating the light. Under the microscope, a dark ring showed around the corpuscle, whose thickness depended on the magnification. It was not clear whether the ring arose from total refraction of light at the edge of the particle or was an optical illusion caused by the diffraction of light. Some observers measured the whole of the ring, some none, and some tried for half of it. As amplification increased, so did the discrepancy in the results. Another related problem was created by the light apparatus. Different lighting angles gave different impressions of the three-dimensional corpuscle. Last, but certainly not least, came the optical trade-offs between amplification, definition, and flatness of the visual field. Flatness provided an opening for many mistakes, since it compelled the observer to relocate each measured corpuscle in the center of the field. These trade-offs impeded optical progress. Hence, Ewell's contention that "there is no advantage gained in the use of high powers, as claimed by Dr. Richardson and others; and especially is this so, where, as in the paper of Dr. Formad, amplification is pushed to such an extent as to destroy definition and leave the margin of the corpuscle to be measured a nebulous patch, whose beginning and ending it is impossible to determine."[69]

We are not done with difficulties. Microscopy suffers from the so-called *personal equation*, an irremediable error, peculiar to the individual, that accompanied every human measurement. Investigated in its astronomical context already in the 1820s, by Friedrich Wilhelm Bessel, the effect was still under investigation in the 1860s and 1870s. It was found that the personal equation varies according to the objects seen; it differs with the sun, the moon, and the stars; the magnitude, brightness, direction, and movement of a star; and with many other factors as well. The discovery of so many conditions of variations made the attempt to take them all into account hopeless.[70] The notion of the *personal equation* soon spread from astronomy to other fields of measurements, among them forensic microscopy, where the discrepancies in the measurements made by different observers were notorious.

In order to assess the relative accuracy of micrometric measurements with different apparatus and different observers, Ewell ruled two series of lines onto a glass plate, one set of eleven lines equally spaced about 0.004 inch apart and another set of five lines spaced about 0.008 inch apart. Then he asked six microscopists to measure the spaces at least five times with the same standard stage micrometer. The means of the different series of measurements differed by more than $\frac{1}{9000}$ of an inch, a discrepancy "greater than the greatest difference between the different measurements of blood corpuscles of man and some common domestic animals, e.g., dogs." Additional experimentation convinced Ewell "that identical results are not to be expected even by the same observer when the measurements are made at different sittings," and that "even with the best apparatus varying results are to be expected."[71]

Ewell's research resulted in a warning. "The very utmost the present state of science enables us to state," he told the members of the American Microscopical Society as its president in 1892, "is that the blood in question is or is not that of a mammal, a bird, fish, or reptile. To say more than this, is, in my judgment, without warrant to imperil human life, and, in the words of the late Dr. Woodward, to make scientific experts more dangerous to society than the very criminals they are called upon to convict."[72] Ewell's warning was timely. Their use spreading into civil cases, the microscopists found themselves increasingly deployed in cases involving large sums of money, such as commercial forgeries and contested wills. These cases tended to tarnish the public images of all parties involved, especially of the hired experts. "There is one interesting difference between astronomy and microscopy," noted a sarcastic editor of a Chicago newspaper, after a particularly scandalous case:

> The astronomer is never a humbug, but the same cannot be said of the microscopist. The astronomer is never a hired witness in a court of justice, and that is the highest ambition of the microscopist . . . The astronomer is generally an honest old fellow with seedy clothes and an empty pocket book, while the microscopist is sleek and prosperous and is strongly tempted to be a charlatan. His evidence is very unreliable, from circumstances beyond his control, such as the peculiarities of his own eyes and the blundering unscientific qualities of his own mind; but it is to be feared that his vision is also sometimes distorted by something else than the

inequalities of his crystalline lens and the undue refractive power of his vitreous humor.[73]

Ewell's research on the "personal equation" attracted much attention, because it seemed to add a new argument to the old debate on the micrometric blood test. "The consequences of Prof. Ewell's startling proposition," Professor Moses White told the members of the American Microscopical Society in 1894, "are too great to allow it to pass over without careful criticism." Fifteen years after doing so poorly in the Hayden trial, the young and conscientious professor of pathology of Yale College had become a court veteran, a prominent microscopist, and a celebrated inventor whose improved micro-spectroscope had overcome the sticky problem of keeping the eyes steady while looking for spectral bands. White took it upon himself to vindicate the micrometric blood test. He faulted the low powers (100 to 250 diameters) used by Ewell's observers. He cited the results of a parallel research by Professor Theodor Wormley, in which eleven lines (ten spaces), measured by three observers with different instruments, produced a difference of no more than $\frac{1}{200,000}$ of an inch. White: "With the above demonstrations, I think I may dismiss the objections of Professor Ewell." Reviewing the latest results in the field, White declared triumphantly that a testimony by a skilled microscopist remained "of untold importance in saving the lives of the innocent, and often in overthrowing the plea of those who are guilty."[74]

Plus ca change. "The question of distinguishing human blood from that of other animals," confessed Dr. Henry Tolman in January 1894 before the Microscopical Section of the Chicago Academy of Sciences, "is by no means settled, and in fact the study is yet in its infancy. There are so many factors involved, so many nicely balanced questions to be considered, that the subject is one great difficulty." Dr. William S. Thorne of San Francisco concurred. "Volumes, innumerable brochures and monographs have been written on this subject, and yet the position of the red blood corpuscles in judicial proceedings is underdetermined, the medical mind is unsettled and the opinions of experts conflicting." The mounting difficulties stopped neither Tolman nor Thorne from having a fixed opinion. Tolman held that "the determination of human blood, as against all other animals except perhaps a dog, can be made with such a degree of certainty, as to entitle the evidence of a well qualified expert on that subject to much weight." Thorne claimed that "we have experts among us who boast of having sent persons charged with crimes to the gallows upon the certainty of a microscopical distinction between human blood and that of lower animals. When we consider that microscopical opinion as to the source of the blood is usually, if not wholly, based upon measurements of the red blood corpuscles, that measurements of the same corpuscles vary with different observers . . . a difference that is often greater than that which exists between blood of man and that of the lower animals, such statements appear appalling."[75]

Losing Interest: 1900s

Nineteenth-century microscopists dreamed of becoming "a true amicus curia . . . a terror to the guilty and a joy to the innocent."[76] They found, however, that the production of scientific knowledge for legal purposes involved much more than technical progress and disinterested measurements. The participants in the debate over the micrometric blood test all agreed about the scientific facts, that is, that there are a few mammals whose red corpuscles were so similar in size to those of man as to render distinguishing among them doubtful. They disagreed, however, about what should be derived from these facts for judicial purposes. White, Richardson, Treadwell, Piper, and Tolman thought that in most cases the microscopist could and should be able to pronounce reliable judgment. Gulliver, Woodward, Ewell, and Thorne believed that as long as the experts could not be certain, they should withhold their opinion.

Both camps justified their view by appealing to the moral responsibility of science "to secure for humanity, by our researches, the greatest benefit with the least injury." For one group, this meant the responsibility to take part "in saving the lives of the innocent, and in overthrowing the plea of those who are guilty." For the second group, it meant vigilance, "without which scientific experts may become more dangerous to society than the very criminals they are called upon to convict."[77] These considerations were not mere rhetoric. They shaped research programs, affected the content and tenor of publications, and influenced the politics of the microscopists' community. Thus, Richardson attempted to support his views by experimenting with the gigantic blood-disks of reptiles, Ewell bolstered his position by experimenting with the relative accuracy of micrometric measurements, Richardson sidestepped the fact that similiarities in the corpuscles of certain animals made his conclusions doubtful, and Woodward dedicated a large portion of his extensive research and publications to confirming this very fact.

In spite of the deep and lingering scientific disagreement, legal convention remained unwavering throughout the second half of the nineteenth century that the micrometric blood test, "although not infallibly correct, is worthy of the greatest consideration by court and jury as being the best of opinion evidence."[78] Why? A clue to the answer may be found in the 1879 case that opened this chapter. The officers of the law had good reason to believe that William Young had murdered all five members of the Spencer family in their sleep. Yet, they could not prove it. It seemed within the power of microscopy to translate the little dark spots found on Young's shirt and pants into an unmistakable proof of guilt, illuminated by the "colorless light of scientific truth." What court could ignore such powerful agency? Thus, despite the tenacious debate that polarized the scientific community and despite the recurring scandals in highly publicized murder trials, the results of the micrometric blood test were received in evidence in many cases throughout the second half of the nineteenth century. As the *Central Law Journal* reported in 1896, the test "is now received without objection."[79]

Leaders of the microscopical community feared their colleagues' recourse to the law. The microscopists' business was dispassionately to weigh and measure the smallest visible objects of creation and the uniformity of their results was crucial to their credibility. Alas, once the results of the micrometric blood test entered the courtroom, the lawyers wasted no time taking advantage of the many uncertainties and contingencies related to the micrometric blood test to generate strong interpretive disagreements among the scientific witnesses and challenge their reliability.[80] Worried about the public image of their young profession, leaders of the microscopical community kept pressing for conformity. This, however, was not easy to achieve. Late-nineteenth-century microscopists had only limited employment opportunities, usually in hospitals or universities, and the legal system offered a most lucrative market for their expertise. Thus, as Richardson had predicted, "bloody dog tales" kept appearing, providing a good livelihood for the scientific experts, who continued to appear in criminal cases and testify under oath to the results of their blood micrometric tests.[81]

In 1900, gathered at the grand ballroom of a New York Hotel to look back on a glorious century of achievement and to speculate about an even greater one ahead, the members of the American Microscopical

Society listened to their president, Dr. Bleile, as he rewrote their history. There was a time, Bleile told his audience, when

> men were found enthusiastic enough to declare even under oath that the results [of the micrometric blood test] were positive and trustworthy, thus taking upon themselves the greatest responsibility in vouching for the correctness of the statement that the blood in hand was or was not human blood, [but] newer and better knowledge soon led to a more conservative position. For we know now that the differences in size do exist in the corpuscles of one individual and in individuals of the same species, differences so great that they may easily overlap the average measurements given for another species . . . This method of recognition, then, as between mammalian bloods has been generally given up as untrustworthy; and while it is easy to distinguish between the oval and nucleated corpuscles of the ovipara and the circular non-nucleated of mammals, it is to the highest degree unsafe according to more conservative view to attempt more.[82]

The debate, however, was far from being decided. Microscopists would have continued to oppose each other until the end of time had it not been for a new development that rendered the whole debate otiose. During the 1890s, Europe was preoccupied with the development of antitoxic sera for infectious diseases. By 1900, the relevant body of knowledge included the fact that blood possesses the remarkable property of developing distinct defense mechanisms against each foreign protein it encounters. In 1901, Paul Uhlenhuth, a young *docent* at the Hygienic Institute of the University of Greifswald, suggested that this property could be utilized to identify the blood of any given species. If the blood of animal A were injected into animal B of a different species, Uhlenhuth suggested, the blood serum of B would develop a particular defense mechanism that precipitated the specific proteins of A. "I have taken blood of men, horses, and cows," he wrote in a small article in 1901, "dried them on a board for four weeks, then re-moistened them in NaCl solution, and was able to identify the human blood at once with the help of my serum—this should be of great importance to forensic medicine."[83]

Indeed, it was. The new testing procedure, which came to be known as the serological test, quickly dethroned the older microscopical-method after its sixty-year reign. The long and winding scientific debate concerning the differentiation of human from animal blood by microscopical measurements was, therefore, never resolved. Instead, it was dropped altogether, rendered useless by a new method.[84] Falling from grace, the micrometric blood test has never made it into the chronicles of nineteenth-century microscopy. Concentrating on the establishment of microscopy as an educational and research program in various scientific and medical fields, histories of nineteenth-century microscopy tend to present a linear success story, with a "take-off" in the 1830s and its eventual establishment in the 1870s and 1880s. Incorporating forensic microscopy, a major player, into the picture counterbalances this view and reveals a far more complex story.[85]

Attempts to distinguish human from animal blood constituted an important chapter in the histories of nineteenth-century law and microscopy in America. For the legal system, the micrometric blood test remained the only credible scientific technique for the crucial task of identifying human blood. For the fledgling American microscopical community, it provided not only legal patronage, but also important loci for technical pressures and theoretical discussions concerning central problems such as standardization, accuracy, and reliability, and for an intensive ethical discourse about the character of the microscopic community and its function in society.

Microscopy is not the only science whose historians could benefit from a visit to the legal archives. The biggest story of fin-de-siècle science, the discovery of x-rays, is another example. The discovery of x-rays affected legal practice no less than it affected medical practice and, as the next chapter suggests, the reactions of the two learned professions to the mysterious rays need to be understood each within the context of the other.

5

The Authority of Shadows: The Law and X-Rays

A document purporting to be a map, picture or diagram,
is, for evidential purposes simply nothing.
Except so far as it has a human being's credit to support it.
It is mere waste paper – a testimonial nonentity. . .
It is somebody's testimony, or it is nothing.

~John Henry Wigmore, *A Treatise on a System of Evidence
in Trials at Common Law (1904)*

THE EARLY YEARS OF X-RAY PRACTICE in America
were described as "a piebald proceeding, a sort of Joseph's coat of many
colors, which fitted no one."[1] Indeed, commencing early in 1896, x-ray
practice in America remained for more than two decades an unregulated territory, inhabited by a welter of photographers, electrical engineers, physicists, medical novices and other speculative souls. It was
only in the 1920s that an organized community of medical specialists,
the radiologists, successfully claimed monopoly over x-ray practice.
The slow warming of the American medical community to the new
discovery has posed a puzzle to historians. Some explained this as being
the results of the difficult theoretical and technical challenges that faced
the infant technology.[2] Others concentrated on the social and institutional intricacies involved in the molding of the expertise of the photographer, the electrician, and the doctor into a new profession.[3] Still
others pondered over the great epistemological barriers involved in
shifting from the tactile and the verbal to the visual.[4] This chapter
reveals this "piebald proceeding" as a coherent plot impelled by a distinct legal impetus.

The late nineteenth-century discovery of the x-ray baffled the medical and the legal worlds. The possibility of looking into the human body as if through an open window challenged the time-honored medical monopoly over the inner cavities of the human body. Likewise, the possibility of putting on view objects unavailable to the naked eye challenged the established legal theories and practices of illustration and proof. In recounting the reactions of the medical and the legal professions in America to these challenges, this chapter treats the two professions as two deeply connected social institutions, carrying ongoing negotiations through which legal doctrines affect medicine no less than scientific discoveries and medical applications affect the law. The joint analysis rewards us with a rich story about an early and overlooked chapter in x-ray history; the professionalization of radiology, the origins of defensive medicine, and the evolution of the legal theory and practice of visual evidence.

Malpractice

On June 5, 1895, James Smith of Denver, Colorado, had a great fall. Some say that he injured his left thigh. Others mention his right hip. Being a young man of modest means, Smith tried to outlast the pain, but after three weeks with no improvement he gave up and sought medical help. In 1895, patients still expected their doctors to make house calls, the cost of which ranged from two to five dollars. If the doctor had to travel more than a mile, an extra dollar was added for each mile, two dollars at night. Some doctors had a meter-like device attached to the wheel of their buggy to measure the mileage. Thus, on June 29, 1895, Dr. William W. Grant, a surgeon of national reputation, who headed the surgical staff at St. Joseph, Denver's first private hospital, paid Smith a visit to check on his injury.[5]

Late-nineteenth-century orthopedic diagnosis was a trying task for both the surgeon and the patient. In order to determine the position of the bones, the doctors had to manually probe the injury. Deformity, due to displacement or swelling, was the most important diagnostic clue. However, this did not preclude the possibility of congenital asymmetries or of fractures that caused no deformity. Abnormal mobility was doctors' second clue. However, in cases of a fissure, or incomplete or intra-articular fractures, it was not always recognizable. Finally, there was the "forlorn hope of the surgeon"—crepitus, the sound generated by grating bones. But again, not all fragments could be brought against each other, and not all injuries generated sounds. In addition, the traveling of the sound through the bone could lead to afalse impression as to the seat of the fracture. In many cases, late-nineteenth-century surgeons had no choice but to resort to the patient's own account of pain and loss of function for their orthopedic diagnoses.[6]

Dr. Grant examined Smith thoroughly but found no evidence of a fracture. Thus, he did not attempt to immobilize Smith's limb but, on the contrary, advised various exercises as though he were treating a mere bruise or a strain. A week later, when Smith did not improve, Dr. Grant again examined him and again found no evidence of a fracture. That was the last Dr. Grant heard from Smith, until April 1896, when Smith brought a malpractice suit against him for the lofty amount of $10,000. Presumably, when Smith's health failed to improve, he sought the advice of other physicians who diagnosed his injury as a fracture of the femur. Smith then sued Dr. Grant, arguing that Grant's mistaken diagnosis and improper treatment constituted negligence and want of skill, which prevented his recovery, caused him much suffering, and damaged him for life.[7]

Smith's malpractice suit was not unusual. Heightened expectations and demand created by advancing medical technology, coupled with a rising preoccupation with health and physical appearance, created an epidemic of malpractice litigation throughout the second half of the nineteenth century. The majority of these malpractice lawsuits were orthopedic cases, usually involving compound fractures, where patients often found themselves with demonstrable problems and sued the doctors who had taken care of them. Disgruntled patients found it easy and cheap to seek legal remedy. Filing a malpractice suit cost a nominal fee; a growing population of lawyers, eager to take such suits for contingency fees as high as 50 percent, provided victims with easy access to legal services, while a lack of professional standards and abundant intra-professional rivalry assured them of plenty of medical expert testimony to support their claims in court. Finally, if they were poor enough, they were usually exempt from paying costs in case of defeat.[8]

While malpractice suitors had usually little to lose and much to gain, the defendants had everything to lose and nothing to gain. "To be a defendant in a malpractice case and win the verdict is only a little less disastrous than to have been beaten," complained one Colorado surgeon. In either case, costs mounted and notoriety was unavoidable. The patient, concurred a bitter Maine surgeon, "sues upon the principle of flipping the pennies, heads I win, tails you lose. The suit depletes the surgeon's pocket and ruins his reputation. To pay is ruinous, to defend is ruinous, and to live in constant dread is ruinous." Ironically, it was the most successful surgeons who were also the most frequent target of malpractice suits because they offered the best incentive for suing and the best chance to collect. The more prosperous the practitioner was, the more he had to lose, and the more likely he could be frightened into cash settlement.[9]

Dr. William W. Grant's practice was prosperous and his reputation nationwide. Ten years earlier, in 1885, he had performed the first successful appendectomy in the United States. He was also the first to operate for facial paralysis by anastomosis of facial and spinal accessory nerves, and his operations for disease and deformity of the mouth nerves were considered standards in the profession. Nor did his public career lag behind. He was among the founders of the American College of Surgeons and the Western Surgical Association. At various times, he had served as the president of the Western Surgical Association, the Colorado State Medical Society, and the Denver City and County Medical Society. He had served also as a trustee of the American Medical Association for sixteen years, being nominataed twice for its presidency. Grant, therefore, had much to lose. Still, known for his feisty nature, he had no intention of being frightened into a cash settlement by a suit he conceived of as simple blackmail. Instead, he chose to follow the advice of his Philadelphia colleague, Dr. Mordecai Price: "Such cases should under no circumstances be compromised—not a penny for tribute, but all we have for self-defense."[10]

Thus, on Wednesday, December 2, 1896, both parties appeared at the District Court of Arapahoe County, now known as the District Court of Denver, Colorado, ready to do battle before Judge Owen E. Lefevre and his jury. Dr. William W. Grant, tall and well dressed, was represented by three well-known lawyers: two former judges and a U.S. Senator, Charles J. Hughes, who was considered by many to be the most brilliant lawyer Denver ever had. James Smith, ailing, his legs hanging useless on crutches for all to see, was represented by two attorneys in their early twenties. However, having made his meager living typing legal documents and depositions for various law offices, Smith must have had access to excellent legal advice, for he had chosen his legal team extremely well. Practically unknown at the time, both his counsels, Ben B. Lindsey and Fred W. Parks, had brilliant careers lying ahead of them—Parkes, as the youngest senator in Colorado's history and Lindsey as a well-known judge and founder of the Denver Juvenile Court.[11]

Producing the Evidence

What made *Smith v. Grant* extraordinary was not the assembly of famous names but rather the attempt made by the young prosecution team to use x-ray photographs as proof of James Smith's injury.[12] Professor Wilhelm Roentgen of the Bavarian University of Würzburg had first made the discovery of x-rays public less than a year earlier, in December 1895. The notion of a new kind of ray, unrefractable and indifferent to electromagnetic fields, befuddled the scientific world and precipitated feverish research into their nature and implications for the long-standing theories of light and matter. Popular culture was equally mesmerized. The notion of a "dark light" that could penetrate flesh as easily as glass and produce photographic images of the skeleton was intoxicating. Overnight, the mysterious rays became popular icons constantly encountered in advertisements, prose, songs, and cartoons. More than one thousand articles and fifty books were published on the subject in 1896 alone.[13]

The doctors were as excited as everyone else by the new discovery. A series of diagnostic mechanisms, such as the stethoscope, the ophthalmoscope, and the laryngoscope, had already allowed late-nineteenth-century practitioners to extend the power of their senses into the body's inner cavities. Still, the notion that they could inspect, not to mention photograph, the most recondite structures and functions of the body seemed almost surreal. "The surgical imagination can pleasurably lose itself in devising endless application of this wonderful process," meditated one doctor in the *Medical News*. "If it becomes possible to drive these mysterious rays through the entire body as clearly as they now penetrate the hand, the realm of utility will be practically boundless."[14] Others preferred to reserve judgment. The weaknesses of the infant x-ray technology were many. Moreover, there was neither a body of anatomical knowledge to confer precise meaning on the images produced nor a medical language by which such meaning could be communicated.[15] The majority of medical practitioners, therefore, considered the new invention as "more an interesting toy than a weapon of value in medicine," and preferred to stick to their time-tested methods of manually probing the injury in order to diagnose the position of the bones. "It may not be unreasonable to hope for much more important results in the near or remote future," commented the influential *Journal of the American Medical Association.* "At present however, the limitations of the methods are too great and the medical nature of the discovery is, as yet, a largely unknown quantity."[16]

In February 1896, a few weeks after Roentgen's announcement of his discovery, two professors from Colorado College traveled from Colorado Springs to Denver to demonstrate the new marvel before the Denver Medical Society. The grand ballroom at the Brown Palace Hotel in downtown Denver was packed with more than three hundred physicians who came to view the grand display of glass tubes and huge batteries, to hear the crackling sounds and see the eerie lights, and finally to witness for themselves the "power of the mystifying and mysterious X-rays."[17] Two months later, the grand ballroom was packed again "with physicians and their ladies, with scientists, and with the curious." This time the guest of honor was Colonel Charles F. Lacombe, president of the local Mountain Electric Company. Lacombe reviewed the history of x-rays and reported on the latest improvements in the apparatus and methods of operation used to produce them. The evening's highlight came when Lacombe x-rayed the hand of a physician in the audience who had suffered a gun shot wound several years earlier and had always wanted to know where the shot had lodged. After three minutes of exposure, the photographic plate was developed, and the resultant picture clearly demonstrated a piece of buckshot deeply embedded between the third and forth metacarpals. The audience was spellbound.[18]

During the ensuing discussion, Colonel Lacombe was surprised to find that his medical audience was troubled with the question "whether the X-rays had been used in courts as evidence, and if they would be accepted as such" He did not know the answer, nor did anyone else in the audience.[19] Soon, however, Colonel Lacombe found out why the question was so important to his audience. When he offered to make x-ray plates without charge as a public service, his laboratory was "besieged by persons who were certain that their physicians were wrong and wanted X-ray photographs to prove it."[20] The dismayed Colonel began to allow in only patients accompanied by their physician. However, his x-ray machine was not the only one in town. All that was needed was a partially evacuated gas tube, a power source (batteries and induction coils, or a static generator), and regular photographic dry plates—all easily available from any of the growing number of photographic and electrical machine shops. In Denver, at least two other people were also experimenting with x-ray apparatus—H. H. Buckwalter, a professional photographer who was asked by the local *Rocky Mountain Daily News* to take up the x-ray issue for the benefit of its readers, and Dr. Chauncey Tennants Jr. of the Homeopathic Hospital.[21] It was to these two people that Smith's attorneys turned when they planned Smith's suit. Tennants and Buckwalter agreed to attempt together to make an x-ray photograph of Smith's injured limb.

Although no one was sure what x-rays were, it was clear how to produce photographs with them. That explains much of the unprecedented enthusiasm and widespread activity. Still, securing a good x-ray image was far from simple in 1896.[22] The complex relations between the electrical characteristics of the tube, its gas pressure, and the properties of the rays it emitted were far from being understood. X-ray operators were aware that the tubes could emit x-rays of varying degrees of penetration. Some rays were "soft" and could barely pass through the skin. Others were so "hard" that they passed through the thickest bones, and produced little contrast on the photographic plate.[23] The x-ray operators were also vaguely aware that the properties of the rays were mainly determined by the gas pressure in the tube. The higher the vacuum in the tube, the "harder" the x-rays it emitted. The degree of the vacuum, however, varied from tube to tube, and from minute to minute within the same tube. Worse still, a tube might have a low vacuum but still deliver rays of high penetration, and high-vacuum tubes sometimes refused to emit any rays at all. Bewildered, x-ray operators expounded on the "seasoning" of the tubes. If properly used with suitable intervals of rest, a given tube might become "seasoned" within a few weeks, or it might take several years. No generally accepted explanation existed for this.[24] A great part of the operator's expertise was to know his tubes by heart and choose the right one for the specific task. No precise measurements of the voltage across the tube's terminals or of the tube's output were possible in 1896. The operators had to estimate the amount and the quality of the x-rays given off by examining the colors of the tube's fluorescence.[25] This, however, did not always work. Sometimes the tube would fluoresce in a manner similar to a normal tube, but no x-rays of sufficient penetration would be given off. "It seemed, at times," one operator later recollected, "as though gas tubes had been invented for the specific purpose of trying man's soul."[26]

On November 7, 11, 21, and 28, 1896, Dr. Tennants and Mr. Buckwalter made four attempts to take x-ray photographs of Smith's injury.[27] Securing a good x-ray image of Smith's hip and thigh was especially challenging. The thickness of the bones necessitated the generation of highly penetrating x-rays for a very long time. To achieve high penetrability, a high-vacuum tube had to be used. The higher the vacuum, the higher the voltage that was needed to excite the tube. The higher the voltage, the hotter the tube became and the more quickly it broke down. The longer the exposure, the more likely other things would also go wrong. The heat developed tended to liberate adsorbed gas from the glass and electrodes, and so to raise the gas pressure and "soften" the rays. The size and location of the tube's focal spot also depended on the gas pressure and could vary appreciably during the same exposure, significantly degrading the image's definition.[28] A great deal of ingenuity was also needed to place the subject in such position that he would keep perfectly still during the long exposure. Sometimes, the only solution was anesthesia. Tennants and Buckwalter got lucky, however. In one of the sessions, after an exposure of eighty minutes, they secured a satisfactory photograph that showed a well-defined fracture of the neck of Smith's femur. The outline of the bone appeared a bit hazy, but "the irregular dark line which ran across the bone was quite apparent and left no possible room for doubt that a fracture did really exist."[29] Lindsey and Parks were thrilled. They were determined to persuade Judge Owen E. Lefevre to do what no U.S. court had yet agreed to do—to allow them to use their photograph in evidence before the jury.

Evidence of Things Seen

A picture may be worth a thousand words, but for centuries words ruled the legal domain. Rhetoric, the art of using language, has always been the trademark of lawyers, and trials, especially in Common Law, have been widely understood as battles by words. Alas, all glory is doomed to pass and the second half of the nineteenth century saw a new mode of persuasion rising to dominance, driven by a new class of machine-made testimonies that threatened to turn words into an inferior mode of communicating facts. Ever alert and never involved, machines such as microscopes, telescopes, high-speed cameras and x-ray tubes purported to communicate richer, better, and truer evidence, often inaccessible otherwise to human beings. The emblem for this new type of mechanical objectivity was visual evidence. "Let nature speak for itself," became the watchword, and nature's language seemed to be that of photographs and mechanically generated curves.[30]

Nineteenth-century American courts, with their processes of fact-finding and proof traditionally geared to the reception of *viva voce* testimony by human witnesses, found it hard to adapt to this new logic of mechanical objectivity. Machines may have promised to succeed where human beings failed, but they could not be put under oath, interrogated and cross-examined. The incipient conflict came to a head in the 1870s and 80s, with the deployment of photography in a constantly broadening range of legal contexts. Was the photograph an especially privileged form of evidence, nature's unmediated testimony? Or was it an especially repugnant form of evidence, a human artifice claiming to be mechanically objective? Late nineteenth-century lawyers fought bitterly over these questions and the result was a fascinating legal discourse concerning the nature and status of photographic evidence.

The historian Jennifer Mnookin described two sorts of attitudes that governed the nineteenth-century legal discourse over photography. The first took photographs to be an especially privileged form of evidence. Those with this sort of attitude took one of two positions. The first emphasized the unmediated nature of photography; that is, its ability to mechanically transcribe nature as if it were a process in which nature reproduced itself directly. The photograph, this argument ran, was an equivalent of what it represented. Thus, it was the best proof of what was in it. As the Georgia Supreme Court put it in 1882, "We cannot conceive of a more impartial and truthful witness than the sun, as its light stamps and seals the similitude of the wound on the photograph put before the jury."[31]

The second position emphasized the unprecedented lifelike fidelity of the photograph. "If a difference exists between a photograph and an eye-witness testimony," asked one writer, "should we not give the greater credence to the photograph, whose testimony, we know, is perfectly truthful and generally commensurate with the fact, while that of a vouching witness, and also of the witness to speak to the question of identity, may be mistaken or perjure?" The same argument was applied also to the realm of objects. "Until photography was discovered nothing in nature was exactly like any other thing," argued another lawyer in an attempt to admit a photograph of a document instead of the document itself. "Until this discovery there was, therefore, reason in the rule that required the production of the original paper writing as the best evidence of its appearance. Science now steps forward and relieves the difficulty." Understood this way, as the outcome of a mechanical and deterministic process, the photograph was projected as the highest form of evidence, capable of offering uniquely objective knowledge of the world. As one judicial opinion put it in 1871, "as well we might deny the use of the compass to the surveyor or mariner, the mirror to the truthful reflection of images; or spectacles to aid the failing sight, as to deny, in this day of advanced science, the correctness . . . to the photographic instrument, in its power to produce likeness."[32]

The second sort of attitude viewed the photograph not as a mechanical and unmediated replication of nature but as an artificial and constructed representation of it. Here too, people with this view tended to take one of two positions. The first pointed to the inherent distortions involved in photography, while the second pointed to the significant role that human intervention played in the construction of the photographic image. Photographers, for example, did not tire of emphasizing that well-constructed photographs were not a product of the camera itself, but rather of the skills of the photographer. For others, however, this argument only served as a further reminder not only of the photograph's fallibility but also of its ability to be manipulated. "Can the sun lie?" asked the Virginia Law Journal in 1886. "The question is supposed to carry its own answer. Perhaps we may say that though the sun does not lie, the liar may use the sun as a tool. Let us, then, beware of the liar who lies in the name of truth."[33]

As Mnookin emphasized, the two competing attitudes described here were not necessarily mutually exclusive. Indeed, many nineteenth-century commentators on photography developed multifaceted conceptions of the photograph as offering both truth and artifice. In most cultural settings it was not terribly problematic to accept these competing notions and even enjoy their internal contradictions. Not in the courts, though. There, judges confronting photographs did not have the luxury of relishing the paradox. Understood the first way, as nature's unmediated testimony, the photograph was primary evidence, which, like a signed document, afforded the greatest certainty of what was inside it. Understood the second way, as a human construct prone to deceit and manipulation, it was, at best, low-grade evidence that could be resorted to only out of necessity, in cases where better evidence was not available. Understood both ways, as a human artifice claiming to be mechanically objective, as second-grade evidence holding special powers of persuasion, it was repugnant evidence, too powerful to be given to the credulous jury without being subjected to constant checks and balances.[34]

No clear judicial strategy for handling the photograph emerged until the 1880s. By then, however, all conceded that no matter how one understood photography, it was so generally accepted in society that the court must recognize it.[35] Various judicial solutions were suggested, and the one finally adopted followed a characteristic Common Law strategy—it tried to avoid the horns of the dilemma by denying its existence. Photographic evidence, the courts asserted, was neither new nor problematic.

To see how this was done, let us look at the 1881 opinion of Chief Judge Charles J. Folger, in *Cowley v. People*, which became the principal precedent concerning the admissibility of the photograph. This opinion, which came to be known as the Cowley opinion, did not treat the photograph as a new sort of evidence. Instead, it equated it with a more traditional sort of picture—man-made pictures:

> Photographic pictures do not differ in kind of proof from pictures of a painter. They are the product of natural law and a scientific process. It is true that in the hands of a bungler, who is not apt in the use of the process, the result may not be satisfactory. Somewhat depends for exact likeness upon the nice adjustment of machinery, upon atmospheric conditions, upon the position of the subject, the intensity of the light, the length of the sitting. It is the skill of the operator that takes care of these, as it is the skill of the artist that makes correct drawing of features and nice mingling of tints for the portrait.[36]

Photographs, Judge Folger maintained, were not different in essence from other, man-made, images that had been used in court for centuries. The portrait, accepted in court on questions of identity and appearance, was only one among various kinds of visual representations at work in the nineteenth-century courtroom. Other kinds included maps and surveys that described the terrain and its boundaries in trespass cases and land disputes; paintings and drawings that described property in civil cases and wounds and scenes of the crime in criminal cases; diagrams, plans, and models that described inventions in patent cases, and so forth. Dependent by their nature on human observation and unavailable for cross-examination, these visual representations were never regarded as a proper basis for the fact-finding process. Instead, if proven to be correct representations by witnesses personally familiar with the scene or object depicted, these visual representations were allowed into the courtroom, where they served as explanatory devices, to help the verifying witnesses to communicate their oral testimony to the jury.[37]

Equating the photograph with one of these types of man-made visual representation legitimized the photograph and diminished its value at the same time. It debunked the myth of the mechanically objective camera. At the same time, it also suggested that there was no reason why machine-made pictures should not fill the same illustrative functions that man-made pictures had filled for so long. After all, as Judge Folger put it:

Most of evidence is but the signs of things . . . A witness who speaks to personal appearance or identity, tells in more or less detail the minutiae thereof as taken in by his eye. What he says is a description thereof, by one mode of signs, by words orally uttered. If his testimony be written instead of spoken, and is offered as a deposition, it is a description in another mode of signs, by words written, and the value of that mode, the deposition, depends upon the accuracy with which his words uttered are put into words written. Now if he has before him a portrait or a photograph of the person, and it shows to him a correct copy of that person, if it produces to his view a correct description, which he testifies is a likeness, why may not that be given to the jury, as a description of the person by the witness in another mode of signs? The portrait and the photograph may err, and so may the witness. That is an infirmity to which all human testimony is lamentably liable.[38]

The difference between the new machine-made pictures and the older man-made pictures, Judge Folger suggested, was no greater than the difference between words uttered and words written. The analogy to verbal evidence was not incidental. All evidence, the Cowley opinion implied, verbal or visual, was but different modes of representation by which witnesses communicated their knowledge to the tribunal. As such, none of them had any evidentiary value independent of the witness whose perceptions and knowledge it purported to represent and communicate. Thus, all types of evidence were legitimate, but only when sponsored by the perceiving witness who could be examined and cross-examined about their content in the presence of the jury.[39]

Building on the Cowley rationale, later judicial opinions constructed a new broad legal category termed "illustrative" or "demonstrative" evidence, into which they stuffed photographs together with other visual representations such as maps, plans, models, diagrams, portraits, and paintings. Officially, illustrative evidence was secondary evidence, admissible only for illustrative purposes, to aid the witness in explaining and the jury in comprehending the testimony given in court. Arguably, this was a fiction, at least as far as photographs were concerned. Photographs, after all, seemed to remember more accurately, describe in more detail, and communicate their observations more clearly than any human testimony. Equally important, they appeal directly to the senses, allowing the judge and the jury to derive their own perceptions rather than satisfy themselves with the perceptions of others. Thus, photographs possess immediacy and reality that render them particularly persuasive. They document rather than illustrate, verify rather than explain.[40]

Legal fictions have their purpose, though, and crowding a new phenomenon into an already familiar category has long been a popular strategy within Common Law. It has allowed the law to escape the stifling effects of its own formalism, while preserving the facade of uniformity. The legal procedures to deal with existing categories are well known, and Common Law has its ways to compensate for any growing incompatibilities created by new additions. In time, if needed, the law can carefully develop new tools to deal with the, by then, already familiar phenomenon. In this way, nothing is ever really new in the legal universe. Either the phenomena or the procedures to deal with them are always well rehearsed.

The case of photographic evidence followed the same pattern. Crowding photographs, together with models, maps, and pictures, into the category of illustrative evidence allowed the courts to integrate photography into the existing evidentiary terrain, while keeping its contradictory notions in the desired balance. Treating photographs as illustrative evidence enabled the courts to argue that photographic evidence was neither new nor problematic. The conventional test used to establish the veracity of maps, plans, models, and paintings—having a witness familiar with the represented object testifying under an oath as to the accuracy of the representation—was readily applied as the principal test for the admissibility of photographs. The demand that photographs be verified by eyewitness testimony provided a powerful check on photographic evidence and secured the status of the traditional eyewitness, whose mediating presence seemed to be threatened by this new species of evidence. All together then, the doctrine of illustrative evidence seemed an attractive solution that gave photographic evidence a recognizable identity, an uncontroversial past, and hopefully a predictable future.

Still, containing the suggestive force of the photograph by treating it as just another species of illustrative evidence was only partially successful. It worked well on the formal admissibility level, which was controlled by the judiciary. However, the judges had far less control over how evidence was used inside the courtroom.[41] The judges may have conceptualized photographic evidence as a nonverbal expression of an eyewitness, but the jury often treated it as independent evidence that represented its subject matter directly. Consequently, once the photograph was allowed into the courtroom, it was no longer clear where illustration ended and proof started, or what was illustrating what—the photograph illustrating the testimony or the testimony illustrating the photograph.

Thus, by the end of the nineteenth century, photography was already established in the American courtroom as a powerful evidentiary tool, routinely applied in a constantly broadening range of cases. Still, its theory and practice took on a double character. Formally, as John Wigmore, the legendary dean of the Northwestern Law School and the leading authority on evidence put it, the photograph was "Simply nothing. Except so far as it has a human being's credit to support it. It is mere waste paper—a testimonial nonentity. It speaks to us no more than a stick or a stone. It can of itself tell us no more as to the existence of the thing portrayed upon it than can a tree or ox. We must somehow put a testimonial human being behind it (as it were) before it can be treated as having any testimonial standing in court. It is somebody's testimony, or it is nothing."[42]

Yet, inside the courtroom, properly verified photographs often functioned as independent, sometimes central, pieces of primary evidence. The discrepancy was not intolerable. The trial judges, certainly, did not seem troubled by it. They allowed the jury to carry photographs with them into the jury room to be used in their deliberation as if photographs were independent evidence. They also allowed witnesses, including expert witnesses, to use photographic evidence to explain their testimony, but usually forbade them to instruct the jury what was in the photographs. The final interpretation of the content and credibility of photographic evidence was left to the discretion of the jury, and through this uncensored discretion the popular understanding of the photograph as a privileged form of evidence continued to exert its powerful influence in the courtroom.[43]

This rather intricate judicial arrangement worked well as long as two conditions were met. The first condition was explicit—a testifying human being had to stand behind the photograph. The second was more tacit—the jury was expected to use the intuitive information photographic evidence conveyed to better understand the rest of the evidence. Verified by eyewitnesses the photographs were then admitted into evidence as an illustration of the witness's testimony, to be considered by the jury. But what about photographs that depicted evidence that was not available to the naked eye? How could a human being be provided who could vouch from personal knowledge as to their accuracy? And what about photographs that did not convey intuitive information easily accessible to the jury? How could the jury be trusted to read them properly? Smith's lawyers, Lindsey and Parks, had to come up with good answers to these questions if they wanted to persuade Judge Lefevre to allow them to present their precious x-ray photographs of Smith's fractured femur as evidence before the jury.

Evidence of Things Unseen

Early roentgenography was considered to be a new specialty in the field of photography. After all, the new images were produced inside the same dark room, by the same materials and chemical processes used in regular photography. Consequently, many of the workers actively engaged in making x-ray images were photographers, and those physicians "desiring to acquire a knowledge of X-ray work," were advised to "first serve an apprenticeship in a photographic gallery."[44] The result was that the anatomical images produced by the new x-ray technology were generally thought of as photographs, which were admissible in court. Lindsey's and Parks' strategy was built on this popular notion. If x-ray images are photographs, and photographs are illustrative evidence, they argued, then x-ray images are illustrative evidence, admissible as secondary evidence to aid the witness in explaining his or her testimony. This strategy strained the already uneasy foundations of the legal category of illustrative evidence. Its main weakness was that the standard legal test used to establish the veracity of illustrative evidence was not applicable to x-ray images. Whatever it was that they depicted, it was hidden from the naked eye. Thus, no witness could be provided who could vouch to the accuracy of the representation. Without the standard eyewitness testimony, one judge explained to Ben Lindsey to clarify why he had forbidden x-ray evidence in his court, "there is no proof that such a thing is possible. It is like offering the photograph of a ghost when there is no proof that there is any such thing as a ghost."[45]

Lindsey and Parks planned to bypass this difficulty by suggesting that x-ray plates should be admitted upon verification not of the accuracy of their representation but of the reliability of the process that produced them. Such a test seemed to have satisfied some British courts. In February 1896, the Court of the Queen's Bench of Montreal, Canada, was the first on record to admit an x-ray photograph; it was made at the McGill University physics lab by Professor John Cox and showed the location of a bullet in the leg of a young man who had been shot during a pub brawl on Christmas Eve. Shortly after, in Nottingham, England, an actress, Miss Gladys Ffoliett, sued the Nottingham Theater Company for damages incurred during a fall in the course of her work. The judge allowed Miss Ffoliett's lawyer to present to the jury a Roentgen plate of her left foot, made at the local University College Hospital, showing a displacement of the cubic bone.[46] The American judges, however, were not as receptive as their royal counterparts, and for good reason. The dominant British judges could afford to be flexible with the evidence they admitted, trusting that they could later guide the jury's assessment of it during the trial. The American judges, however, were generally forbidden to comment on the weight of the evidence during the trial. Thus, they were more guarded about the evidence they allowed before the jury.[47]

Young and inexperienced as they were, Lindsey and Parks turned out to be excellent instigators. On the morning of the trial, they produced all the x-ray equipment (electrical apparatus, batteries, tubes, and so forth) in the courtroom, together with actual x-ray photographs of various bones in the human body. They offered to show the jury, right then and there, the bones in their hands—an offer that "created such terrific excitement about the courthouse that extra bailiffs were called in to keep the court in order during the argument." Next, Lindsey and Parks introduced the photographer, Buckwalter, (rather than the doctor, Tennants Jr., who was also present in court) as their leading expert witness. Buckwalter carefully explained to the judge and the jury the use of the rays and the method of operating the apparatus that produced them. In order to convey the similarity between roentgenography and photography, Buckwalter contrived a box with a small hole at one end through which candlelight could cast shadows upon a screen at the opposite end of the box. He first showed the jury the shadow of a hand, then an x-ray photograph of the hand. Next, the shadow of a normal femur was cast on the box's screen, followed by the x-ray photograph of that same femur. Next, he described the manner in which the plates of Smith's injured thigh were taken and how the negatives and final photos were developed from them. Finally, Lindsey and Parks displayed the x-ray images of Smith's left femur, and proposed that they be submitted to the jury as evidence that there had been a fracture of the neck of Smith's femur.[48]

Senator Hughes, the leading defense counsel, objected. In a deeply learned speech that lasted more than three hours, Hughes provided a long line of authorities for the well-settled legal principle that for a photograph to be accepted in evidence there should be a witness familiar with the object in the photograph to verify under oath that the object had been accurately portrayed. But how could the photograph of an object unseen by the naked eye be verified? Hughes asked the court. There was no way to ascertain that this was Smith's femur in the photograph. And even should it be admitted that this was Smith's femur, there was no way to verify that the photograph represented it truthfully. Being impossible to verify, Hughes concluded, x-ray photographs were not competent evidence.

Expecting this line of argument, Mr. Lindsey replied in an equally spirited manner. Senator Hughes had emphasized the subjective aspect of photographic evidence—presenting it as a fallible human artifact that bestowed no reliability. Lindsey replied by emphasizing the objective aspect of x-ray photographs—presenting them as the deterministic product of the immutable laws of nature. The Senator had argued that having a witness who was familiar with the content of the photograph was the only test of its veracity. Lindsey argued that the law recognizes a second test based on the assumption that photographs were the product of a scientific process. Quoting from numerous recent scientific journals, he maintained that the accuracy of the x-ray process had been fully proven and that whenever an incision had followed, the condition had always been found to be as the photograph had shown. Exhausted, Judge Lefevre finally adjourned the session, informing the parties that he would rule the following morning on the admissibility of the x-ray photographs.[49]

The x-ray paraphernalia, the offer to roentgenize the jury, and the attempt to admit the mysterious x-ray photographs in evidence had made a profound impression upon the reporters, who published the whole story in the morning papers. When the court convened at nine the next morning, the courtroom was packed. "The excitement was intense," recollected Lindsey, "the gallery all on my side, restrained from breaking into applause on several occasions because of their anxiety to have this 'miracle' demonstrated and actually recognized by the court." Lindsey's powers of recollection may be questioned, however. Many physicians present in the audience were anxious to see the x-ray evidence rejected. Their reasons were deeper than mere sympathy for one of their own. "Should the court hold the evidence was good," explained the local newspaper to its readers what was at stake, "a physician would be almost compelled to take up the study of the rays as part of his education or pass doubtful and serious cases to someone who understood the new and mystic way of penetrating the patient in'ards."[50]

Judge Lefevre, a sporting gentleman who had made a large fortune in the risky business of mining, was well aware of the importance of his ruling and did not fail to rise to the occasion:

> [We] have been presented with a photograph taken by means of a new scientific discovery, the same being acknowledged in the arts and in the science. It knocks for admission at the temple of learning and what shall we do or say? Close fast the doors or open wide the portals? These photographs are offered in evidence to show the present condition of the head and neck of the femur bone which is entirely hidden from the eye of the surgeon. Nature has surrounded it with tissues for its protection and there it lies hidden; it can not, by any possibility, be removed or exposed that it may be compared with its shadow as developed by this new scientific process. In addition to these exhibits in evidence, we have nothing to do or say as to what they purport to represent; that will, without doubt, be explained by eminent surgeons. These exhibits are only pictures or maps, to be used in explanation of a present condition, and therefore are secondary evidence and not primary. They may be shown to the jury as illustrating or making clear the testimony of experts.[51]

Judge Lefevre, therefore, accepted the equation between x-ray and regular photography suggested by the prosecution. He also adopted their line of reasoning—that the x-ray images could be admitted as secondary evidence on the strength of their reliability. The idea was not entirely new. Courts had referred to the deterministic nature of the photographic process as a central constituent of the photograph's credibility. The Cowley opinion, for example, had suggested that verifying the reliability of the photographic process (for example, the skills of the photographer and the circumstances in which the particular photograph was taken) should be part of the admissibility test. However, such a verification procedure was difficult to perform, and most courts had satisfied themselves with having an eyewitness testify to the accuracy of the representation.[52] But if such a witness could not be provided, Judge Lefevre now ruled, verifying the reliability of the photographic process could and should suffice. To this end, Lefevre shifted the accepted balance between the interwoven understandings of the photographic evidence and emphasized the hitherto suppressed view of photographic evidence as a the product of a scientific process capable of offering reliable representations, indeed even of things unseen:

> During the last decade at least, no science has made such mighty strides forward as surgery. It is eminently a scientific profession, alike interesting to the learned and to the unlearned. It makes use of all science and learning. It has been of inestimable benefit to mankind. It must not be said of the Law that it is wedded to precedent; that it will not lend a helping hand. Rather, let the court throw open the door to all well considered scientific discoveries. Modern science has made it possible to look beneath the tissues of the human body, and has aided surgery in telling of the hidden mysteries. We believe it to be our duty, if you please, to be the first to so consider it in admitting in evidence a process known and acknowledged as a determinate science. The exhibits will be admitted in evidence.[53]

With the x-ray photographs admitted into evidence, the trial finally commenced. For the next two weeks a crowd of physicians arrayed itself enthusiastically along the two sides of the slippery $10,000 question—whether Dr. Grant had used a "reasonable" degree of care, skill, and diligence in his treatment of Smith's injury. By the end of these two weeks, the jury failed to agree upon a verdict. This meant that the whole case had to be gone over again.[54] There are no records of a second trial, however. No longer certain of his imminent victory and dreading the prospect of a longer and more expensive retrial, Dr. Grant probably agreed to settle the case. Denver's medical community, on the other hand, was up in arms. Predictions were made that the suit against

Dr. Grant was but a forerunner of other similar suits; the legal fraternity was castigated for the growing malpractice fad, and the various
local medical societies called special meetings to discuss means of protection "against unjust and malicious suits for malpractice, instituted by
irresponsible parties, and conducted ordinarily by impecunious attorneys upon shares."[55]

The courts were not impressed. Medical testimony had proved a
serious bottleneck in malpractice suits throughout the nineteenth century, as the experts constantly failed to agree with each other about their
findings. The courts had little to lose therefore, and the benefits, it
seemed, were well worth the risks. True, x-ray images may not have
been more reliable then palpation, but at least they seemed to allow
judges and jurors, for the first time, direct access to the disputed facts.
Having such access via properly verified x-ray plates, the courts hoped,
would enable them to reduce malpractice litigation and shorten its
course, and would allow the lay jurors to draw better-informed decisions. Thus, despite the fact that the infant x-ray technology was widely
viewed as unreliable, and even though fundamental questions of what
x-ray images conveyed and who should be considered a competent
x-ray expert were still under discussion, other American judges soon
began to follow Lefevre's ruling and admit x-ray photographs on the
strength of the argument that photographic evidence could be admissible on the proof not of its intrinsic accuracy but of the reliability of the
processes that produced it. "New as this process is," reasoned the
Tennessee Supreme Court in September 1897, "experiments made by
scientific men, as shown by this record, have demonstrated its power to
reveal to the natural eye the understructure of the human body, and that
its various parts can be photographed as its exterior surface has been
and now is."[56] Being the first supreme court to deliberate the admissibility of x-ray evidence, the opinion of the Tennessee Supreme Court
carried great weight all over the country. Within a month, in October
1897, x-ray images were admitted for the first time in a criminal trial. In
1899, the Supreme Court of Maine declared that the judge presiding at
a trial could admit x-ray images at his discretion. In 1901, the supreme
courts of Massachusetts, Wisconsin, and Washington, D.C., reached
similar conclusions, and in 1902, 1904, and 1905, Nebraska, Illinois,
and Arkansas also fell into line.[57]

Thus, by the start of the twentieth century, a growing agreement had
emerged among American courts concerning the admissibility of x-ray
images. Understood as a variant of photographic evidence, x-ray images
were crowded together with regular photography in the category of

illustrative evidence. The courts were well aware of the fine line they were walking on. "It is not to be understood," the Tennessee Supreme Court cautioned, "that every picture taken by the cathode or x-ray process would be admissible. Its competency, to be first determined by the trial judge, depends upon the science, skill and intelligence of the party taking the picture and testifying with regard to it, and that lacking these important qualifications it should not be admitted."[58] Such caution fell well within the established doctrine of photographic evidence, which treated the photograph as a fallible human artifice that needed to be carefully verified. True, the methods for verifying regular photographs and x-ray images differed, but the courts discounted the difference. "There would seem to be no reason for making a distinction between an X-ray and a common photograph," proclaimed the Supreme Court of Washington, D.C. "That is, either is admissible as evidence when verified by proof that it is a true representation of an object which is the subject of inquiry."[59]

But while a certain agreement was achieved concerning the admissibility of x-ray images, a considerable amount of disagreement remained about what to do with them after they were admitted into the court. Underlying the application of photographic evidence, indeed the whole legal category of illustrative evidence, was the presumption that the average juror could easily comprehend it and use the intuitive information it conveyed to better understand the rest of the evidence. However, unlike regular photography, the information conveyed by x-ray images was not intuitive. The human body, an object of varying densities, produced under the x-rays an image of superimposed shadows of varying opacity. The information conveyed in the image was contours and densities without real perspective. The only perspective information lay hidden in the sharper shadow's outlines of nearer objects. To corectly read the lesson in any x-ray picture, one had also to make a mental adjustment for the disproportion created by the obliqueness of the rays caused by the fact that they were not parallel but emanated from a point. Thus, once the courts began to admit x-ray images, a demanding problem emerged: Could the jurors interpret the images correctly by themselves or should experts be allowed to explain to the jury what they saw in the image?

The judicial solutions to this problem varied, exposing in the process some of the potential interpretive conflicts buried within the doctrine of illustrative evidence. The rules of evidence dictated that expert witnesses be allowed to advise the trial jury on the evidence only when it was clear that the members of the jury did not have sufficient knowledge to enable them to draw an informed decision from the evidence. That, however, was clearly not the case with photographic evidence, which was assumed to be freely accessible to the jury, and was tacitly treated in court as independent evidence that speaks best for what it shows. Recognizing the difficulty, Judge Lefevre tried to resolve it by allowing experts to tell the jury what they thought they saw in the x-ray plates, while still saving the last word for the jurors who were allowed to take the x-ray plates into the jury room, examine them, and interpret what their meaning was. However, many of Judge Lefevre's colleagues, who accepted his invitation to think of x-ray images in terms of photographic evidence, rejected his solution. Photographic evidence, they insisted, was designed to explicate oral testimony, not the other way around. Thus, they forbade experts from telling the lay jury what, in their opinion, was in the images. "No argument is required to show," explained in 1907 the Iowa Appellate Court why allowing a doctor to testify to what appeared in an x-ray image was an error, that photographs "are the best evidence of what appeared in them."[60] Thus, experts were only allowed to describe in general terms what a certain bodily condition would look like on the x-ray plate, and they had to leave it to the jury to decide whether that was also the case in the specific image before them.

The Reactions of the Medical Profession

The medical community watched the legal developments with a growing sense of alarm. It was common knowledge among physicians that juries in malpractice trials were strongly biased toward the usually poorer and indisposed plaintiffs. The possibility of having their clinical judgment subordinated in malpractice suits to the authority of the x-ray image as interpreted by the jury, seemed, therefore, like a nightmare.[61] Here is how Denver's *Daily News* explained to its readers the implications of Judge Lefevre's decision as to the future of surgical practice:

> It means that unless a physician is absolutely certain that he knows what is the matter with one of his patients, he must shadowgraph them and make a correct diagnosis with the x-ray aids, or the first thing he knows someone will attack him in court and by shadowgraphs prove that the doctor was a long way off from the correct treatment. Whether this will mean that each doctor will have to carry an x-ray laboratory and dark room around with him when he

makes his calls, or will be obliged to cart each of his patients to his office and there do the shadowgraph act, does not appear, but it seems certain that some easy method of knowing that he is all right will be a necessity to the physician with reputation and property to protect.[62]

The deployment of x-ray images as legal evidence, thus, became a central topic in all of the major American medical meetings of 1897. The two general concerns among the doctors were that it would stir up a new outbreak of malpractice litigation and that the new technology, which seemingly allowed the jury and the lawyers for the first time direct access to the disputed facts, would undermine the authority that medical witnesses had so far enjoyed in the courts. "There could be no doubt that skiagraphs will figure largely in suits for damages," William White, professor of clinical surgery at the University of Pennsylvania, warned the members of the American Surgical Association at their 1897 annual meeting in Chicago. "They have already been admitted in evidence in such cases, and it is probable that juries will with increasing frequency have to decide whether to place greater weight on deformity as shown by skiagraphs or on expert evidence as to the absence of genuine disability."[63]

Professor White had two suggestions how to counter the looming danger. First, members of the profession had to modify their practice and make sure "that in all obscure, complicated, and unusually difficult cases the help afforded by the Roentgen rays shall be secured by the surgeon, even if it is done chiefly with the view to his own protection." At the same time, the medical profession had to fortify its courtroom authority in face of the new technology. In particular, the profession had to make sure that no one would assume "that the clinical experience and the judgment based upon it should be subordinated to the pictorial testimony of the skiagraph."[64]

The discussion that followed White's talk illustrated the depth of the medical profession's anxieties about the legal implications of x-rays. "I have talked the medico-legal side of this question over with several of our judges, and some of them are going to admit these skiagraphs," reported Dr. Richardson of Harvard and Massachusetts General Hospital. "We could not bring up a more important subject at this time . . . A great deal of trouble is going to result from these pictures." Dr. Fowler of Brooklyn concurred. "Personal damage lawyers will enter into collusion with those in charge of the X-ray studios, and patients who from curiosity or other motives apply for skiagraphs will be encouraged to institute legal proceedings to recover damages for alleged improper treatment." Agreeing that "there is no doubt that many suits for malpractice will result," the participants decided to appoint Professor White chairman of a committee that would look into this troubling issue and recommend further actions.[65]

The powerful railway companies also felt threatened by the new rays. They had long considered their trains to be filled with "men and women who long for a wreck, a jolt, a fall, something on which to base a claim."[66] To counter such claims, the railway companies employed armies of surgeons in towns along their tracks in order to provide medical assistance to both trainmen and passengers "in the manner which is most conductive to the interests of the Company."[67] They also paid leading physicians lavish salaries to represent them in court. The concern with a new outbreak of "x-ray litigation," therefore, became the central topic at the 1897 meeting of the American Academy of Railway Surgeons. In his opening address, Dr. Harvey Reed, a professor of surgery at Ohio Medical University and chief surgeon of the University Hospital, who was just about to be nominated surgeon-general of Wyoming, declared that: "As long as it is possible to take radiographs of the normal limb, which gives the appearance of being fractured, and on the other hand, when skiagraphs are taken of fractured limbs that fail to show the existence of the fracture, we are compelled by the circumstances to distrust its accuracy, and for that reason if for no other, we consider it a dangerous source of evidence which should not yet be admitted by our courts."[68]

Once again, the discussion following Reed's opening address was most revealing. "If the profession treats it as it usually does everything new," warned Dr. Grant of Denver, the x-ray "will prove an instrument of danger. It will be an expensive thing for the surgeon."[69] Dr. Lacey of Iowa, agreed. "It seems to me that if a skiagraph were placed before the jury it would injure the case in which the surgeon might be deeply interested. So I feel that the Academy, as well as other association of this character, should be guarded in pronouncing the X-ray, or skiagraphs, of such vast importance until their actual value is demonstrated." Dr. Galbraith of Omaha concurred: "It would be a grave mistake for us to place ourselves in a position to state that it is necessary for the X-ray to assist us . . . for we will receive much abuse in times of tribulation from such recommendation." Dr. James Burry of Chicago, one of the few American physicians who had mastered the new x-ray technology, tried to be more practical. "We can not keep it out of court," he reminded the participants. "Supposing we are all on the defense side, the plaintiff will bring it into the court." Dr. Cole of Helena, Montana, backed him up. "It is useful. It has come to stay and will unquestionably be used hereafter to an extent perhaps which we do not anticipate." The conclusion reached by the members of the American Academy of Railway Surgeons was similar to that reached by the members of the American Surgical Association—that the best that could be done was to have x-ray technology's "proper limitations defined," so that "no one should think that the X-ray is a referee or is the final arbiter in the matter." No one who could serve as a juror, that is.[70]

Meanwhile, the ominous predictions about an outbreak of "x-ray litigation" quickly materialized. "Ever since its discovery, especially in the last year," complained in 1898 Professor Pratt of Chicago,[71] "every malicious person who can scrape up enough money to pay for a shadowgraph, is having one taken for the purpose of bringing a damage suit for personal injuries or malpractice. It is coming to this: That a surgeon is not safe unless he has a shadowgraph taken before and after each operation. It is surprising to see the number of damage suits now pending against corporations, individuals and especially surgeons for supposed injuries sustained or for malpractice, depending entirely on the shadowgraph as evidence." The participants of the 1898 annual meeting of the American Association of Physicians and Surgeons were warned that "negligence will be claimed for failure to use it [x-ray] in all cases where bad results follow." The lecturer, Dr. Frank Ross of New York, cautioned all practitioners "to take unusual care in giving opinion of a case without the consent and knowledge of the attending physician, particularly where a suit for malpractice is likely to ensue," and urged them to "shape our examinations in this line as to make them conform to the rules of legal evidence in all its technicalities, or our work may not only be productive of no good, but may be an actual source of menace to us personally and professionally."[72]

The Professionalization of X-Ray Practice

The last years of the nineteenth century were, therefore, characterized by a seemingly incongruous medical reaction to the new x-ray technology—while the reliability of the new x-ray technology was relentlessly attacked, its practice was constantly recommended. The orchestrated effort to undermine the reliability of the new technology reached its peak in the 1900 meeting of the American Surgical Association with the publication of the official report of the Committee on the

Medico-Legal Relations of the X-rays, appointed at the 1897 meeting. Read by the chair of the committee, Professor William White, the conclusions of the long and detailed report were that "the routine employment of the X-ray in cases of fracture is not at present of sufficient definite advantage to justify the teaching that it should be used in every case," and that "skiagraphs alone, without expert surgical interpretation, are generally useless and frequently misleading."[73] Following the publication of the report, the members of the American Surgical Association unanimously voted for a resolution that the skiagraph be inadmissible as evidence in court, because "being a picture of a shadow and not of the object, it is inaccurate and unreliable."[74]

By 1900, however, the tide was already beginning to turn. Improved equipment, accumulated expertise, better recording media, more refined techniques, and a mushrooming body of literature were producing not only better images but also an increasingly organized and confident group of experienced medical practitioners who could speak for the significance of the new technology. In December 1900, the first organized meeting of x-ray operators attracted 150 medical practitioners to New York, and the national organization they founded, the American Roentgen Ray Society, worked hard during the following decades to obtain recognition of the x-ray practice as a medical specialty.

Dr. Carl Beck of New York was one of the more active voices among this growing group of x-ray boosters. By 1900, he had carried out more than 3000 x-ray sessions, and his 1904 publication list included seventy-six articles and a book on the rays. For Beck, "the proofs of the immense usefulness of Roentgen rays in surgery are so overwhelming, indeed, that to discuss them would be carrying owls to Athens."[75] Still, the initial reaction of the profession did not surprise him: "The Roentgen rays have brought about a revolution. They show the conditions as they are, and are impolite to do so without the slightest regard for great authorities. No wonder such brusque information was received with a feeling of uneasiness, often by the same men who should have been but too glad to learn of their diagnostic errors in order to correct them."[76]

The source of the roentgenology infamy within the medical profession, according to the proponents of the new specialty, was the unfortunate combination of the strangeness of its subject and the simplicity of its apparatus. The mystique of the new rays attracted many speculative minds, and the simplicity of the apparatus allowed many of them to enter the exciting new field. The result was an unregulated field, full of ignorant laymen and medical novices, and last but not least, shyster lawyers who succeeded in discrediting it among the members of the medical profession. The key, they argued, to the rehabilitation of the field was its professionalization, that is, the replacement of the individual x-ray operator who developed and marketed his own expertise by an organized community of medical specialists who shared and were defined by medical standards for education, training, competence, and ethics. Here is how a prominent member of this new group of x-ray professionals, Dr. Charles Leonard, put it in 1901:

> The reasons that so-called fallacies and errors have been found in this method and make it necessary to discuss its accuracy is that as a method of diagnosis, it has been entrusted to persons deficient in that professional education that is essential to the accurate employment of any method of diagnosis. The pictures made by lay operators have been accepted as valuable clinical data, and read, often incorrectly, by those who did not know how they were made. Can the medical or lay public be blamed for doubting their accuracy or for believing that they can tell as much by this new method of examination as the members of the profession? . . . The profession must see to it that this evidence is only given by experts, reliable professional men. Its true value and accuracy will then be evident.[77]

The constant threat of malpractice suits, which was a major reason for the medical campaign against x-ray technology, was turned by the proponents of the new specialty into an effective argument for embracing their specialty. Whether the medical profession likes it or not, they asserted, the pictorial testimony of the x-ray image will play a central role in malpractice and other cases. Thus, instead of fighting the new images, the medical profession should learn to use them to its benefit. Instead of undermining their reliability, it should strive to monopolize their interpretation. And instead of castigating the legal fraternity for deploying them, the doctors should welcome it as a partner in their effort to regulate x-ray practice. Collaboration, Dr. Mihran Kassabian, chairman of the Medico-Legal Committee of the newly founded American Roentgen Ray Society, appealed to both professions, would be of unquestionable value to all sides: "First to the physician, in sustaining double diagnosis; secondly to the patient, who is suing for damages, and thirdly to the judge and jurymen, to whom medical nomenclature is unintelligible, and to whom 'seeing is believing.' . . . The physician by this means will avoid the possibilities of wrong diagnosis, and if they should be brought, he has a means of protecting himself. In addition, many suits may be discouraged, or compromises effected outside the court, by the ability to give a clear and definite statement of the condition of the injury."[78]

The first two decades of the twentieth century indeed saw the rise of the medical specialty that came to be known as radiology.[79] The radiologists struggled long and hard to differentiate themselves from the welter of individual x-ray operators—photographers, physicists, engineers, electricians, and others—who were producing x-ray images. To that end, the medical practitioners differentiated between the production of the images and their interpretation. Everyone could produce the images, they maintained, but only a few could read them. "The language of an X-ray picture," as Dr. Samuel Monell, the author of an important medical book on the rays put it in 1902, "is intelligible only to those who speak it themselves. Behind the picture must be the trained understanding of what it ought to represent." Mastering the technology, the radiologists emphasized time and again, was not enough. To read the images with any reasonable degree of accuracy, to be able to distinguish between normal and pathological appearances and between essential and accidental details, one needed to know anatomy, histology, and pathology in detail and to be familiar with the various ways that both normal and abnormal conditions appear on the x-ray plate. Without such knowledge, no meaningful reading of the images was possible. An x-ray image, as Dr. Sidney Lange, a physician and a radiologist to the Cincinnati Hospital, put it in his 1907 summary of the first decade of radiology in America, "has no intrinsic value. It is worthless, even dangerous, unless in the hands of one able to correctly interpret it."[80]

The differentiation between production and interpretation allowed the radiologists to shift the focus of their growing discourse from the technology to its human agents. Instead of blaming the "failure" of the rays on the deficient technology, which was shared by all operators regardless of their training, the radiologists were now blaming it on the incompetence of the nonmedical operators. "The Skiagraph is never wrong," the slogan ran. "When error exists it lies in its interpretation."[81] "There are fallacies and errors and distortions referable to this means of diagnosis," Kassabian explained to the members of the prestigious New York Medico-Legal Society. "But it is not the science that is at fault, but the incompetence of the persons entrusted with the work of making the X-ray examinations."[82] "When this point of view has become more prevalent," another radiologist concurred in 1908, "the

pictorially excellent skiagraph of a hand or a foot made by some enthusiastic amateur will no longer excite wonder, and the photographers,
electricians, and janitors who now make the so called X-ray photographs in many hospitals, will have their activities transferred to other
fields where they will be less menace to the public health."[83]

The growing specialty attempted to monopolize not only the interpretation of the images but also the physical prints themselves. "It is not
well for you to send your patient to the Röntgenologist for the purpose
of 'having an X-ray picture taken,'" George Stover, professor of radiology and Dean of the Denver Medical College, instructed the general
practitioners in 1910: "It is not the picture you wish; it is the opinion of
this consultant you desire; it happens that he bases his opinion largely
on the information he gains by means of Röntgenographic examination. His opinion should be given to you, not to the patient . . . The
records, notes, and plates of the Röntgenologist are in the nature of
privileged communication, and should be regarded technically as being
as confidential as the information acquired in the consulting room."[84]
Led by Stover, the American Roentgen Ray Society passed in its
1910 annual meeting an official resolution that "endorses the
views . . . vesting a property right in the radiogram with the radiographer, inasmuch as the radiogram is a part of his clinical record of
examination." According to the Society, the service that the paying
patient was entitled to was the radiologist's opinion, not the image. The
image was to be kept away from the inquiring eyes of the patient,
carefully filed with the complete case record, available for inspection on
proper demand. Furthermore, since a lawyer could compel the radiologist to present the image in court, the radiologists were advised that "it
is necessary to see that the written reports on X-ray examination are not
burdened with too much gratuitous surgical information."[85]

Thus, by the second decade of the twentieth century, an organized
community of medical specialists had emerged that made increasingly
successful claims of monopoly over the ownership, control, and the
interpretation of x-ray images. By that time, x-rays no longer enchanted
the public. The rays were integrated into the scientific theories of
radiation and matter, and the growing awareness of their dangerous
nature diminished lay interest in them. The popular press stopped
printing the images alongside explanations of what they contained, and
the advertisers and the poets moved on to new icons. The increasingly
specialized apparatus also was no longer a bargain, and the nonspecial-
ists found it too costly to keep up with the pace of technological change
that quickly consigned their equipment to the scrap heap.[86] In short,

the x-ray image ceased to be a part of the layman's universe and became relocated within the private province of radiology. The increasingly specialized x-ray technology was gradually black-boxed. When not totally ignored by the radiological discourse, it was treated as an unproblematic and controllable factor. At the same time, the images it produced were presented both as a human construct prone to manipulation and a code that needed to be cracked. The differentiation between the stable technology and the problematic images, between the routine production of the images and their artistic interpretation, and between the medical and nonmedical actors was institutionalized via a division of labor and status. The American Roentgen Ray Society closed its ranks to nonmedical practitioners. The nonmedical operator, who had once dominated the field, was turned into a radiographer—an invisible technician in charge of the mechanical stage of the actual image making. The medical operator became a radiologist—a keen-eyed expert, whose specialized training made him competent to interpret the enigmatic images and whose professional ethics made him suitable to represent them before the visually impaired public.

The Silent Witness

The early decades of the twentieth century saw the American courts struggling to adapt to the changes in the cultural and professional status of x-ray images. What they had embraced as a new species of evidence that would allow lay judges and juries direct access to the disputed facts was turning into exclusive domain, accessible to experts alone. The judges were ready perhaps to sanction medical monopoly over the images outside the courtroom but not inside it. X-ray images were photographic evidence, and as such, they were claimed inside the courtroom by a mightier monopoly—the institution of the lay jury, the ultimate trier of facts. Still, the judges were well aware that if x-ray images were to appear authoritative in the courtroom, they needed the sanction of the medical community. Thus, while holding the jury to be the final authority on x-ray images, the judges nevertheless were willing to recognize radiography as a medical specialty and searched for ways to carve a more significant role for the radiologist in the courtroom.

The courts found little difficulty in carving a meaningful role for radiologists in the admissibility stage, where photographic evidence was treated as a fallible human artifice that needed to be carefully verified. The illustrative evidence doctrine permitted the witness to adopt a visual representation instead of a verbal description, as long as a witness could verify that the evidence was an accurate representation of what he had seen. Unfortunately, x-ray images, by their nature, could not be verified by such eyewitness testimony. In the absence of such testimony, the courts looked for other ways to establish the reliability of the images. Some judges allowed the substitution of expert testimony by radiologists for eyewitness testimony, stating that the anatomical images reproduced the conditions sought to be represented with reasonable accuracy.[87] Most courts, however, opted to rely on circumstantial evidence in order to establish the reliability of the x-ray images. The procedure generally adopted was the one suggested by Judge Lefevre, that is, summoning expert witnesses who were familiar with the various stages of the process by which the particular image had been produced to testify that the process had been applied in a way that was known to produce satisfactory results.[88] Once it was established that a skilled operator had produced the particular image under the proper conditions and with adequate equipment, the image was admitted into the courtroom as a graphic expression of the expert witness's testimony to be weighed by the jury. That is where the troubles began.

Inside the courtroom, the conception of visual images as mute objects that needed someone to speak for them ran counter to the implicit but powerful legal understanding that properly verified photographic evidence spoke for itself. Thus, in spite of the growing authority of radiologists, many judges continued to forbid them to tell the jury, except in general terms, what, in their opinion, the meaning of the images was. "Skiagraphs or X-ray photographs are the best evidence of what appears thereby," explained an Iowa court in 1919, expressing why it found a medical testimony to be a reversible error, "and physicians should not be permitted in a personal injury action in which such evidence is introduced, to interpret and explain same."[89]

The radiologists repeatedly protested. The x-ray image they maintained, is "as far from being a photograph as possible for two things to differ"; its lay interpretation was "worthless, or worse than worthless," and "there is really no more reason why a jury should be allowed to see the skiagram than that there should be exhibited to them the clinical thermometer, stethoscope, measuring tape, and chemical apparatus, etc."[90] Still, the courts never seriously entertained the possibility of shifting the responsibility for intrepreting the images from the lay jury to the radiologists. The judges did become increasingly aware, however, of the need to ensure that lay jurors have access to reliable expert guidance before they decided on x-ray evidence. This led some of them to rethink their tacit practice of treating properly verified x-ray plates as independent evidence that could speak for itself. "Nobody but an x-ray expert could tell anything from the plates," declared an often quoted decision by a New York appellate judge in 1915: "I do not think that the doctrine that an ordinary photograph is the best evidence of what it contains should be applied to X-ray pictures. They constitute an exception to the rule concerning ordinary documents and photographs, for the X-ray pictures are not, in fact the best evidence to laymen of what they contain . . . The opinion of the expert is the best evidence of what they contain—the only evidence."[91]

The disunity of judicial opinion continued well into the 1920s. Eventually, however, the courts came to endorse expert testimony as an essential element in the process that made the x-ray image reliable evidence. In time, the courts took judicial notice of both the reliability of the x-ray process and the difficulties of nonexperts in reading its images.[92] That greatly simplified the admittance of x-rays in evidence, which now required only a testimony that the particular x-ray image offered in evidence was properly produced and identified. At the same time, the courts did not explicitly deviate from their illustrative evidence doctrine. Instead, they treated x-ray evidence as an exception to the doctrine, necessitated by the unavailability of direct eyewitness testimony. Experts were increasingly allowed to interpret the x-ray images for the jury. And once the first expert had finished, it was common and general practice to permit other witnesses—plaintiffs and defendants—to comment on and draw inferences from the images too. In short, the courts treated x-ray images as substantive evidence of the conditions revealed by them, while still discounting all other photographs by admitting them only as illustrations.

The judicial attempts to sustain the authority of the illustrative evidence doctrine by treating x-ray evidence as an exception were only partially successful. If x-ray images were admitted on proof of the reliability of the process by which they were produced, why shouldn't regular photographs be admitted according to the same criteria? If the legal procedures governing regular photography were applicable to x-ray photography, why not vice versa? Indeed, in the absence of clear doctrinal distinctions between regular photographs and x-ray images, the x-ray admissibility procedure was gradually recognized as being applicable to regular photography too. By the end of the second decade of the twentieth century, it was already generally acknowledged that regular photographs could be authenticated not just by an eyewitness but by any witness judged competent by the court to speak for the reliability of the process that had produced them. The practical implications were minor at first. Litigation tactics and economics still dictated that photographs be authenticated by eyewitnesses and x-ray images by radiographers and radiologists. Nevertheless, the theoretical implications were far-reaching. The basis on which machine-made images could now be admitted was not only the observing powers of the verifying witness, but also the mechanical reliability of the process that produced them. The essential relationship underlying the doctrine of illustrative evidence—the association between the visual evidence and the witness whose perceptions and knowledge it purported to represent—was severed. Thus, it was no longer clear why properly verified x-ray evidence should be treated in court merely as illustrative of the expert's testimony and not as independent evidence illustrated by expert testimony. Henceforth, visual evidence was no longer just another mechanism by which human witnesses communicated their knowledge to the tribunal. Machine-made pictures have become, potentially at least, independent evidence that could serve as direct evidence of things that were depicted in them.

The courts were not eager to advance an alternative theory under which photographic evidence could be admitted independently of the illustrative evidence doctrine. Without such a theory, the issue was left to the broad discretion of the trial judges. These indeed sporadically recognized the photograph's independence but without expressly recognizing that they were deviating from the illustrative theory. Instead, they resorted to various kinds of specific factual elements as a sufficient foundation for admission.[93] Legal commentators recognized the growing anomaly.[94] In 1944, a popular handbook on evidence warned its readers that "some photographs, even though offered as explanatory of condition, take on a double character and are both illustrative of what the witness describes and mediums through which original evidence reach the jury."[95] Two years later, in 1946, Dillard Gardner, librarian of the North Carolina Supreme Court, made a strong appeal for a new theory of visual evidence that would reflect the way photographic evidence was actually used in court:

> The x-ray, which is a photograph whose accuracy cannot be checked by human vision, is welcomed in court, while the photograph, whose accuracy can be independently verified, is at times coldly turned aside. An x-ray neither verified nor accepted as competent is admitted and the court takes judicial notice that what it represents is accurate, while a competent, verified photograph

has, at times, been limited to the doubtful status that it may be used only to "illustrate testimony" . . . We have drifted into this strange anomaly in our law by losing sight of this significant fact: photographs may, under proper safeguards, not only be used to illustrate testimony, but also as photographic or silent witnesses who speak for themselves.[96]

The conservative approach of the illustrative evidence doctrine, Gardner contended, may have been justified when photography was not yet well developed, so that the sponsoring witness was also testifying to the accuracy of the photographic process. However, this accuracy was now universally recognized. Thus, the time had come for the courts to explicitly recognize that photographs were not only illustrative of other evidence but could also tell their own stories.

Gardner's plea for a new theory of admissibility failed to impress the courts. New visual technologies were introduced into the twentieth-century courtroom, expanding the realms of visual evidence and increasingly turning judge and jury into virtual witnesses who could judge the facts as if they saw them directly with their own eyes. Some of these technologies—such as 16-mm motion pictures, already at work since the early 1920s; color photography, which was introduced in the early 1940s; and videotapes, which appeared in the late 1950s—posed little difficulty for the courts, which continued to muddle along within their established evidentiary framework.[97] But in the late 1960s, with the rise of the surveillance camera, the watchdog of the modern public sphere, the issue of whether photographs could be admitted as independent evidence that tells its own story rather than illustrates the testimony of others was finally put squarely before the court.

Surveillance cameras, just like x-ray machines, provided valuable images for which no verifying eyewitness could be provided. However, unlike x-ray machines, the surveillance cameras needed no one to speak for them in court. They produced traditional photographic evidence that conveyed intuitive information easily accessible by the jury. Thus, for the first time, the courts faced machine-made visual evidence that was no longer required to be coupled with a human agency in order to express what it contained.

The status of this evidence was gradually resolved during the 1970s in a series of cases involving bank robberies and check fraud.[98] In these cases, the courts admitted surveillance-camera photographs on the proof of the reliability of the process that had produced them. Then, the courts allowed the lay jury to extract the identity of the perpetrators from the photographs—a finding that was not supported in any way by verifying witnesses, but was clearly based on the photographs alone. This, of course, was vigorously contested as a violation of the illustrative evidence doctrine, and in order to justify the decision appellate courts attempted for the first time to advance a general theoretical approach that explicitly recognized machine-made pictures as reliable representations of what they depict. Under this approach, there was no longer a need for a witness to testify that the photograph offered in evidence accurately represents what he or she had observed. This approach came to be known as the "silent witness" doctrine because it recognizes the photograph as one that "speaks for itself" and not for a human patron. This doctrine was readily embraced and has quickly grown into a recognizable alternative to the illustrative evidence doctrine for the deployment of visual evidence in court.[99]

Does that mean that the deep judicial ambivalence toward machine-made evidence has been finally resolved? Not necessarily. As commentators keep noticing, machine-made images continue to challenge the boundaries we so laboriously erect between nature and artifice, between truth and fiction, between lay persons and experts, and between human beings and machines.[100] Meanwhile, the growing presence of technologies purporting to provide true evidence of facts otherwise inaccessible to human beings and of authoritative claims made by experts in the name of these technologies, have been a mark of the twentieth century. So have the judicial efforts to check the exceptional powers of persuasion of machine-made evidence and to keep judicial processes as public and accessible as possible. The tension created by these conflicting tendencies is the subject of the next chapter. This time, however, the battleground was not the human body but the human soul.

6

Science Unwanted: The Law and Psychology

[Soon] there will be no jury, No hordes of detectives and witnesses, no charges and countercharges, and no attorney for the defense. These impedimenta of our courts will be unnecessary. The State will merely submit all suspects in a case to the tests of scientific Instruments, and as these instruments cannot be made to make mistakes nor tell lies, their evidence would be conclusive of guilt or innocence.

↝"Electrical Machines to Tell Guilt of Criminals," *The New York Times* (1911)

B Y THE TURN OF THE TWENTIETH CENTURY, science had permeated the American courtroom.[1] Serving as consultants and witnesses for the interested parties, American men of science played an increasingly pivotal role in the mounting litigious activities in modern matters such as energy (first gas and then electricity), the environment (pollution and contamination), public health (food and drug adulteration, water supply, sewage treatment), communication, transportation, agriculture, mining, industry, insurance, malpractice, and, of course, patents.[2] The forensic sciences also prospered. At the beginning of the nineteenth century, the forensic sciences had employed only basic microscopy and toxicology; by its end, they had not only pushed microscopy and toxicology to a much higher level, but had also incorporated physics, mineralogy, zoology, botany, fingerprinting, and what used to be called anthropometry.[3] By the turn of the twentieth century, forensic experts were able to detect stains and forgeries using infrared and ultraviolet light, identify people by the shape of their head and fingerprints, trace minute quantities of arsenic by its organic chemical print and inorganic substances by their line spectra, differentiate between the various species by the shape of their blood cells, reconstruct important characteristics of a corpse from partial clusters of its bones, and photograph the insides of things with the mysterious x-rays.[4]

By the early years of the twentieth century, some of the social sciences, which had gone through a rapid process of "professionalization" in the last quarter of the nineteenth century, were also allowed into the courtroom.[5] "The black-letter man may be the man of the present," declared Justice Holmes in 1897, "but the man of the future is the man of statistics and the master of economics." Holmes' maxim reflected the growing acceptance by the legal community of the idea that the law is an organic part of the greater society and should reflect its mores. Consequently, the legal profession came to accept those social sciences that helped it to determine these mores.[6]

Still, not all new domains mastered by science were allowed into this brave new legal world. Experimental psychology, the turn-of-the-century leading American social science, was excluded. Pounding harder than any other social science on the courtroom door, experimental psychology offered the court an analysis of its own central processes of fact-finding, persuasion, and proof. The court, however, was determined to keep these processes out of the reach of the long arm of science. Although it developed a deep interest in the insights that human science had to offer, it was unwilling to accept the deterministic implications that went along with these insights. It was ready to accept the expert interpretation of human processes outside its walls but not inside them. The legal subject was purposive, and its data was not to be comprehended by way of any mechanical formulae. Thus, at bottom, the court insisted on explaining human and social phenomena in terms of their own underlying ontological substance—an autonomous causal agency endowed with a spontaneous nature and a free will—that is, in the language of values and purposes and not of numbers and forces.

This chapter unfolds the story of experimental psychology's pounding on the courtroom doors. Describing it within the wider and increasingly problematic context of scientific expert testimony, the chapter describes how the court, in its efforts to keep its doors shut before experimental psychology, abandoned its traditional neutral approach toward the processes of scientific proof and became, for the first time, an active participant in the development of scientific proof, thereby inaugurating a new judicial state of mind that has dominated American legal thought ever since.

The Case of Richard Ivens

In late May of 1906, Harvard professors William James and Hugo Münsterberg, two of America's leading psychologists, received a letter from a Chicago psychologist, John Christison. The letter asked for Münsterberg's and James' support in the struggle to save the life of one Richard Ivens, a retarded young Chicagoan, who had been convicted on his own confession for the brutal murder of a young housewife. "I am absolutely convinced," wrote Christison, "that he [Ivens] is innocent of the crime charged against him and that his 'confessions' are the product of hypnotic suggestion."[7]

The Ivens case began on a mild winter morning, January 12, 1906, when young Mrs. Bessie M. Hollister left her home in high spirits, taking a wall clock to the jewelers for repair. Around 9:30 A.M. she bought two carnations, white and pink, and fish to be delivered to her home, then she disappeared. The next morning, around seven, her body was found in a barn, half a block from her home, lying face down on a manure pile. According to the coroner's physicians, Mrs. Hollister had been raped and then choked to death after a hard struggle.

Mrs. Hollister was the fourth woman to be murdered in Chicago within five months. None of the first three cases had been solved. The public was agitated. Women, the newspapers complained, did not feel safe any more walking the streets as "groups of young men stand on the streets corners, at the car barns, and public places, accosting the women as they pass." Thus, once the rumor spread about the brutal rape and murder of a righteous housewife right next to her doorstep in the middle of a lively neighborhood, public outrage erupted. A large crowd gathered next to the barn and demanded satisfaction. "Only the power of the gospel makes possible our marvelous self-restraint at this time," Mrs. Hollister's priest, Dr. Hall, warned at her memorial service. "While we do not cry for revenge, we do demand justice." The call was echoed from other pulpits as churchmen decried the new Sodom where "right mindedness, and decency, and propriety cannot go on the street without danger of murder" and demanded that the police protect "the virtue, the honor, and the life of our mothers and sisters," and that the mayor restore "the streets to the public by closing the saloons at night and on Sundays."[8]

Mayor Dunne, returning hastily from Denver on hearing the news, declared that "it is notorious that Chicago has not enough patrolmen to watch its streets," and blamed it on the city council whose duty it was to provide the means for an appropriate police force. Chief of Police Collins rose to the occasion and promised that "if the police force were increased by one thousand men, I would undertake to keep the streets clear of all hoodlums and degenerates." The city council, however, rejected the demand and blamed the police for being "signally ineffi- cient." Charges were openly made that the police force, including Chief Collins, was corrupt, and the council resolved to create a committee to investigate why the police were unable to provide the people "the protection to which they, as tax-payers, are entitled." Mayor Dunne was quick to declare that he had also ordered an independent investigation of the police to be conducted by the state attorney. The police, no doubt, were under stress. To appease the angry public and its anxious representatives, they needed to quickly solve the Hollister murder.[9]

Inspector Lavin, the chief investigator on the murder scene, wasted little time before finding a decent suspect—Richard Ivens, a feeble- minded youth in his early twenties who had been the first to find the body, when he went to the barn to feed his father's horse. One quick talk was all it took for the experienced inspector to be convinced that Rich- ard Ivens was his man. Ivens, Lavin later explained, "did not give straight answers," and looked as if he had not slept the night. That was all Lavin needed to tell the young lad right there and then that "he was the guilty man." He arrested Richard Ivens and took him to the local police station for interrogation.[10]

Richard Ivens was the elder of two sons of a poor but respected family. He had never been convicted of any offence, nor had he been seen in bad company. If he was known for anything, it was for his obliging disposition. He was a timid boy, pushed around by his friends, who "thought him rather stupid and decidedly trusting and credulous and absent minded."[11] In short, he was what Henry Goddard, the influential American psychologist, would categorize a few years later as a "moron"—a term Goddard coined from the Greek word for "stupid." The moron's estimated mental age was between eight and twelve, and he topped Goddard's chain of feeblemindedness. Below him were the "imbecile," whose mental age was between three and seven, and the "idiot," whose mental age was between one and two. Mental deficiency was considered a sure sign of criminal tendencies at the turn of the twentieth century.[12] "Not all criminals are feeble minded," explained one well-known psychologist, "but all feeble-minded persons are at least potential criminals . . . Moral judgment, like business judgment, social judgment, or any other kind of higher thought process, is a function of intelligence. Morality cannot flower and fruit if intelligence remains infantile." Of all feebleminded, the moron was considered the most dangerous to society. The idiot is perhaps more loathsome, explained Goddard, but "he lives his life and is done. He does not continue the race with a line of children like himself . . . It is the moron type that makes for us our great problem." It was also the hardest to identify and, therefore, to control.[13]

By 10 A.M., three hours after the body was found, Richard Ivens was brought to the local police station, which was already flooded with reporters. Once there, the customary "sweating" routine began. Ivens was stripped, searched, and pressed by four officers, including the local chief of police and his assistant, to confess his guilt. Nervous and afraid to deny or admit the allegations directed at him, Ivens looked sullenly at the floor and said nothing. Finally, the officers left the room, leaving Richard to be interrogated in private by the local expert, Assistant Chief of Police Schuettler. It is not clear what happened in that closed room, but within thirty minutes Richard had confessed to the brutal murder.

During the next hour or so, Ivens was made to repeat his confession twice before an anxious crowd of officers. Later, around 3 P.M., he made another confession and signed written documentation of it at the official coroner's inquest. There, in the presence of a large crowd of officials, William Hollister, the murdered woman's brother-in-law, tried to shoot him. Hollister was detained but was soon released. "No one could blame him," the newspapers reported, for trying to get the "brute." It is not clear how many more confessions the brute made. According to one of the newspapers, he made fifteen of them within the first three days of his arrest. During all that time he saw neither his parents nor a lawyer. Only after a week in the police station, after all ranks were satisfied with his multiple confessions, was Ivens finally moved to the local jail. Shortly after that he retracted all his confessions. "I have no knowledge of having made them, and I am innocent of that crime," he claimed. "From the time I was arrested I do not believe that I was myself for a moment until after I was here in the jail. Everything about that time was a blur to me . . . I can see, through this blur, the times in the station when the police would bring me up every little while and tell me that I had done it."[14]

At the trial, Ivens pleaded not guilty. According to his confessions the murder had occurred between 7 and 7:30 P.M., beyond the closed gate of the barn, only inches from the busy street. Still, no one had heard anything, including three witnesses whose businesses were located a few feet from the alleged scene of "hollering" and "screaming," and the officers who inspected the premises found no evidence of the confessed scuffle. Ivens also presented an alibi, supported by twelve witnesses, that excluded the possibility of his having committed the murder at that time. He had little chance to prevail, however. The newspapers campaigned relentlessly against him for weeks. "Every husband," the *Chicago Chronicle* recounted later, "perceived that the fate which fell upon this hapless woman could have fallen upon his wife; every father shuddered with the comprehension that his daughter might as easily have been assailed." In his closing statement, Mr. Olson, the public prosecutor, rolled his eyes upward and reminded the jury that Mrs. Hollister was looking down at them at that moment from heaven, crying "I was murdered! murdered! murdered! Gentlemen, do your duty!" The jurors did their duty. They convicted Richard Ivens, who was then sentenced to death by the trial judge. The defense appealed, and Christison's letter, sent to more than twenty notable psychological authorities worldwide, was a side effort designed to add much-needed respectability to the efforts to reverse the verdict.[15]

Christison's claim that Ivens' confessions were the product of hypnotic suggestion was far from being scientifically outrageous. Once a source of public amusement and scientific disdain, hypnotism had become, by the early twentieth century, a legitimate research subject and therapeutic technique among university psychologists, medical practitioners, and many others. It was widely recognized that most people could be hypnotized, many of them with surprising ease. Following studies that showed that there are as many hypnotic states as there are persons, hypnotism came to represent a mode of suggestion that could be applied in states varying from normal consciousness to a hypnotic trance.[16]

Reading the protocols of the police interrogations enclosed with Christison's letter, William James was inclined to agree with him that Ivens "was probably 'hypnotized' by the police treatment." But having only Christison's records to rely upon, James was careful to emphasize in his answer only the need to reopen the case. "Whether guilty or not guilty," he wrote, "Ivens must have been in a state of dissociated personality so exceptional that only experts could be expected to treat it as credible. If guilty, he must have lapsed into that state shortly before doing the crime . . . Or, if not guilty, the shock of the experience of the first morning must have thrown him into that state . . . In any case it is a foul scandal that he was not submitted to a hypnotic examination by experts . . . I think him a victim of popular ignorance of morbid psychology." Münsterberg, by his nature, was more confident. "I feel sure," he wrote, "that the so-called confessions of Ivens are untrue, and that he had nothing to do with the crime . . . It is an interesting yet rather clear case of dissociation and auto-suggestion." Münsterberg blamed the conviction on the highly charged public atmosphere. "The witches of the seventeenth century were burned on the account of similar confessions and the popular understanding of mental aberrations has not made much progress since that time."[17]

Made public in Chicago by Christison, James' and Münsterberg's replies were met with a hostile reaction. The city prosecutor mocked the hypnosis charge, and the local headlines cried "Harvard's Contempt of Court." The editorials pictured psychology as the newest scientific fad for "cheating justice," "emasculating court procedures," and "discouraging and disgusting every faithful officer of the law." Ivens, one editorial declared, "is guilty as hell, whatever vagaries 'experts' and theorizing university professors be led into . . . and the long distance impudence of profs. James and Münsterberg can have no effect except to make themselves and their science ridiculous." "Illinois has quite enough people with an itching mania for attending to other people's business without importing impertinence from Massachusetts," repeated another editorial. "We do not want any directions from Harvard University irresponsibles for paltering still further."[18]

The Supreme Court of Illinois recognized the legal merit of the defense's argument that Ivens' confession, coaxed from him by the police without a warning that it would be used against him, was inadmissible. Still, it rejected the request to retract the confession and refused to postpone Ivens' hanging until the legal question could be decided. On June 22, 1906, with record crowds cheering outside the jail walls, Ivens was executed.[19] Shortly thereafter, the city saloon license fee was doubled to pay for the one thousand men added to the Chicago police force. A year later, *The Third Degree*, a dramatic play based on the Ivens case, but with a happy ending, became a huge blockbuster on Broadway. Münsterberg was shaken. The Ivens' case, he wrote, "was a tragedy not only of crime but also of human error and miscarriedjustice, and my scientific conscience as a psychologist compels me to speak of it because the tragedy of yesterday may come up again, in some other form, tomorrow."[20]

Hugo Münsterberg and the Rise of Experimental Psychology

Hugo Münsterberg was born in 1863 in the Baltic port city of Danzig, the third of four sons of a prosperous Jewish family that was involved in the expanding lumber trade between the rural East and the industrialized West. Like so many other middle-class Jews who grew up in Germany at the height of the emancipation period, education became the backbone of Münsterberg's identity, the key to his assimilation in a society that defined itself through *Kultur*. At the age of six, little Hugo started learning music. At seven, he wrote his first poem; by eight, he was studying French; by twelve, he was already conducting a full intellectual life, attending to an extensive botanical collection, publishing stories in *Kinderlaube*, a journal he started with a friend, and cultivating an avid enthusiasm for the theater and the opera. An interest in physics soon followed and Münsterberg explored the mysteries of electrical machines. Then came philology and Münsterberg ventured into Sanskrit and Arabic. In his spare time, when not working on his demanding gymnasium drills, Münsterberg also played the cello in an amateur orchestra, frequented literary clubs where drama was read aloud, and wrote long epic poems on romantic German themes that had sparked his fertile imagination.[21]

In the fall of 1882, Münsterberg started to study medicine at the University of Leipzig, one of the best academic institutions in Germany and a leading center of the German medical reform, which emphasized a science-based medicine. As a first-year medical student, Münsterberg was required to spend much time in the laboratory, where he was to be armed with the experimental skills needed for his future war with disease. In addition, he attended physiology and anatomy lectures, where authorities such as the biochemist Karl Ludwig, the anatomist Rudolf Leukart, and the embryologist Wilhelm His initiated him into the mechanistic approach to natural science.[22]

In his second year, Münsterberg was able to exercise his *Lernfreiheit* to satisfy some of his natural inclination towards philosophy. By 1883, speculative philosophy, the powerful German academic field that once held all truth within its four corners, had been consistently losing ground in the lecture halls for over three decades. Non-Euclidean geometry frustrated Kantian epistemology, and the theory of evolution was threatening to make it incoherent. The law of conservation of energy and the extension of the reflex arc to the cerebral cortex, the putative office of the soul, were doing the same to idealism and its central notions of free and active will. Finally, the most impressive reductionist victory of them all had been scored by experimental physiology in its holy war against the vitalism of *Naturphilosophie*.

Psychology, a branch of speculative philosophy by tradition, was able to connect its prestige to that of the natural sciences, physiology in particular, and thus enjoy new heights of popularity. By the 1870s, psychology was a growing field whose courses were outnumbering those in metaphysics and ethics by a ratio of four to one, competing with logic for second place in popularity after the history of philosophy, which ordered the great philosophical systems in their succession and treated them critically as things of the past.[23]

Nineteenth-century German philosophers maintained a deep interest in the insights that the prospering natural sciences had to offer about the laws of the inner world of the soul. They paid particular attention to the new discipline of physiology and its successful research program, which relied heavily on novel instrumental techniques and mathematical analysis to reduce physiological knowledge to physical and chemical laws. Physiologists, it seemed, were able to overcome some of the obstacles that had long frustrated the eagerly awaited empirical science of the soul. In particular, they had figured out a way to quantify some mental phenomena, such as sensations, and experiment with them indirectly by measuring the stimuli that created them and the time it took to perceive them.[24]

Much of physiology's exponentially growing body of knowledge concerned the mind's agents: the sense organs and the nervous system.[25] The wave of experimental findings started with Magendie's 1822 demonstration of the fundamental distinction between the function of the motor nerves and the sensory nerves. This was followed by Johannes Müller's 1826 researches on vision, E. H. Weber's 1834 researches on the sense of touch, Ohm's 1843 researches on hearing, Du Bois-Reymond's 1848–1849 findings on the electrical nature of nerve impulses, and the 1850 measurements of their velocity by Helmholtz. The peak of this wave came with Helmholtz's seminal works on vision and hearing, which were published between 1856 and 1866.[26] These findings, and the methodological and theoretical preferences that accompanied them, created a wealth of important insights into the central problem of post-Cartesian philosophy—the relations between the external and the inner worlds.[27]

In February 1875, in a daring move, the Leipzig philosophical faculty offered a philosophy chair, which had been kept vacant by internal disagreements for more than a decade, to Wilhelm Wundt, a physician who made his name in the academy as an experimental physiologist and had no formal qualifications in philosophy. The minutes of the faculty decision noted that thousands of students in Leipzig would benefit from lectures on the relations between the material and the mental realms, and that the successful candidate should have a solid scientific background in order to avoid dilettantish coverage of this important topic. Wundt had spent seventeen productive years in the field of experimental physiology, much of them in close collaboration with its highest authorities. Formally moving into the field of philosophy, Wundt brought the best that *Naturwissenschaften* had to offer in ethos, institutional forms, and experimental methods to the rich philosophical tradition of the psychophysical problem.[28]

As a philosopher, Wundt's interests turned out to be remarkably broad and his output of staggering proportions. His unique academic contribution, however, was to create a coherent psychological discourse grounded in experimental evidence, aimed at doing to philosophy what physiology had done to medicine, that is, to supply it with a scientific base.[29] Like most of his generation, Wundt believed that a true German soul was endowed with a spontaneous nature and a free will. This, in his analysis, dictated the existence of a distinct and independent mode of psychic causality. The natural sciences ignored it, since it was not reducible to physical causality. Philosophy could not study it because the observation of mental processes was tricky and insufficient. It was the task of psychology, he argued, to take the middle ground between scientific reductionism and speculative philosophy, and to apply the experimental method to the ultimate goal—unraveling the laws of psychic causality. To that end in 1879, Wundt founded in Leipzig what is considered to be the first psychological laboratory in the world. There, equipped with their master's conceptual framework, Wundt's many students toiled over a well-defined corpus of problems, with a distinct set of experimental techniques and apparatus, in an attempt to provide psychology with a solid base on which to place the principles of psychic causality.[30]

Münsterberg chose to attend Wilhelm Wundt's popular lecture course on experimental psychology. There, in the new field, he found his true destiny. Three years later, in the spring of 1885, Münsterberg received his Ph.D. in physiological psychology from Wundt's laboratory. After adding a medical doctorate from Heidelberg in 1887, Münsterberg became a *Dozent* in Freiburg, a small but prestigious university in Baden. There, besides teaching courses on epistemology, ethics, Schopenhauer, and the philosophy of the natural sciences, he also developed a psychological laboratory with his own money, which soon attracted considerable attention with a series of highly original experimental and polemical publications.[31]

Münsterberg considered himself a champion of Fichtean idealism, whose view of reality as the creation of the self-determining ego was enjoying renewed popularity in the early 1880s.[32] For Münsterberg, the dichotomy between the physical and mental realms was, much like everything else, an artifact created by willing subjects for their own purposes. In the real world, there were no physical or mental facts, only willing subjects and the objects they created to fulfill their needs and desires. Nature and consciousness were, therefore, nothing but abstract constructions—hypothesized loci of all physical and mental "facts," respectively. Science studied the first, psychology the second. Both were distinct and inexplicable in terms of each other. However, since they referred to the same underlying reality, there was a perfect parallelism between them.[33]

Psychology could not study mental phenomena directly, since these were private. It could rely though on the perfect parallelism between mental and physical phenomena and study the former indirectly through the latter. Wundt, according to Münsterberg, understood that but failed to understand that the two should not be mixed. Once the physical point of view was taken, there was no place in it for such entities as an autonomous will, a spontaneous soul, and psychic causalities. Like all psychological process and constructs, these too must be accounted for by their physiological counterparts. Thus, while still holding extreme idealistic metaphysical views, Münsterberg nevertheless subscribed in his experimental work to a program that considered the mind as a purely reactive mechanism.

Münsterberg's theory of consciousness conceived of each mental sensation as a bodily function of two variables: the original stimuli and the initial condition of the perceiving bodily mechanism, which colored the perception of the original stimuli. These bodily functions were the fundamental ingredients of Münsterberg's theory of consciousness. We perceive their looped interactions as feelings, their anticipation and recollection as will, and their succession as time and space. The idea that the initial condition of the perceiving body determined perception allowed Münsterberg also to maintain, on a pure mechanical ground, the idealistic view that the active self created its phenomena.

Münsterberg's theory of consciousness was an ambitious attempt to bridge the deepest schism of late-nineteenth-century German intellectual life—that between philosophical idealism and mechanical science. It was also the boldest attempt yet to objectify psychology. By the early 1890s, Münsterberg's laboratory was attracting students from all over Germany, and he was internationally recognized as the leader of "the Freiburg School of Psycho-Physics," a bastion of the anti-Wundtian positivistic wing of experimental psychology.[34] At the same time, Münsterberg drew heavy criticism from the established leaders of German psychology, who found his work to be unsuitably polemical, presumptuous, and worst of all, scientifically flawed. "Whether the theories of the Beiträge *[zur experimentellen Psychologie]* stand or fall," wrote Edward Tichener, one of Wundt's disciples, "their experimental foundations have very little positive worth. Dr. Münsterberg has the fatal gift of writing easily—fatal especially in science, and most of all in a young science, where accuracy is the one thing needful."[35]

Münsterberg's work was more esteemed in the Anglo-American world. His mechanistic approach accorded with much of the British tradition of empirical psychology, which conceived of the human mind as a passive entity whose concepts were formed by external objects and its values by utilitarian arithmetic.[36] "What distinguishes Münsterberg," wrote the editor of the prestigious *Mind*, "from any others who in this country have conceived of the physical series of nervous events as bearing the whole causal strain of the checkered play of mental life, is just the experimental art which he brings to bear upon special questions of psychology."[37]

One of Münsterberg's greatest enthusiasts in the Anglo-American world was William James of Harvard, whose theory of the mind coincided to a great extent with Münsterberg's.[38] The two had first met in the summer of 1889, at the International Congress of Physiological Psychology in Paris and continued to correspond afterward. James called Münsterberg's 1888 *Willenshandlung* "a little masterpiece," and considered his evolving *Beiträge* to be a work that promised more to psychology "than the work of any one man who has yet appeared."[39]

By the time he met Münsterberg, James was already weary of experimental work and was looking to move on to philosophy. This was reflected in the psychology program he headed in Harvard. The program, which pioneered the new science of experimental psychology in the United States in the late 1870s, found itself by the late 1880s lagging behind the newer and better-financed programs at Johns Hopkins, Clark, and Pennsylvania. James was looking for ways both to revitalize the program and to rid himself of its experimental duties. By 1891, he had already raised money and used it to update Harvard's outdated laboratory. Now, he wanted Münsterberg, "the ablest experimental psychologist in Germany," to run it. In February 1892, he wrote to Münsterberg and invited him to come to Harvard for three years and become the new head of the psychological laboratory. "The situation is this," James wrote. "We are the best university in America and we must lead in psychology . . . We could get men here who would be safe enough, but we need more than a safe man, we need a man of genius if possible."[40]

Münsterberg found the unexpected invitation intriguing. He loved to travel, and the New World, which was shedding its reputation for intellectual backwardness by such appointments as von Holtz (the former rector of Freiburg) to the new University of Chicago, had attracted him. He was well aware of the advance of experimental psychology in the New World, and James, whose recently published *Principles of Psychology* had scored an immediate triumph, was the rising new star of the New World in psychology. Finally, Münsterberg had just secured an *Extraordinarius* (associate professor) position in Freiburg, so he had little to lose by temporarily changing his academic environment. Thus, after some indecision, Münsterberg decided to accept James' invitation.[41]

Kulturpolitiker in the New World

The incipient field of experimental psychology constructed its tender identity around Wundt's laboratory at Leipzig and around the *Philosophische Studien*, the first journal of experimental psychology, which Wundt had founded in 1881. Still, the growing group of self-conscious young researchers found it impossible to synthesize a distinct autonomous discipline out of their massive experimental work and publications. Late comers to Germany's highly structured academic scene, the experimental psychologists found themselves up against well-entrenched interests, especially those of philosophy, which claimed much of the psychological subject matter for itself. Wundt himself spiritedly objected to any attempt to separate psychology from philosophy. Such a development, he felt, would turn psychology from a serious intellectual discipline into a mere craft dominated by practical considerations. Thus, well into the twentieth century, practically all those in

Germany who researched and taught psychology held appointments in philosophy, and their choice of problems, methodology, and rhetoric remained dominated by philosophical agenda.[42]

Wundt's stellar career happened to coincide with the period of Germany's undisputed ascendancy in the field of university education. Almost all of the first generation of experimental psychologists in the United States had spent significant time at Wundt's laboratory before 1900. They returned home in time to get in on the ground floor of the modern American university system and to couple the creation of their discipline with the overall academic expansion and professionalization that was taking place. Thus, if psychology as a distinct field of knowledge was a German novelty, psychology as an autonomous and influential discipline was an American one.[43]

American experimental psychologists had no philosopher-mandarins looking over their shoulders. While their German companions had to guard their philosophical respectability, the Americans psychologists had to answer to totally different masters—the businessmen and politicians who controlled university funds and appointments and were interested in techniques that enhanced social power and performance. It was to this tribunal that one of Wundt's students, the Yale psychologist E. W. Scripture, appealed when he wrote in 1895 that "it is to the introduction of experiment that we owe our electric cars and lights, our bridges and tall buildings, our steam-power and factories, in fact, every particle of our modern civilization that depends on material goods. It is to the lack of experiment that we must attribute the medieval condition of the mental sciences."[44] Experimentation was seen in that period as the key to progress, and it was the laboratories, first the independents and later the industrials, that bore the wonderful fruits of technological progress. The laboratory also became the engine of the American enterprise of psychology, allowing it to differentiate itself from philosophy and the other social sciences and to present itself as a superior "scientific" discipline. Stanley Hall, who worked with Wundt in the first year of his laboratory, founded the first American laboratory at John Hopkins in 1883. Then, from 1888 to 1895 a wave of laboratories swept over America, lagging only a little in phase behind a similar wave in Germany. By 1891, seventeen American colleges and universities had experimental laboratories, seven of them primarily for research.[45]

Münsterberg arrived at Harvard in the fall of 1892 and took immediate control of the laboratory, establishing the German model that emphasized direct control, original experimentation, and introspective methods. Success was rapid. The following year James referred to the laboratory as "a bower of delight," and it was not long before authorities such as Princeton's James Baldwin and Columbia's James Cattell recognized Münsterberg's laboratory as the most important one in America. "Both for original research and for demonstration," reported *McClure's Magazine* in 1895, "this laboratory is the most unique, the richest, and the most complete in any country; and in witness of the fame and genius of its present director, and of the rapidly spreading interest in experimental psychology, particularly in America, there are already gathered here, under Professor Münsterberg's administration, a larger number of students specially devoted to mental science than ever previously studied together in any one place."[46] In 1898, Münsterberg reported unprecedented attendance (365 students) in his lecture course on empirical psychology. In 1899, he was elected president of the American Psychological Association and vice president of the American Academy of the Arts and Sciences. In 1900, he published the first part of his intended *magnum opus* in psychology and was elected chair of the philosophy division at Harvard. In 1903, he established the *Harvard Psychological Studies* as his laboratory organ, and a year later he played a leading role in the organization of the International Congress of the Arts and Sciences at the World's Fair in St. Louis.[47]

Despite his brilliant American academic career Münsterberg was still hoping for a chair at a major German university. Finding one was not easy. The rapid growth of experimental psychology had already peaked in Germany. All available chairs were occupied, and the growing objection of philosophers uncomfortable with psychology's agenda and success stifled further growth. Being a Jew and a reductionist did not help either. Nonetheless, thirteen years after coming to America, Münsterberg's hopes seemed to come true. In 1905, the University of Königsberg in East Prussia offered him the prestigious chair of philosophy that had once belonged to Immanuel Kant. Alas, the offer came too late. Harvard had just given him a brand-new twenty-four-room laboratory that he had meticulously designed and equipped with the best equipment money could buy. Additional motivation for staying was provided by the multiple emotional pleas of his American colleagues not to leave. "The loss," urged Cattell of Columbia, "not only to Harvard, but to the whole country would be irreparable." Münsterberg declined Königsberg's offer. His long-standing dream of returning to the Fatherland was gone. He had found a new destiny as a leading figure in the German-American community and as a self-nominated *Kulturpolitiker* in the New World.[48]

Unlike many educated Jews who had drifted away from their Jewish identity toward a cosmopolitan one, Münsterberg was an earnest German patriot, who internalized the German values of duty and a life dedicated to ideal ends. Finding himself in America, Münsterberg solved his identity problem by undertaking to infuse American culture with the eternal values of truth and order. In 1901, he began to fulfill his "duty to measure critically the culture of the one country by the ideals of the other," by writing a book that informed the American public of "the German ideals [that] can be serviceable to American culture." In 1904 he fulfilled his complimentary duty to "awaken a better understanding of Americans in the German nation" by writing a second book that informed the German public of the "lasting forces and tendencies of American life." Then, in 1906, the Ivens Case came his way and set his public career on a new path that eventually made him the best-known psychologist in America.[49]

Interpreting the public and official hostile reactions to his intervention in the Ivens Case as an attack on him as a foreigner and a German, Münsterberg was determined to fight back. He decided to carry out an educational campaign from the pages of the popular Sunday magazines on the importance of legal psychology. The first article appeared in January 1907, followed by two more in March and April. In these articles, Münsterberg charged the officers of the law with brutality and ignorance, and announced that "a new special science has grown . . . It started in Germany and . . . spread rapidly over the psychological laboratories of the world." In America, "practical jurisprudence is, on the whole, still unaware of it," but he was determined to change that. Analyzed by modern experimental methods, held to standards of proof, and judged by criteria of evidence, the new science, psychology of testimony, would soon become an indispensable auxiliary of American justice as well.[50]

Psychology of Testimony

If the promise of the experimental method was responsible for the creation of the field of experimental psychology, the difficulties in fulfilling this promise were, to a great degree, responsible for the creation of its subfield, *Psychologie der Aussage*, or the psychology of testimony. That was especially true for introspection, psychology's main observational method. At its basis, introspection is a very simple empirical procedure that has served the students of the inner world since the beginning of time. An individual submits himself to a given experience and then renders an account of it. As a scientific procedure, however, it was considered by many philosophers to be a vestige of the prescientific era. Asking the mind to interrogate itself, they argued, was an endeavor worthy of Baron Münchhausen, who could pull himself by the hair out of any trouble. The introspective act alters and distorts the state of the phenomena it aims to observe. The mind cannot think about its own thoughts without changing them. Moreover, as Kant had already pointed out, introspective evidence was subjective and could not be repeated, compared, or tested by others. Thus, it was not reliable enough to build a solid science on. According to John Herbart, Kant's successor at Königsberg, all that the attempts to force these pseudo-empirical methods of observation upon mental phenomena were able to do was to transform psychology "into a mythology in which no one will confess a serious belief."[51] Nevertheless, as long as systematic introspection seemed to be the only avenue to the higher levels of human consciousness, and as long as these were considered to be the real subject matter of psychology, introspection remained the principle method of psychology. "Without introspection and trust in the introspective method," wrote George Ladd, a Yale philosopher turned psychologist, "experiment gives us no psychical data or strictly psychological laws. And much of what has been discovered in this [introspective] way belongs to physics and biology."[52]

Well aware of introspection's problematic nature, experimental psychologists invested much effort in domesticating it. Massive control procedures, mostly adapted from the experimental protocols developed by physiologists who studied the senses, were used to furnish the act of inner mental observation with properties enjoyed by the outer physical one—standardization, quantification, and repeatability. Chronoscopes, kymographs, and tachistoscopes were employed to measure and control stimuli, and to ensure identical conscious experiences for the various introspectionists. Standardizing the introspectionists by a long training process was another popular procedure, and the time interval between the original perception and its report was minimized to avoid reflection and self-consciousness as much as possible. Generating mounds of data that could be treated statistically to find means and variances was another way to give decisive meaning to the introspectionists' reports on their qualitative comparisons of sensations. But in spite of the industrious effort, introspective evidence remained highly problematic, especially that which was produced by experimental investigation of the complex mental processes of feeling, thinking, and judging, where the various control procedures were virtually useless.[53]

Deeper analysis of the technical achievements and limitations of the scientified method of introspection need not concern us here. Our interest lies elsewhere, with the growing recognition by experimental psychologists that even under the best circumstances, reports of the same phenomena, under similar conditions, by the same person (not to mention by different observers) could vary significantly. This recognition was not limited to experimental psychology. As Kant had already established a century earlier, the main instrument that produced phenomena was the human apparatus. And while Kant was interested in the transcendental analysis of this apparatus, other scientists soon followed who had an interest in the human apparatus as a source of observational errors, and saw this apparatus as one whose limitations had to be understood according to physical laws.[54]

The human apparatus became the heart of various experimental programs that involved introspection. The physicist Wilhelm Weber, for example, had to grapple with the introspective method in the late 1820s, in his experimental efforts to establish a standard measure for sound as a tool for his research program into the molecular structure of matter. A more pronounced example came three decades later when Helmholtz used introspection extensively in his experimental work in physiology, which gave the will a decisive role in his theories of vision and hearing.[55] Considerations of the human apparatus turned out to be relevant even to nonintrospective experimental practices that seemed at first sight to be detached from human bias. Indeed, in the first part of the nineteenth century, it was the astronomers who found involuntary cognitive differences among observers to be the Achilles heel of their observational procedures. Consequently, they carried on a most comprehensive investigation into the human sensory apparatus. Finally, there was another significant avenue, which was opened by the studies of the French psychiatric schools of Salpêtrière and Nancy that disclosed the ease with which emotion and suggestion could lead perception astray.[56]

It was the psychologists who turned this set of singular issues scattered over varying domains into a distinct and well-defined experimental investigation. Driven by their general interest in the human apparatus and by their attempts to objectify their experimental procedures, they invested in the identification and factorization of the various sources of bias in the human apparatus. The imperfectabilities of each sense organ, memory, attention, and perception; the suggestive influence of various personal particularities, of habits, emotions, and preconceptions; the effect of the various communication channels and the chosen narrative on the report; all were researched and registered. It was only a question of time before their applicability to the problems of legal testimony was recognized.

"When I began in 1901 to examine the correctness of recollection among my students," William Stern, the Breslau psychologist, recalled 36 years later: "I was determined by theoretical interests in the realm of memory rather than by any practical considerations. Yet, once confronted with the results, I realized the importance of this research beyond the borders of mere academic psychology. For the crucial outcome demonstrated that a perfectly correct remembrance is not the rule but the exception; that even the most favorable conditions for witnessing and remembering fail to protect people against illusions of a more or less serious kind." Many others—doctors, alienists, neurologists, psychologists, artists, poets, and jurists—fully shared Stern's insight. The modern self, they all concurred, was as erratic and indeterminate as the world of political, social, and cultural flux it inhabited. Driven by inner unconscious motives and by outer suggestions, it could not be trusted even when truthful. This widely shared sentiment was but a part of a larger framework brought about by the great European fascination with the unconscious. In clinics, in psychiatric offices, in the arts, and in the courtroom—words, dreams, and memories became a code to be deciphered, and intentions, perceptions, and expressions, a guise to be interrogated.[57]

In 1896 in Munich, Johan Berchthold was put on trial for murdering three women. A well-known psychiatrist, Albert von Schrenck-Notzing, testified as to the effect that the sensational press campaign against the accused had on the witnesses, leading them, as it were, to "retroactive memory-falsification." In the life of a normal man, Schrenck-Notzing explained to the court, there are "very particular forms of error and deception, to which we are all exposed," and "one definitely does not have to be hysterical or a pathological liar to fall prey [to them]."[58] Berchthold was convicted, but the sensational trial attracted much public and judicial attention to the debate over the power of suggestion. "The development of this trial," wrote the Austrian jurist and criminologist Hans Gross, "showed us the enormous influence of suggestion on witnesses, and again how contradictory are the opinions concerning the determination of its value." Gross' 1897 classic *Kriminalpsychologie* offered a new direction to the rich nineteenth-century European tradition of criminal psychology and anthropology, which had dealt so far exclusively with the psycho pathology and natural history of the criminal mind. "No doubt crime is an objective thing," Gross wrote:

> But, for us, each crime exists only as we perceive it,—as we learn to know it through all those media established for us in criminal procedures. But these media are based on sense perception, upon the perception of the judge and his assistants, i.e.: upon witnesses, accused and experts. Such perceptions must be psychologically validated . . . We must know how all of us—we ourselves, witnesses, experts, and accused, observe and perceive; we must know how they think, and how they demonstrate—we must take into account how variously mankind infers and perceives, what mistakes and illusions may ensue; how people recall and bear in mind; how everything varies with age, sex, nature and cultivation.[59]

Gross' plea for a new psycho-legal science was echoed in France by Alfred Binet, the director of the psychological laboratory at the Sorbonne. In 1900, Binet presented the results of his experiments on the influence of suggestion on observation and memory, called attention to their legal application, and pointed to "the advantage that would occur from the creation of a practical science of testimony." Shortly thereafter, William Stern took up the project in Breslau. Investigating the question of what kind of people are best for testifying, Stern defined a measure of "suggestibility" and tried to measure its dependent variables—sex, age, education, and other characteristics—in a systematic set of experiments. To this end, he developed what came to be the paradigmatic *Aussage* (memory) experiments. Subjects were shown pictures or introduced into a simulated event and were subsequently asked to recall the details through different forms of questioning in a variety of suggestive contexts.[60]

A second experimental category that evolved from the *Aussage* experiments combined the "association test," first introduced by Sir Francis Galton in 1879, with the reaction-time techniques developed in the 1880s. The association test was developed by Galton to eliminate the distortion created by conscious introspection in the investigation of the mental activity of association. It presented the individual tested with a group of words and asked him or her to utter the first thought generated by each word. The reaction experiments, in which mental processes such as will, attention, and discrimination were timed in thousandths of a second, seemed in the 1880s to be the great new tool of the "new psychology," which would give rise to a chronometry of the mind. Two decades later, young researchers such as Carl Jung and Max Wertheimer combined the two procedures, hoping to determine guilt by the chronometry of associations. The underlying theory, also shared by psychoanalysis, was that emotionally charged complexes of ideas disclosed themselves by the involuntary disturbances they create in consciousness. In the *Aussage* application, the tested person went through a carefully prepared list of words, the time needed to call up the first suitable response was measured, and delays, irregularities, or hesitations were looked for in associations that had bearing upon the crime.[61]

Thus, by the turn of the twentieth century, alongside the booming psychologies of education and therapy, there had developed in Europe the applied field of the psychology of testimony. Its chief purpose was to determine the accuracy of human testimony and the conditions on which this accuracy depends, and to educate police and judicial officials in the fallibilities of the cognitive processes. In 1903, Stern founded the journal *Beiträge zur Psychology der Aussage*, which quickly became the official mouthpiece for the new field.[62] By 1906, Binet and Édouard Claparède, the director of the psychological laboratory at the University of Geneva, introduced the German results in the pages of the *Année Psychologique* and repeated the call for a *"science psycho-judiciare."*[63] The feeling in Germany, Austria, and France was that systematic experimentation would allow psychologists, in the near future, to detect guilt and to rank testimonial reports on a scale of reliability.[64]

While legal psychology was making progress on the Continent, it remained uncommon in Common Law countries. As early as 1878, George Miller Beard, a physician from New York, published a manuscript as three successive articles in *Popular Science Monthly*, in which he discussed the limitations of the senses and memory and complained that "the subject of human testimony has never been scientifically studied." The existing legal rules, he noted, "are based on incorrect assumptions in regard to the value of human testimony, [and] they frequently lead to serious error." Beard's articles failed to precipitate much interest. Two decades later, in 1895, James Cattell, Wundt's first assistant in his Leipzig laboratory, who became head of psychology at Columbia University, suggested again that "the probable accuracy of a witness could be measured and his testimony weighted accordingly." A decade later, in 1906, the year of Ivens' trial, George Arnold, a retired Crown official, who had served as a judge in Burma and India, complained, in the first book on legal psychology written in English, that the legal professionals "obtained their conclusions concerning the human mind in the days before psychology existed as a science and they do not appear to realize that those ancient notions require revising now." On the whole, as an American expert on psychological testing observed in 1909, "English and American investigators are conspicuous by their absence [from the field of legal psychology.]" The only notable experimentalist who took up the field of the psychology of testimony was Hugo Münsterberg, who did so in his Harvard laboratory.[65]

Law and Psychology—The Struggle for Authority

Münsterberg's three popular articles on the psychology of testimony attracted much attention, and in June 1907, *McClure's Magazine* commissioned him to write an essay on a sensational Idaho murder trial. Earlier that year, Frank Steunenberg, a former Governor of Idaho and a leading opponent of organized labor, had been assassinated. The authorities arrested Harry Orchard, who confessed not only to this murder but also to a series of others ordered by Big Bill Haywood, a leader of the newly founded Industrial Workers of the World union, and others from the "inner circle" of organized labor radicals. The confession initiated a trial that brought to a climax almost a decade of industrial violence and immediately became a symbol for international labor protests. Thus, at the end of June 1907, Münsterberg defied the heat and traveled to Boise, Idaho, where, enjoying the hospitality of the authorities, he spent four days and six courtroom sessions observing how "the twelve jurymen sat, each in his own rhythm, in twelve rocking chairs." Then, the Governor of Idaho, Frank Gooding, drove him with a trunk full of scientific instruments to the State penitentiary where he was allowed, off the record, to conduct for seven hours various tests on Harry Orchard in an attempt to evaluate the truth of his testimony. Orchard passed the tests with flying colors. "As far as the objective facts are concerned," Münsterberg summarized the results, "my few hours of experimenting were more convincing than anything which in all those weeks of the trial became demonstrated."[66]

Back in Boise, where the trial was still going on, Münsterberg had the good sense to keep his sensational scoop private. But on the train back to his cool summer cottage in Clifton, Massachusetts, he could not resist the temptation any longer and revealed his findings to a lucky reporter who happened to meet him on the train. By the next day, he was quoted all over the country as saying that "every word in Orchard's confession is true." Crowds of reporters descended on him, eager to know how the professor had reached his conclusion. The possibility of a scientific test that could penetrate the mind and detect deception caught their imagination, and they played it for all its worth. An immediate sensation was created, and the myth of the lie detector was born.[67]

A long article in the *New York Herald*, written by a colleague of Münsterberg's, described, with illustrations, the apparatus that psychologists used to detect deception, including the automatograph, which recorded involuntary movement; the sphygmograph, which recorded heart beats; and the pneumograph that recorded rate and rhythm of breathing. The article reappeared in other newspapers and was also cabled to London, where it was depicted as the "crowning life work" of the Harvard psychologist who had invented new machinery to detect lies. The momentous news traveled to France, where the implications of the new technology for the discovery of love affairs were discussed in detail. The word "wrist" was mistakenly translated as "back" giving the impression that the demonic sphygmograph was strapped to the victim's back to test his truthfulness.[68]

Back in America, the author of a fictitious interview with "Prof. Hugo Monsterwork of Harvard" found the commotion farcical. Others were not so amused. The serious newspapers began to question Münsterberg's Svengali-like abilities. The legal profession, in particular, found the sensational public affair, which occurred before the verdict had been delivered, to reaffirm its "little confidence in the opinions of experts and professors, who often have more knowledge than judgment."[69] Orchard's lawyer, Clarence Darrow, pointed to the fact that Münsterberg was a guest of the prosecution and accused him of graft. Fearing a lawsuit, Münsterberg asked *McClure's Magazine* to postpone his Orchard paper until the verdict was known. Meanwhile, he defended his scientific integrity. "To deny that the experimental psychologist has indeed possibilities of determining the 'truth-telling powers,'" he answered his critics, "is just as absurd as to deny that the chemical expert can find out whether there is arsenic in a stomach or whether blood spots are of human or animal nature."[70]

At the end of July, the trial jury, distrusting Orchard's confession and testimony, acquitted Haywood. Münsterberg decided to retract his Orchard paper completely. Instead, in the September issue of *McClure's*, he published a passionate article on the importance of psychology to the litigation process. Warning against "the blind confidence in the observation of the average normal man," he argued that the jury and the judge did not possess the knowledge needed to clear out the chaos prevailing in the witnesses' observations. "Does the court take sufficient trouble to examine the capacities and habits with which the witness moves through the world which he believes he observes?" Münsterberg asked rhetorically.[71] "The study of these powers no longer lies outside of the realm of science. The progress of experimental psychology makes it an absurd incongruity that the state should devote its fullest energy to the clearing up of all the physical happenings, but should never ask the psychological expert to determine the value of the factor which becomes most influential—the mind of the witness."

Charles Moore, a legal scholar, resented the didactic tone of the psychology professor. True, Moore agreed, legal scholarship had so far failed to systemize and make public the knowledge it had accumulated concerning the psychological side of litigation, but that did not mean that the knowledge was not there. In fact, for years he himself had been engaged in the formidable task of going through something like 2000 thick volumes of legal reports, unearthing the psychological treasures hidden deep within the opinions of experienced judges. With the results of his laborious efforts about to emerge from the printer in the shape of a two-volume *Treatise on Facts*, Moore took it upon himself to represent the legal profession against the presumptuous intruder and his unjust accusation that the court possesses insufficient psychological knowledge.[72]

"A Northwest Passage to truth has been discovered by no less a personage than Dr. Hugo Münsterberg, professor of psychology in Harvard University," Moore announced in an article titled "Yellow Psychology," in the legal journal *Law Notes*. "He found it in his class room, and has mapped and charted it." Alas, Münsterberg did not have "the remotest conception of the process by which judges are accustomed to reach their conclusions," and that caused him to overrate "the value that the courts attach to direct testimony." In fact, Moore explained, the judges had taken into account all the problems Münsterberg alluded to and more. After all, the trial judge has the best psychological laboratory imaginable—the courtroom itself, where "year after year, he conducts or presides over experiments of living and momentous consequence." Therefore, being the true experimental psychologists, "on almost every topic that has a proximate and practical relation to the trustworthiness of testimony delivered in court, the judges have the psychologist 'beaten a mile.'" Moore provided an example: "a psychologist will tell you, peradventure, that 4.1144 per cent of eleven negro women, whom he catechized, recollected such and such a class of incidents in their childhood. But a native Southern judge knows a thing or two about the memories and mental character of the negro race, and says so when occasion arises."[73]

Münsterberg responded immediately. He would not be scared off, he wrote in a letter to the *Law Notes*, "by lawyers that suddenly point to my psychological radicalism and my yellow demagoguery." Addressing the yellowish nature of his public campaign, he explained that he "knew too well that any reforms could come only through a widespread public interest. But to stir up such interest would be impossible if the presentation was to be confined to the technical statement which is suited to the legal journal."[74]

The articles, thus, kept coming. By March 1908, Münsterberg had published four more articles in which he presented the American public with the lessons of the *Psychologie der Aussage*. Although "there is no one who desires to increase the number of experts in our criminal courts," he wrote, "the idea of the psychological expert in court cannot be withdrawn from public discussion. The mental life—perception and memory, attention and thought, feeling and will—plays too important a role in the court to reject the advice of those who devoted their work to the study of these functions." Unfortunately, "while the court makes the fullest use of all modern scientific methods when, for instance, a drop of dried blood is to be examined in a murder case, the same court is completely satisfied with the most unscientific and haphazard methods of common prejudice and ignorance when a mental product . . . is to be examined." Münsterberg was very clear about both the root of and the remedy for the problem. While all quarters of society "are ready to see that certain chapters of applied psychology are a source of help and strength to them, the law alone is obdurate. The lawyer and the judge and the juryman are sure that they do not need the experimental psychologist . . . They go on thinking that their legal instinct and their common sense supplies them with all that is needed and somewhat more. If the time is ever to come when the jurist is to show some concessions to the spirit of modern psychology, public opinion will have to exert some pressure."[75]

The public campaign seemed to be successful. Suddenly legal psychology was in the air, "psychology and the law" seemed to be the topic of the day, and Münsterberg, "who reduced a knowledge of the truth to exact science," was much in demand as a lecturer by all the clubs and societies eager to hear of the most advanced ideas of science.[76] In December 1907, Münsterberg was made president of the American Philosophical Association. In March 1908, his eight popular articles on legal psychology were edited and reprinted in one volume titled *On the*

Witness Stand, which, thanks to the previous controversies, became a popular success and went through numerous editions in both the United States and Great Britain. The reactions were mixed. Most critics agreed on the importance of the subject matter and on the book's timely appearance but rejected Münsterberg's suggestion to replace what they considered to be the court's practical wisdom with arbitrary psychological guidelines and stressed the need for further inquiries. Everyone's interests would be ill served, suggested the *Nation*, by placing the scientific frontier in the courtroom too soon, while it was still open to question and dispute. The reviewer of Moore's *Treatise on Fact* in the *American Journal of Psychology* held a different view. He took the opportunity to hail "Münsterberg's courage in attempting to bring into general notice a growing and practically unknown branch of psychological research," and appealed to Münsterberg to take Moore's criticism with "a grain of tolerant salt [for] he knew the true character of the original *Aussage* work, its laborious, cautious experimentation and its hesitant, qualified conclusions."[77]

But it was not Moore that Münsterberg had to worry about. There were bigger fish in the legal pond, and many of them considered Münsterberg's campaign, carried out on the pages of Sunday magazines, to be a mortal sin. Professor John Henry Wigmore, the legendary dean of the Northwestern Law school, whose monumental multivolume *Treatise on the System of Evidence* made him the leading authority on evidence of his generation, had been watching the unfolding of the affair from afar with increasing annoyance. Since early in his career, Wigmore had had a lively interest in the emerging forensic sciences and took special pride in his progressive attitude toward science. Münsterberg's book, which charged the legal community with ignorance and presented it as an obstruction to the advance of science, infuriated him. Highly skeptical of Münsterberg's claims, Wigmore's legal mind was especially irritated by Münsterberg's constant failure to cite the authorities for all the "easily available data" he quoted. Wigmore expected other psychologists to examine Münsterberg's claims. Nothing, however, was forthcoming. Meanwhile, students and colleagues frequently asked him about the facts concerning the psychology of testimony. Tired of saying he did not know, Wigmore finally decided to take on the task himself. Having a working knowledge of several European languages, Wigmore spent much of his spare time in 1908 in Chicago's libraries, surveying the voluminous Continental psychological literature and familiarizing himself with much of the work that Münsterberg constantly referred to. By the end of 1908 he was ready, as he notified Münsterberg in advance, "to poke some fun at your indictment of us."[78]

In February 1909, Wigmore published an acerbic article, subtitled "A Report on the Case of Cockstone v. Münsterberg," reporting an imaginary libel suit against Münsterberg. The action was entered in the Supreme Court of Wundt County, on April Fool's day. The plaintiff's name, Cockstone, was the combination of Coke and Blackstone, two giants of Common Law.[79] The defense counsel was Mr. R. E. Search, and his assistants were Mr. X. Perry Ment and Mr. Si Kist. The defendant himself was a foreigner (Wigmore's unkindest cut), who came to Windyville (Chicago) to give two lectures. The first to the Ambitious Affratellation of Office Boys concerning the uniform psychological connection between the personality of the office chief and the number of times the letter M appeared on the scrap paper in the wastebasket. The second, titled "Studies in Domestic Psy-collar-gy," was addressed to the Honorific Order of Suburban Dames, concerning a method of locating lost collar buttons.[80]

Putting the author of *On the Witness Stand* on the witness stand gave the last, but not the least touch of satire to the article. However, this was also where the article turned serious. Declaring his will to vindicate the honor of the legal profession, the plaintiff's counsel, Simplicissimus Tyro (simplest beginner), cross-examined the defendant on each of his assertions and nullified them one by one. Providing an extensive bibliography of 127 relevant psychological titles in five languages, Wigmore demonstrated, in a grand display of scholarship and rhetorical wit, psychology's tender age and infantile character as revealed by its hasty generalizations, inner conflicts, and fanciful thinking. Stressing psychology's infancy enabled Wigmore to suggest that it was too soon for psychology to be used in court, without committing himself to any final conclusion. It allowed him also to present the legal system not only as innocent of the neglect of psychology, but on the contrary, as eager for the services that psychology promised, but did not deliver.

Following their erudite champion, the legal profession continued to nibble at Münsterberg's heels. In June 1910, for example, an editorial comment in the *Law Notes* described how "It has been found that a person's view of objects while standing on his head is curiously incorrect, and it cannot be doubted that in cases where witnesses were upside down when they observed the facts to which they testify, a psych-physiologist would be a star witness."[81] Belittling Münsterberg, however, was not enough for the serious minded. True, his style was presumptuous and his solutions probably premature. Nevertheless, he seemed to open a Pandora's box, and the first demon to come out of it—in the shape of a witness who could not perceive, report, or judge accurately the world around him or the world inside of him—genuinely disturbed the legal community. As the leading nineteenth-century American text on legal evidence had explained: "Evidence rests upon our faith in human testimony as sanctioned by experience, that is, upon the general experience of the truth of the statements of men of integrity, having capacity and opportunity for observation, and without apparent influence from passion, or interest to pervert the truth . . . It is upon this faith in the credibility of human testimony that the court or jury in a trial are largely obliged to rely."[82] Thus, if the psychology of testimony was correct that "even the most favorable conditions for witnessing and remembering fail to protect people against illusions of a more or less serious kind," then it unearthed a fundamental weakness in the central legal dogma according to which the external world could be accurately reproduced inside the courtroom through eyewitness testimony.[83] Trying to address this problem in his satirical attack on Münsterberg, Wigmore echoed Moore when he argued that the psychologists failed to address the critical legal issue, which was not the frequency of testimonial errors, but their judicial impact. The judge with his instructions, and the jury in the course of its deliberations, were able, according to Wigmore, to compensate for testimonial errors and to come up with accurate verdicts.

But there were other demons in the box, and the problem was deeper even than the crusader of the psychology of testimony himself had suggested. For too long the Anglo-American legal community had been neglecting systematic investigation of the psychological bases of litigation. The legal community had developed high scientific aspirations during the nineteenth century, but these were directed elsewhere. Nineteenth-century legal scholars had dreamed of reducing the law to first principles from which, as in geometry, all necessary consequences could be deduced. The late nineteenth century saw a move towards the experimental model. Langdell's case method, which had made Harvard's Law School famous and had dominated legal education since the 1870s, conceived of the courtroom as its laboratory and of judicial opinions as its sole data. The legal scientists were to empirically research their facts in the library, in the law reports (preferably of appellate cases that dealt with unadulterated questions of law), trying to generalize from them first principles, and from them to deduce an entire rational body of laws.[84] Others still hoped that like geology and biology, the law also could be explained evolutionarily through the history of its empirical content.[85] But, whatever the scientific enterprise of the day was, experimental investigation of the law's own processes of fact-finding and proof was never among its targets.

In defending the legal profession, Wigmore found himself encountering the same faults for which he had ridiculed Münsterberg. Common Law's doctrines of evidence were heavy with inner conflicts, fictions, and unsubstantiated generalizations about human nature. Worse yet, the psychology of testimony might not yet have succeeded in demonstrating the practical advantages that could arise from its studies of human cognition, but at least it had a well-developed experimental program to base its future efforts on. The Anglo-American legal system, on the other hand, had never even tried to develop the experimental capabilities needed to test the efficacy of its own testimonial processes. Captured by adversarial ideology, its illustrious tradition of authors writing on legal evidence (with the great but outdated exception of Bentham) occupied itself exclusively with the procedural side of the law, virtually ignoring the "real world"—the one in which people were constantly involved in the processes of persuasion and proof. These were left to be consolidated under the weight of tradition and precedent, and streamlined by the experience of judges and the common sense of juries.

During the nineteenth century, the inner workings of the law were largely shielded from the uninitiated, and the working assumptions of the law concerning the autonomous, willful, and responsible individual who chooses between lawful and unlawful conduct, coincided with the larger liberal paradigm of psychological individualism.[86] This allowed the legal profession to argue that, in its own way, common law represented a systematic analysis of human nature, comparable to other disciplines. By the end of the nineteenth century, however, the tide was turning. The individualistic paradigm was wearing away, the new social sciences were presenting increasingly differentiated bodies of organized and standardized knowledge, and its professionals were making increasingly successful claims of authority in different areas of American culture such as education, industry, therapeutics, and now also the law.

Psychology may "hardly hope to become an exact and applied science," conceded Josiah Morse, professor of psychology in the University of South Carolina in 1913. But "all law is psychological. It is human product, drawn up to direct and control human conduct. The trail of human is over every inch of it." True, many judges and lawyers may have gained by their wide experience a deep knowledge of human nature. But that was exactly where the problem lay—the legal practitioner had to go through a lifetime of experience and pay the costly price of numerous errors in order to secure that knowledge. "Therein lies the value of psychology," argued Morse, for by the second decade of the twentieth century, a legal practitioner could have "profited by the labors of at least two generations of scientists who have devoted their lives to the accurate study of consciousness and human behavior, the results of which had been boiled down in books and monographs, enabling him to learn in a year or so what it took others a lifetime to discover."[87]

The legal community had no good answer to such a challenge. But it needed one. Its machinery, which had replaced religion as the public authority in American society, had suggested instrumental instead of sacred justifications. Its law was "made," not "given," and its authority lay in the "rationality" of its processes, which were now challenged by external experts in the holy name of science. Wigmore was well aware of this. "Our profession should awaken," he told the members of the Iowa Bar Association in 1909, shortly after his rebuttal of Münsterberg. "As our fellow professions are not standing still, we shall be left entirely behind, unless we endeavor to catch up. The people will take from us the one talent which we have unworthily preserved and give it to other servants who have been more diligent."[88]

Four years later, Wigmore took it upon himself, again, to reestablish legal dominion over human affairs. In 1913, he published a twelve-hundred-page book called *The Principles of Judicial Proof as Given by Logic, Psychology, and General Experience and Illustrated in Judicial Trials.* "This book aspires," Wigmore declared on its opening page, "to offer, though in tentative form only, a *Novum Organum* for the study of judicial evidence." The Baconian aspiration was not accidental or disproportional. The book proposed a completely new legal science that would redefine the relations between law and the social sciences to create a synthesis adequate for litigation in the twentieth century. "There is," Wigmore asserted, "and there must be, a probative science—the principles of proof—independent of the artificial rules of procedure; hence, it can be and should be studied:

> This science, to be sure, may as yet be imperfectly formulated. But, all the more need is there to begin in earnest to investigate and develop it. . . . [It] is the ultimate purpose of every judicial investigation. The procedural rules for admissibility are merely a preliminary aid to the main activity, viz. the persuasion of the

tribunal's mind to a correct conclusion by safe materials. This main process is that for which the jury are there, and on which the counsel's duty is focused. Vital as it is, its principles surely demand study.

"We must seek," Wigmore warned, "to acquire a scientific understanding of the principles of what may be called 'natural' proof—the hitherto neglected process." Failing to do so, we are bound to continue to carry on judicial trials "by uncomprehended, unguided, and therefore unsafe mental processes."[89]

Dedicated to Hans Gross, "who has done more than any other man in modern times to encourage the application of science to judicial proof," this ambitious book tried to close the Pandora's box that Münsterberg had opened, with a lid made of a peculiar synthesis of all relevant fields of human knowledge. Designed as a textbook with sources in law, logic, various forensic sciences, psychology, and even literature, the *Principles* aimed at supplying the legal student and practitioner with basic scientific knowledge about the logical, psychological, scientific, and judicial dimensions of the cognitive processes of proof and persuasion. The book also included Wigmore's original work. One part consisted of an experimental program that extended the *Aussage* experiments to include a jury, in addition to witnesses. In these experiments, Wigmore demonstrated how the jury, during the course of deliberation, was able to compensate for testimonial errors and arrive at accurate verdicts. Another part of the book was devoted to a stunningly original method Wigmore had developed to quantify the complex processes of legal proof. "The mind is moved," he posed, "then can we not explain why it is moved? If we can set down and work a mathematical equation, why can we not set down and work out a mental probative equation?" The graphic method he developed allowed one to analyze a mass of evidence, break it down into its elements, express the logical relations between the elements and the factual issues to be proved, and calculate their probative value.[90]

The *Principles of Judicial Proof* received enthusiastic reviews but left no visible impact on American legal thought. From 1913 until shortly before his death in 1943, Wigmore offered a compulsory course, based on his book, that was known as "Evidence I," and preceded the standard course on the rules of evidence, "Evidence II." But the book was never adopted outside Northwestern's law school. The work of a legal mind that had tried to master the field through a huge inductive enterprise, Wigmore's book turned out to be an impossible crossbreed between a legal textbook, a popular anthology, and a mathematical logic treatise. Its author's reputation carried it through three editions and thirty years in print. But the *Principles of Judicial Proof* remained estranged from the legal as well as scientific mainstream in title, form, and program. Today, it is a forgotten book, a mute reminder of the existing gulf between the law and the science of man.[91]

The Search for the Lie-Detector

Wigmore's scathing 1909 attack on Münsterberg had a powerful effect on legal psychology. According to a report from 1927, "since the rebuff met by Professor Münsterberg psychologists have left the law rather severely alone." Ironically, because it was an excellent survey of the field, Wigmore's article became popular among psychologists and university instructors, who used it in their reading lists. In 1935, it was again suggested that Wigmore's article had discouraged a whole generation of American psychologists from pursuing further research in the field of legal psychology.[92] Psychology of testimony did not disappear, however. The experimental psychologists may have been inclined now to be ultra-cautious when asserting the applicability of their techniques to the real-life processes of litigation, but their attempts to enter the courtroom did not come to an end; they only changed course. During the 1910s, the field of experimental psychology seemed to be shifting from its traditional emphasis on the measurement and interpretation of mental processes to the measurement and interpretation of the more definite physiological changes. Those experimentalists who were interested in the psychology of testimonial processes followed the same course.

The notion that emotions produce bodily phenomena is an ancient one. Many accounts exist of archaic procedures that took advantage of what they assumed were the bodily manifestations of guilt and fear to detect deception. Experimental curiosity had to wait however until suitable measurement equipment was developed. Galileo is credited with developing the first scientific pulse counter in 1581; it measured pulse by means of a synchronized pendulum. The first blood-pressure measurements are ascribed to Stephen Hales, an English clergyman, who, in 1733, inserted tubes into the left crural artery of horses and dogs and measured the rising of the blood in the tubes. Less direct methods soon followed, and by the turn of the twentieth century, the modern method, consisting of a rubber cuff that applies calibrated pressure, was already in use.[93]

With the development of suitable measurement apparatus, experimental investigation became possible, and interest in the problem of detecting deception grew. By the turn of the twentieth century, while some psychologists tried to capitalize on changes in mental reaction times to detect guilt, others were hoping to capitalize on physiological changes in order to detect deception. Already in the late 1870s, while Galton was working on his word-association tests, Musso, an Italian physiologist, was studying the influences of anger and fear on pulse and respiration. During the following decade, Tarchanoff, a Russian psychologist, devised an experimental program to find the relation between the resistance of the human body to electrical current and the intensity of emotions. In 1895, Musso's tutor, the famous Italian criminologist Lombroso, described how he used variations of blood pulsation, caused by fear and anger, to detect deception in criminal interrogations. A decade later Carl Jung, the famous Swiss psychiatrist, combined the word-association test with Tarchanoff's findings in order to detect charged mental complexes by measuring variations in the electrical resistance of the body.[94]

In 1908, while Münsterberg was campaigning for legal psychology, Arthur Macdonald, an American disciple of Lombroso, described a hypothetical apparatus that contained all of the elements of the modern polygraph, in an attempt to convince a Congressional committee to create a federal criminal laboratory. Nothing came out of it. "If somehow the idea becomes real that there is a science of the mind," an expert on legal psychology gloomily wrote a decade later, "if he can be made to admit, and the more unconsciously the better, that he must come and reason with this science and reckon with these laws, a great battle will have been won." By that time, however, some researchers were already convinced that they had developed the means to win this battle.[95]

In March 1913, Vittorio Benussi, an Italian psychologist who worked in Gross's laboratory in Graz, presented a paper at the meeting of the Italian Society for Psychology, in Rome, describing a method to detect deception by tracing changes in the inspiration-expiration time ratio. That same year, a Harvard undergraduate named William Marston started researching the problem of deception in Münsterberg's laboratory. By that time, Münsterberg had already moved on to other aspects of applying psychology to American life, but he was still supervising experimental work on the psychology of testimony in his laboratory.

Marston spent his first two years in barren research on Münsterberg's favorite deception test—the word-association method. In 1915, after the publication of an extensive study that concluded that "the method cannot be depended upon as a means of determining guilty knowledge," Marston decided to change his approach and to try his luck with the blood pressure test. Using only a stethoscope and a standard medical blood pressure cuff, he immediately achieved promising results. After two more years of testing, Marston published his results, concluding that monitoring the changes in systolic blood pressure "constitutes a practically infallible test of the consciousness of an attitude of deception."[96]

Shortly afterwards, the United States entered the First World War, and Marston was summoned to Washington to join a committee of psychologists that evaluated a variety of known deception tests for security purposes. According to Marston's own account, the committee found his test superior, with a 97 percent success rate, and recommended appointing him as a special assistant to the Secretary of War with authority to use his technique in spy investigations. The Army Intelligence Service declined the unconventional suggestion but allowed Marston to participate, as a civilian volunteer, in several investigations of spy cases. Robert Yerkes, chief of the Psychological Division of the Sanitary Corps and one of America's leading psychologists was impressed by Marston's test and asked Marston to train other military psychologists in his techniques. But in the middle of the training the war ended, and Marston's military run ended with it.[97]

With the cessation of the war work on deception tests, Marston looked for new sources of patronage. The obvious place to look was in the legal system. The immediate years after the war witnessed a surge of research and publication on the issue of lie detection, and in February 1921 Marston declared that "a sufficient psychological background probably exists to qualify an expert upon deception in court." The same month, another ex-student of Münsterberg's, Dr. H. E. Burtt of Ohio State University, who had studied the value of the deception tests for the use by the Army, published the results of his laborious studies of Benussi's and Marston's deception tests. According to Burtt, Benussi's breathing test was correct in 73 percent of the cases, while Marston's blood pressure test correctly diagnosed 91 percent of the cases. The combination of the respiration and blood pressure tests, Burtt concluded, would constitute a most effective lie detection procedure. Later the same year, John Larson, a medical student who worked for the Berkeley Police Department in California, combined all three methods deemed reliable for detecting deception. His assembled apparatus secured a continuous graphic record of respiratory and cardiac changes, together with a record of association time reactions.[98]

Marston continued to update and improve his methods, and in the summer of 1922 he finally got his opportunity to seek admission for his deception test in court. Richard Mattingly, a Washington-based attorney, contacted him and asked him to run his deception test on one of his clients. The client, a young African-American named James Alphonso Frye, had been arrested on August 16, 1921, in the District of Columbia, on robbery charges. Six days later, Frye had confessed to an unsolved murder of a wealthy African-American physician named Dr. Robert W. Brown, who had been shot in his office eight months earlier, on November 25, 1920. The late Dr. Brown was a leader of Washington's African-American community, and feelings about his murder ran high. However, although there were two eyewitnesses to the murder—a fellow doctor and a patient—and although the Brown family offered a $1500 reward for the arrest and conviction of the doctor's murderer, the case remained unsolved.[99]

When he confessed to the killing, Frye correctly described many details of Dr. Brown's murder, including the make and caliber of the pistol used. Later, he was also identified by Dr. Jackson, an eyewitness to the murder. Consequently, an indictment for premeditated murder was filed against Frye on March 10, 1922, and his trial was scheduled to begin on July 17—then, Frye retracted his confession. Presenting an alibi for the time of the murder, he claimed that he had confessed only because his investigator, Detective Paul Jones, had told him that he would "squash" the robbery charges on which Frye was then being detained and give him half of the $1500 reward for the conviction of Dr. Brown's murder. According to Frye, Detective Jones promised him "there would be nothing to the murder charge after the reward was paid, as he knew I was able to prove a rock-bottom alibi." As for the eyewitness testimony, Frye claimed that Dr. Jackson did not identify him "until after Paul Jones had called him aside and had a private talk."[100]

Frye's counsel was not impressed with his client's story and advised him to plead guilty. When Frye refused, he dropped the case. Instead, the court assigned to Frye an energetic young attorney, Richard Mattingly, who had no reputation to lose and was willing to represent Frye for the experience and the small fee paid by the District of Columbia. Mattingly first tried to exclude his client's confession, arguing that he had not been advised of his right to remain silent, but the motion was denied. Mattingly then sought the postponement of the trial, arguing that because of her ill health, Mrs. Watson, Frye's alleged alibi, would not be able to testify on her scheduled date. The court refused again. The trial records do not specify the reason for this refusal, nor do they describe what happened to the other three defense witnesses whom the defense had subpoenaed. According to Marston, "someone had frightened all the witnesses who might have given Frye an alibi into silence."[101]

Unable to find a single witness who would support Frye's alleged alibi, or a way to discredit Frye's detailed confession, not to mention the testimony of at least one eyewitness, Mattingly made a desperate though imaginative move and contacted William Marston. By exposing Frye's lies with Marston's deception test, Mattingly was hoping that his stubborn client could be persuaded to plead guilty and ask for the mercy of the court. There was no money in it, but Marston agreed to do the test because he saw the opportunity to gain recognition in the legal market. Imagine Mattingly's surprise, therefore, when the results of Marston's test confirmed "that Frye's final story of innocence was entirely truthful! His confession to the Brown murder was a lie from start to finish."[102]

At the trial Frye pleaded not guilty. The defense offered Marston as an expert witness to testify to the result of his deception test made upon the defendant. Walter Irvin McCoy, chief justice of the District Supreme Court, who presided at the trial, refused to admit the results of the deception test on the curious technical ground that the test, which had been made a month earlier in the District Jail, had no bearing on the truth or falsity of the story told by Frye in court during the trial. Mattingly offered to subject Frye again to the deception test, in court. Alas, by that stage Frye had already told his story to the jury. "It is too late," ruled McCoy. "You ought to have had the test made at the time he was testifying if you wanted it at all." Mattingly next tried to have Detective Jones, who took the confession from Frye, take the deception test while testifying but was blocked again by McCoy. Desperately, Mattingly tried to submit the results of Marston's test in evidence and then have Marston take the stand and answer hypothetical questions based on this evidence. This was overruled too.[103]

Frustrated, Mattingly argued for his expert's admissibility. "This offer to attempt to qualify, of course, is for the purpose of showing that this is not merely theory, that it is generally known among experts of this class, that it has been in practical use, that it is not new, and that it is available." But Chief Justice McCoy was not going to take chances with this new psychological fancy and did not allow the defense to try and qualify its expert witness. Sixty-three years old at the time of the trial, McCoy was, according to his own testimony, "too old and too much inured to general principles in regard to the trial of cases to depart from them rashly. I suppose it depends upon whether you are before a conservative judge or a young one who is willing to take chances."[104]

Mattingly insisted, knowing that the results of Marston's deception test were his client's only chance. "We have proof to offer on this point, that it is a scientifically proven fact that certain results will be accomplished under certain conditions. It seems to me that the very least your Honor can do is to permit us to attempt to qualify the expert. I think we are entitled to it as a matter of law." But McCoy would not listen. As far as he was concerned, deception tests were inadmissible until "there is an infallible instrument for ascertaining whether a person is speaking the truth or not." But, "I shall be dead by that time, probably, and it will bother some other judge, not me." The results of Marston's deception test were, therefore, excluded. Nevertheless, the rumor that Frye had been found innocent by a deception test somehow reached the jury and this "undoubtfully saved Frye from hanging." Without much evidence from the defense, the trial lasted only four days, and on July 20, 1922, after deliberating for less than an hour, the jury returned a guilty verdict of second-degree murder. Frye appealed immediately.[105]

The Frye Dilemma

The peculiar admissibility standard of infallibility put forward by McCoy was clearly an excuse. He must have known that it was an impossible standard, which would have excluded almost every other species of expert evidence already allowed in court. If he had allowed the Frye defense to try and qualify its expert witness, he would probably have heard that, according to even the lowest estimates available in 1922, the reliability of Marston's deception test was around 91 percent and that independent researches had found these probabilities to be far better than those of a lay jury trying to extract conclusions about the witnesses' veracity from their behavior.[106] Being the experienced judge that he was, McCoy would also have known that these probabilities were probably better than those enjoyed by other species of expert evidence, such as handwriting, medical experts, or ballistics.[107] McCoy, however, refused to entertain such considerations. "I had certain pamphlets submitted to me yesterday to look at, of some Dr. Marston—I believe," he remarked. "I am going to read them when I come back from vacation."[108]

Clearly then, McCoy's exclusion of Marston's systolic blood pressure deception test had little to do with the reliability of the test. What, then, made him so determined to prevent the test from having its day in court? Why not allow the jury to consider the results of a deception test along with the other evidence in arriving at a verdict? When the defendant's mental state was at issue, medical experts were allowed to provide the jury with evidence as to the mental capacity of the accused. Why should the law allow men of science to look into the defendant's mind to see if he or she knew the difference between right and wrong, but not if he or she had lied about it? What was it in Marston's evidence that so alarmed Chief Justice McCoy?

The standard legal explanation is that the deception test, or the lie detector or the polygraph, as it came to be known, constitutes a clear invasion of the province of the jury. The deception test, this argument runs, offers the jury an expert assessment of the credibility of testimonies. Such an assessment is a vital function of the jury. *Ipso jure*, the deception test has been excluded.[109] Although this is true, such an explanation is, to a large extent, trivial. The expert witness, giving an opinion based upon facts presented in court, has always invaded the jury's province. All the more so when the opinion has been given on mental issues. That did not prevent alienists, psychiatrists, asylum superintendents, and neurologists from advising the jury on the mental state of the defendant. The difficulty remains then: What differentiates an invited expert from an intruder? According to legal theory, it depends on whether the proposed opinion is on matters outside the common knowledge of the jury and whether it has solid ground to stand upon. Alas, the line between common and uncommon knowledge has never been clear cut, and the domain of common knowledge has been constantly shrinking under the pressure of specialization and the growth of scientific knowledge. In 1782, in the Wells litigation, Smeaton's tidal harbors theory was initially objected to because it usurped the function of the jury. X-ray experts also were forbidden at first from telling the jury what was in the images. Still, the courts have always found ways to allow the jury access to the fruits of science whenever they considered them to be of significant help in its work.[110]

If Chief Justice McCoy's aversion to Marston's deception test cannot be attributed to the unreliability of the deception test or to its invasion of the jury's province, what can we ascribe it to? In 1922, Professor Zechariah Chafee of Harvard Law School put part of the answer forward. In reviewing the progress of the law of evidence. Professor Chafee was skeptical about allowing the jury to incorporate into their assessment of witnesses' credibility the results of the new psychological tests, including Marston's deception and Binet's intelligence tests. "If such tests ever are adopted, it is probable that the jury system will have to be abandoned, unless education will have advanced so far that twelve men picked at random will adequately absorb blood pressures, time reactions, and intelligence quotients, and combine the mass to a just verdict. In other words, the jury might also be subject to an intelligence test."[111]

According to Chaffee, then, the motive for the exclusion of Marston's deception test was the judicial distrust of lay jurors' ability to assess its worth. Still, such distrust was neither new nor limited to psychological tests. As discussed throughout this book, the court's mistrust of the ability of the credulous jury to resist the testimony of the scientific expert grew steadily throughout the nineteenth century. Yet, thus far it had not prevented the courts from admitting new types of scientific evidence, including the micrometric blood test, the accuracy of which was contested as much as that of the psychological tests, and the x-ray image, which constituted no less an interpretive challenge for the average juror than blood pressures and reactions times. Something else must be at work then, something that had made experimental psychology and the physiological deception test particularly repugnant to the legal spirit.

On its face, experimental psychologists did not demand much. All they asked for was to be accepted in court as expert witnesses. Like the alienists, psychiatrists, asylum superintendents, and neurologists, they also wanted to advise judge and the jury on mental facts. Yet, this seemingly modest demand posed a great threat to the working of the legal system. Traditionally, experts were invited to the courtroom to provide judge and jury with an interpretation of the facts of the case. Experimental psychologists, on the other hand, claimed superior expertise over the court's own central processes of fact-finding, persuasion, proof, and, worst of all, judgment. The grandeur of these claims had already given rise to strife between the law and experimental psychology. Now, the remarkable new technology of the physiological deception test seemed to have brought this simmering tension to a boiling point. No longer was it human beings arguing among themselves who was the better expert on the human soul. Now, a new technology was put forward, purporting to succeed where all human beings are bound to fail.

In times long gone, God served as the first and ultimate expert at Common Law. Omniscient, God knew who was guilty and who was innocent. All his representatives on earth had to do was to produce those suspected of serious crimes. God provided, through the various procedures of proof by ordeal, both the evidence and the proof needed to guide the frail human judges. Many centuries later, experimental psychology was threatening to reintroduce similar procedures into the courtroom, with machines playing the part previously allocated to deity. Soon, *The New York Times* predicted in 1911, upon the first rumors of such machines: "there will be no jury, no hordes of detectives and witnesses, no charges and countercharges, and no attorney for the defense. These impedimenta of our courts will be unnecessary. The State will merely submit all suspects in a case to the tests of scientific instruments, and as these instruments cannot be made to make mistakes nor tell lies, their evidence would be conclusive of guilt or innocence."[112] It is this image that needs to be kept in mind when we ponder the reaction of Chief Justice McCoy to the prospect of admitting Marston's systolic blood pressure deception test in evidence. It was not the human expert who threatened the jury's province. It was the machine—and it did not threaten merely to invade the province of the jury; it threatened to obliterate it.

The Court of Appeal of the District of Columbia likewise was not about to allow Marston's deception test into the court. Still, being an appellate court, it needed to furnish a better rationale for its exclusion than McCoy's peculiar infallibility standard. This was not an easy task. In 1923, there was no special rule for the admissibility of scientific evidence. Like all other evidence, scientific evidence was mainly evaluated according to the two traditional evidentiary criteria: the logical relevancy and helpfulness of the evidence and the qualifications of the witness.[113] Both criteria offered the Court of Appeal of the District of Columbia little reason to exclude Marston's testimony. The logical relevance of the deception test and its potential helpfulness to the jury were unquestionable. So were Marston's credentials. He was a well-known research psychologist, who possessed special training and extensive practical experience in the subject in question. Thus qualified, the weight of legal precedent of expert testimony was clearly for his admissibility. Unless, of course, there was some clear risk that "outweighed" the value of his testimony. But what? Traditionally, such a risk was either that the evidence would consume too much time or that it would unnecessarily confuse or perjure the jury. Neither of these reasons, however, had been applied to scientific evidence before.[114]

The Frye Decision

On December 3, 1923, the Court of Appeal of the District of Columbia submitted its written decision on Frye's appeal. The decision ignored McCoy's obvious procedural mistake in not allowing the defense to qualify its expert witness. It also ignored McCoy's impossible admissibility standard of infallibility. The court did not try to argue that Marston's testimony would have invaded the province of the jury. Instead, in a remarkably short opinion that specified no references or precedents, the court chose to exclude Marston's deception test on the strength of an innovative argument that it had not yet gained "general acceptance in the particular field in which it belongs."

> Just when a scientific principle or discovery crosses the line between the experimental and demonstrable stages is difficult to define. Somewhere in this twilight zone the evidential force of the principle must be recognized, and while courts will go a long way in admitting expert testimony deduced from a well recognized scientific principle or discovery, the thing from which the deduction is made must be sufficiently established to have gained general acceptance in the particular field in which it belongs. We think that the systolic blood pressure deception test has not yet gained such standing scientific recognition among physiological and psychological authorities as would justify the courts in admitting expert testimony deduced from the discovery, development, and experiments thus far made. The judgment is affirmed.[115]

The decision of the Court of Appeal of the District of Columbia to exclude Marston's deception test became a subject of discussion in the legal periodicals. Setting the exclusionary nature of the decision against the radical nature of the evidence it excluded, the appellate decision was widely applauded as being reasonably cautious. Citing a comment by Dr. John Larson, one of the leading experts in the field of lie detection, that "there is no test in its present state which is suitable for the positive identification of deception," the *Yale Law Journal* argued at length that "it is trite to abuse the courts for not accepting the advances of science. The condemnation is undeserved unless science is advanced." The *Harvard Law Review* concurred. "The experimenting psychologists themselves admit that a wholly accurate test is yet to be perfected," it pointed out. "Because of this admitted uncertainty the result in the principal case seems sound." Other journals also joined the applause. "The attitude of the court in this case seems beyond all criticism," opined the *Law Notes*. "It will not do to declare dogmatically that there is no good in a new thing, and it will not do, particularly in a capital case, to let scientific theories as yet unproven disturb the scales. The court displayed a proper caution, yet left the door open for the coming light."[116]

One may wonder though, how did the Court of Appeal of the District of Columbia reach its conclusion? How, for a start, did the court decide on the particular field to which Marston belonged? In 1923, after all, one could count the experimentalists who worked on deception tests on the fingers of one hand, and each had his own different technique. Assuming that the court had in mind the wider field of experimental psychology, one could then ask: how did the court establish the lack of "general acceptance"? By consulting experts? By counting noses? Whose noses? The court's opinion gave no specific references, only a general one to "physiological and psychological authorities." And what was this "thing" that was not accepted? The theory behind the test? The specific technique used by Marston? Marston himself? The decision of the court specified none of these.

Some of these questions received partial answers in 1926, when Charles McCormick, a law professor from the University of North Carolina, sent a questionnaire to eighty-eight members of the American Psychological Association deemed to be interested in the psychology of testimony. The questionnaire asked for their opinion of whether deception tests, combining the measurements of reaction time and respiratory and blood pressure changes, furnished results sufficiently accurate as to warrant consideration by judges and jurors in determining the credibility of testimony given in court. Of those who replied, eighteen answered yes, thirteen answered no, and seven answers could not be clearly classified. None, however, seemed to doubt that the deception tests were scientifically sound in theory, and only seven of the answers could probably be interpreted as indicating lack of belief in the substantial value of these tests. All the rest seemed to be troubled not so much by the reliability of the test as with the ability of the jury, the judge, or even nonexpert psychologists to handle such evidence.[117]

Thus, the only "generally accepted" opinion evident among the experimental psychologists in McCormick's 1926 survey was the fear of a credulous jury bewitched by charlatans. Like the microscopists half a century earlier, and the medical community a generation earlier, the psychologists distrusted not the science but its legal application, and they agreed that the best way to secure the reputation of their profession was to discourage its application in court altogether.[118] As Miss Johnston, the sister of Leonard Keeler, the developer and patent owner of the modern polygraph, put it in her memoirs: the experts in the relevant field of psychology were "opposed to any further recognition of the lie detector by the court," not because they doubted its ability to detect deception, but rather because they believed that "if the lie-detector operators are allowed to testify, the greatest crop of fakers ever known in the history of jurisprudence will spring up and completely discredit the subject."[119]

Does this mean that the Court of Appeal of the District of Columbia was right in excluding Marston's deception test? Perhaps. The exclusion, after all, should be understood in the context of the protracted controversy over the psychology of testimony. The intriguing question, however, is not whether the Frye decision was correct but why its demanding standard of "general acceptance" has proved so remarkably durable. It has survived many assaults on its exclusionary nature and blurred concepts, and was accepted for most of the twentieth century as the standard for the admissibility of new scientific evidence in practically all of America's courts.[120] Why has its exclusionary rationale been able to dominate American legal thought ever since? The final part of the book provides an answer to this question.

Epilogue

IN 1894, AT THE OPENING of the meeting of the American Association for the Advancement of Science in Philadelphia, Dr. Daniel Brinton, professor of ethnology and archaeology at the University of Pennsylvania and president of the Association, took the opportunity to muse about the nature of scientific verity:

> The one test of scientific truth is that it shall bear unlimited and untrammeled investigation. It must be not only verified, but always verifiable. It welcomes every trial; it recoils from no criticism, higher or lower; from no analysis, from no skepticism. It challenges them all. It asks no aid from faith; it appeals to no authority; it relies on the dictum of no master. The evidence and only the evidence, to which it appeals or which it admits is that which it is in the power of every one to judge, that which is furnished directly by the senses.[1]

The turn-of-the-century American courtroom, where a welter of scientific experts communicated and interpreted for judge and jury all the fascinating domains of hard scientific facts that could be applied in their decisions, appeared to be a perfect illustration for this prevalent and powerful ideology that portrayed science as the best passage to truth and the scientific expert as the ultimate arbiter of all demonstrable knowledge. Alas, by that time the legal profession already harbored a different opinion about the nature of scientific verity. Invited to speak on expert testimony before the New Hampshire Medical Society at its 1897 annual meeting, Judge William Foster chose to open his address with a joke popular within legal circles. "There are three kinds of liars: the common liar, the damned liar, and the scientific expert." This characterization, Foster assured his audience, is bestowed "not only by defeated lawyers and their enraged clients, but also by eminent members of the legal profession, both lawyers and judges, as well as by worthy and respectable members of the general public outside of the professions involved. It is the voice of the people and of the press, as well as that of the bench and the bar. It is a fashion."[2]

Indeed it was a fashion, and it stood in contrast to the authority that science had gained in wider society. This rupture stood at the center of Judge Gustav Eindlich's 1896 address before the prestigious Law Academy of Philadelphia. "Whoever, in recent times, has followed the trials of cases and the utterances of judicial tribunals," Eindlich told the Academy members, "must be struck with these apparently incongruous facts: first, that the volume of expert testimony is constantly increasing, and second, that the modicum of respect paid to that class of testimony by the court is constantly diminishing." Incongruous as it seems, Eindlich warned, this perilous rupture will continue to grow, for, as things stood, both sides were equally right—lawyers in submitting scientific evidence and courts in denouncing it as untrustworthy:

> In our day, when the sub-division of labor and occupation tends to make every man a specialist in his own circle—when the universal demand is for exactness and precision in the smallest minutia—when the advancement of science, of the arts, of mechanics, and a flood of invention have added to the sum total of the world's knowledge, until now scarce any mortal dare boast the mastery of a single department—when by reason of all this, those who rank among the most intelligent can hardly be looked upon as competent to attack problems belonging to other spheres than those in which they move, and when jurors drawn indiscriminately from the whole population are necessarily unequal to such tasks—in our day, the hope of reaching just conclusions upon multitudes of problems submitted to judicial determination, is plainly becoming more and more dependant upon the aid of competent persons. In other words, it must be true that the legitimate field of expert testimony is in process of being vastly extended, not

by the courts but by the march of human progress and the conse-
quent limitation of individual sufficiency to comprehend their
details.[3]

On the other side, Judge Eindlich put forth the second leg of his
argument, stands the stricture of the courts on the worthlessness
of expert testimony, and they also are correct:

> Indeed, it is difficult to conceive of language within the bounds of
> decent and temperate criticism, which ought to be regarded
> as excessively severe in commenting upon the expert testimony
> nuisance as it has, of late years, been infesting our courts. In the
> way of wasting the public's time, in the way of burdening litigants
> with expense, and in the way of beclouding the real issues to be
> tried and effecting miscarriages of justice, it has grown to the
> proportions of an offensive scandal. Instead of being an aid in the
> administration of the law, it has become a positive hindrance to it.
> Instead of assisting in the approximation of the truth, it has
> become the means of obscuring it . . . expert testimony is today
> discredited and rightly discredited by the courts, and ridiculed by
> the hard common sense of the people.[4]

As Eindlich's choice of words makes clear, the feeling, by the end of
the nineteenth century, was that expert testimony had reached rock
bottom. The debate over the causes of and solution to the expert testimony crisis had been picking up steam for almost half a century, and
the remedies suggested, as another judge put it, were "as numerous as
prescriptions for the cure of rheumatism and generally as about as
useful."[5] The problem was that all reform plans seemed to run counter
to one or more of the fundamental postulates of the adversarial legal
system. Creating a special scientific tribunal or simply getting rid of the
jury was counter to the fundamental right to a trial by a jury of one's
peers. Allowing the court to call in experts independently of the parties
in the case was counter to two other equally fundamental
postulates—the right of the parties and their lawyers to present all
evidence and the neutrality of the court. Such suggestions were, therefore, considered by legislators and lawyers alike to be "remedies far
worse than the disease."

The disease was spreading fast, however, and the pressure for a remedy, any kind of a remedy, was mounting. Something eventually had to
give way in this sacred triangulation of the adversarial system—either

the political postulate of the lay jury, the traditional right of the parties to furnish all evidence, or the neutral position of the court. In 1905, Michigan made the first attempt to accomplish such a reform by an act of the legislature. It passed a statute that embodied the most popular reform suggestion—that of allowing the court to nominate its own experts. The statute contained the mildest possible version of such a reform. It did not preclude the parties in a case from using their own witnesses but provided in criminal cases for the additional appointment by the court of no more than three disinterested persons, whose identity should not be made public, to investigate issues involving expert knowledge and testify as to their findings at the trial. Nevertheless, the Michigan Supreme Court held the statute to be unconstitutional. The Supreme Court considered it no part of the duties of the court to select witnesses. Such activity, it pointed out, transferred the power of choosing witnesses from the prosecutor, an administrative officer, to a member of the judiciary, in violation of the provision of the state constitution for a separation of powers. The Supreme Court also considered the statute a violation of the fundamental right of the accused to a fair and impartial trial. The official sanction of judicial appointment, the Supreme Court pointed out, would give the court experts an extraordinary aura of ability and candor, while the other experts in the case would be judged by the jury according to ordinary standards. Declaring the legislation in question unconstitutional, the Supreme Court expressed the opinion that the only available remedy for the acknowledged evils at which the statute was aimed would have to be found in a livelier sense of responsibility in the professions for the proper and decent administration of justice.[6]

The decision dealt a serious blow to those who had been advocating the reform of expert testimony by means of statutory enactment. Accepting that the experts should still be party chosen and the jury still be considered the final trier of facts, early twentieth-century American legal scholars concentrated their attention on enlisting the professions in the effort to improve the standards of admissibility for the party-chosen expert. Their renewed hope of succeeding where their predecessors had so miserably failed hinged on a clear change in the market for scientific expertise, created by the rising professional culture in America. By the second decade of the twentieth century, the individual expert who developed and marketed his or her own expertise had been already replaced by a community of experts who shared, and were defined by, common standards of competence and ethics. A wide range of professions from chemists, physicists, and engineers to architects, surveyors, actuaries, realtors, insurers, and accountants came to be dominated by professional associations of practitioners. These associations developed codes of ethics, generated standards for education, training, and practice, and defined minimum qualifications for certification either through their own examinations or through those of the various state boards of examiners.[7]

During the second decade of the twentieth century, one can find various legal scholars pondering the ways in which the courts could take advantage of this standardized market of expertise to check the problem of expert testimony. "The remedy is not in the enactment of any new statute," wrote one scholar in 1910. "No act of the legislature will make witnesses learned or honest. The reform must come from the professions themselves." It was not clear, however, how exactly such a reform should be carried out. "There is the logical possibility that no remedy exists, or that any proposed remedy brings in its train new evils worse that the disease which it cures," conceded another scholar in 1915. One thing was clear, though. "In the selection of experts no solution can be considered satisfactory that does not provide for the selection by the profession involved."[8] Eight years later, in 1923, the Court of Appeals of the District of Columbia came up with the first effective formulation of such a solution.[9]

Unable to exclude Dr. Marston and his deception test on the basis of the existing admissibility rules, the Frye court formulated a new rationale that demanded the qualification not only of the expert's credentials but also of the particular scientific knowledge he proposed to present to the court. Addressing "the thing from which the deduction is made," and requiring that it "be sufficiently established to have gained general acceptance in the particular field in which it belongs," the Frye decision created a new degree of freedom for the judicial scrutiny of scientific expert testimony within the accepted adversarial structure of the legal system. The jury was still considered the final trier of facts; the experts would still be party chosen; but the judicial ability to control the market of scientific expert testimony was significantly extended from the realm of the expert to the realm of expertise.

One hundred and forty years earlier, in 1782, George Hardinge, barrister of the Middle Temple, suggested a similar rationale to the King's Bench. John Smeaton was a worthy expert, Hardinge argued, but he propounded in court a theory whose scientific status was yet unsettled. Hence, his evidence should be excluded. Lord Mansfield, chief justice of the King's Bench, rejected the proposed distinction between the expert and his expertise. Mansfield's decision, which maintained that it was not for the bench to qualify the expert's opinion, shaped the nineteenth-century practice of expert testimony.[10] If a person was qualified as an expert, his or her expert opinion came along with him or her. It was the job of the cross-examiner to expose the faults of the testimony and for the jury to be the judge of it. A century and a half later, the Frye court reintroduced Hardinge's exclusionary logic into the rules of evidence. Being an expert was no longer enough.[11]

Adopting an active judicial role in the process of reviewing the scientific evidence seemed plausible in the context of the controversial field of the psychology of testimony. But the Frye opinion went beyond that. Proposing to look for the general acceptance of a procedure in the particular field to which the expert belonged, it offered a potent point of departure from the traditional approach to scientific expert testimony. The portrayal of scientific knowledge as the result of an evolutionary process in which approaches had to advance from the experimental to the demonstrated stage before being accepted in court resonated admirably with the new pragmatic epistemology of American professional culture, which conceived of knowledge as a communal product that exists separately from the individual expert and that ought to be empirically evaluated.[12] In a similar fashion, Frye's search for general acceptance within the relevant scientific community accorded well with the dominant progressive views of the age, which conceived of law as an organic part of the greater society and emphasized coordinating various types of expertise to run society efficiently and uniformly.[13]

Still, originating in an extreme case and containing no precedential citations, the decision of the Court of Appeal of the District of Columbia remained, for a while, an isolated solution to a particular problem. During the next three decades, the courts remained content with applying the requirement for general acceptance only to exorcize from criminal trials evidence derived from various lie detection and truth serum schemes. It was only in the post-WWII years, with the gradual expansion of the role of the trial judge as an active gatekeeper, that the courts began to apply the general acceptance requirement as an exclusive test applied to a continually broadening range of novel scientific evidence.[14]

The expanding role of the trial judge as a gatekeeper charged with the responsibility of preventing unreliable scientific evidence from entering the courtroom can be described in terms of both the scope of the cases to which it has been applied and the depth of the judicial scrutiny of the scientific evidence. By the early 1950s, Frye was already proclaimed in the legal literature as the main criterion for the admissibility of novel types of scientific evidence. The U.S. Supreme Court restricted the acquisition of evidence in criminal cases via traditional interrogation techniques, the federally-sponsored crime laboratories flooded the courts with innovative scientific technologies, and the criminal trial judges used Frye as a ready-made tool to decide the reliability of evidence derived from technologies such as voice prints, neutron activation analysis, gunshot residue tests, bite mark comparisons, scanning electron microscopic analysis, and numerous other techniques.[15] By the 1970s, Frye became the *sine qua non* in practically all of the criminal courts that considered the question of the admissibility of new scientific evidence. By the 1980s, as the trial judges adapted themselves to their new role as active gatekeepers, they expanded the use of Frye from criminal to civil proceedings, completing in the process its transformation from a special judicial device designed to check controversial new technologies to a general judicial device for the pretrial screening of scientific evidence. Finally, by the 1990s, as the judges gained confidence in their ability to measure the proposed expert evidence against the standards of its field, the courts moved from deferring to the judgment of the scientific community to independently finding out things for themselves.[16]

The Death of Frye?

The expanding judicial dominion over scientific expert testimony met with increasing criticism. The earliest attacks considered the judicial screening of the scientific evidence as an unnecessary procedure that deprived the jurors of their right to decide for themselves what facts are valuable. Frye's "general acceptance" criterion was also criticized for being too narrow or too slow, thus depriving the courts of what might be reliable and valuable evidence.[17] Frye's proponents argued that it provided the courts with a uniform method for ensuring the reliability of scientific evidence. However, "the thing from which the deduction is made" has meant different things to different courts at different times. The ambiguities inherent in determining the particular field to which new scientific evidence belongs and in deciding how to measure its "general acceptance," have also left ample room for discretion. Consequently, Frye ended up having not one but many "general acceptance" criteria, which the courts seemed to apply in a selective manner, according to their own views about the reliability of the particular forensic technique before them.[18]

Meanwhile, a new twist occurred. In 1975, the rules that federal judges follow were finally codified. Completely disregarding Frye, the enacted *Federal Rules of Evidence* (FRE) prescribed no special test to ensure the reliability of scientific evidence, new or old. Instead, casting the widest net possible, the FRE provided that:

> If scientific, technical, or other specialized knowledge will assist the trier of fact to understand the evidence or to determine a fact in issue, a witness qualified as an expert by knowledge, skill, experience, training, or education, may testify thereto in the form of opinion or otherwise.[19]

Having left open the question of how one defines "scientific, technical, or other specialized knowledge," the FRE was generally interpreted as prescribing more a flexible judicial consideration of scientific evidence in order to create the opportunity for more types of scientific evidence to be used in court. On the other hand, since the FRE did not state an explicit intent to abandon the Frye rule, some federal and almost all state courts remained committed to the "general acceptance" criterion as an absolute prerequisite for the admissibility of scientific evidence.

The debate concerning what was the proper standard for the admissibility of scientific evidence intensified during the late 1980s and early 1990s, following growing fears of a mass tort litigation explosion. Since the 1970s, accidents, technological breakdowns, dangerous drugs, industrial defects, environmental pollutants, and other toxic substances have become the subject of prolonged litigation with ever-escalating financial stakes.[20] In the great majority of these cases, the central legal questions were those of risk and causation, which invariably turned upon scientific evidence and revealed again the all-too-familiar sight of leading scientific experts producing from the witness stand conflicting data and contradictory conclusions. The customary complaints soon followed. The warning was sounded that America's courts are being swamped by "junk science," produced by an unholy alliance between unscrupulous experts and opportunistic attorneys. The judges were urged to raise the bar and rely on the conservative interpretation of Frye in order to protect the credulous jury from pseudoscientific experts, and the deep-pocketed corporations from greedy lawyers.[21] Others objected. The Frye test, they argued, sanctions a stifling and repressive scientific orthodoxy and prevents the courts from learning of authentic scientific innovations. Hence, they urged the court to adopt the relaxed admissibility requirements of the FRE.[22] In short, the two-centuries-old debate between the Goulds, who maintain that the law should exclude from the courtroom certain expert opinions because they are not scientific enough, and the Mansfields, who maintain that the law has no way to give preference to one kind of science over another, was back with renewed vigor.

The conflict came to a head in 1993, in a civil suit by a minor, Jason Daubert, against the giant pharmaceutical corporation, Merrell Dow. Daubert was born with serious birth defects and blamed it on Merrell Dow's Bendectin, an anti-nausea drug his mother took during her pregnancy. To prove a causal link between the drug and the birth defects, Daubert's experts offered three types of scientific evidence: *in vitro* and animal studies, which found links between Bendectin and malformation; chemical analysis, which pointed to structural similarities between Bendectin and other substances known to cause birth defects; and reanalysis of previously published epidemiological data, which found a link between the drug and birth defects. The judges were not impressed. They maintained that the scientific evidence, which was created especially for the trial, was not subjected to peer review and thus could not be considered under Frye as being generally accepted in the relevant scientific community. Thus, they threw the case out before it even reached a jury.[23]

Daubert's lawyers appealed to the U.S. Supreme Court, asking it to decide that the FRE superseded Frye and that according to the FRE, the jurors should determine the persuasiveness of the scientific evidence, not the judges. The Supreme Court agreed with the petitioners that Frye was superseded by the FRE, but rejected their let-it-all-in interpretation of the FRE. Instead, the Supreme Court read the FRE as requiring that the trial judge must ensure that any scientific evidence admitted into the courtroom is reliable. Addressing the main question left open by the FRE: how one recognizes valid scientific evidence, the Supreme Court acknowledged that a ready-made formula, such as the "general acceptance" criterion, failed to provide the desired answer. Nevertheless, the Supreme Court emphasized, an answer must be provided, and if a general formula could not be furnished then it is the responsibility of the trial judge to make his or her own inquiries in each and every case in order to provide it. To that end, the Supreme Court took a dip into the murky waters of the modern philosophy of science and came up with a flexible recipe of five nonexclusive factors that could be used by the trial judge in determining the quality of the scientific evidence proposed:[24]

1. Testability: whether the theory or technique can be or has been tested
2. Peer review: whether the theory or technique had been subjected to peer review
3. Error rate: the technique's known or potential rate of error
4. Standardization: the existence of standards controlling the technique's operation
5. General acceptance (the Frye test): the degree to which the theory or technique has been generally accepted in the scientific community

This is where the affairs of science and law in America now stand. Good faith has turned to disenchantment and finally to institutionalized mistrust. The Daubert decision has been widely celebrated as a breakthrough in the attitude of the courts towards scientific evidence. The decision was labeled the "death of Frye"; its new set of criteria was referred to as the first attempt "to deal substantively with the problem of expert testimony in the courts," and the seventy years since Frye have been declared as "seven decades of hiding from science."[25]

A perspective longer than seven decades could provide us with a different view. True, Frye's "general acceptance" criterion has been displaced. Still, Frye's significance has been far greater than the specific admissibility standard it inaugurated. Although formulated in the radical context of the lie detector, it embodied a general judicial state of mind, the fruit of two centuries of growing legal dependence on, and frustration with, science. The chronic inability of the courts to bridge the gap between experts and juries, the resultant fear of a credulous jury bewitched in the name of science by charlatans and opportunists, the difficulties of science in adjusting to adversarial procedures, the failure to create a better alliance between law and science adequate for litigation in the twentieth century—all these concerns played important roles in the Frye decision and even more so in turning it into a broad coherent rationale. No longer a passive umpire who watches over the rules of the game and counts the points gained by the parties, the twentieth-century trial judge became an active gatekeeper charged with

the responsibility of screening unreliable scientific evidence. First formulated by the Frye court, this exclusionary spirit has since to dominate
American legal thought, and the Daubert decision only served to affirm
it more vigorously. The court's admissibility criteria might have
changed, but its mindset followed the twentieth-century trend towards
ever-greater judicial scrutiny of scientific evidence.

Notes
Index

Notes

Introduction

1. Alan M. Wood, ed., *Science and Technology in the Eye of the Law* (London: Royal Society, 2000); Lewis Wolpert, "What Lawyers Need to Know about Science," in Helen Reece, ed., *Law and Science* (Oxford: Oxford University Press, 1998), 289–298; Steven Goldberg, *Culture Clash: Law and Science in America* (New York: New York University Press, 1994); K. R. Foster, D. E. Bernstein and P. W. Huber, eds., *Phantom Risk: Scientific Inference and the Law* (Cambridge, MA: MIT Press, 1994).

2. David L. Faigman, *Legal Alchemy: The Use and Misuse of Science in the Law* (New York: Freeman, 1999); Joseph Sanders, *Bendectin on Trial: A Study of Mass Tort Litigation* (Ann Arbor: University of Michigan Press, 1998); Zakaria Erzinclioglu, "British Forensic Science in the Dock," *Nature* 32 (1998), 859–860; Daniel Farber, *Beyond All Reason: The Radical Assault on Truth in American Law* (Oxford: Oxford University Press, 1997); K. R. Foster and P. W. Huber, *Judging Science: Scientific Knowledge and the Federal Courts* (Cambridge, MA: MIT Press, 1997); M. Angell, *Science on Trial: The Clash of Medical Evidence and the Law in the Breast Implant Case* (New York: Norton, 1996); Peter W. Huber, *Galileo's Revenge: Junk Science in the Courtroom* (New York: Basic Books, 1991).

3. Gary Edmond, "Whigs in Court: Historiographical Problems with Expert Evidence," *Yale Journal of Law and the Humanities* 14 (2002), 123–175; Sheila Jasanoff, *Science at the Bar: Law, Science, and Technology in America* (Cambridge, MA: Harvard University Press, 1995); Carol Jones, *Expert Witnesses: Science, Medicine, and the Practice of Law* (Oxford: Oxford University Press, 1994); Dorothy Nelkin, ed., *Controversy: Politics of Technical Decisions* (Newbury Park, Calif.: Sage, 1992); R. Smith and B. Wynne, *Expert Evidence: Interpreting Science in the Law* (New York: Routledge, 1989).

4. Harry Woolf, *Access to Justice: Final Report* (London: HMSO, 1996); Editorial, "Development in the Law—Confronting the New Challenges of Scientific Evidence," *Harvard Law Review* 108 (1995), 1481–1605; Lee Loevinger, "Science as Evidence," *Jurimetrics Journal* 35 (1995), 153–197; Carnegie Commission on Science, Technology, and Government, Task Force on Judicial and Regulatory Decision Making, *Science and Technology in Judicial Decision Making: Creating Opportunities and Meeting Challenges* (New York: Carnegie Commission on Science, Technology and Government, 1993).

1. "Where There's Muck There's Brass:" The Rise of the Modern Expert Witness

1. John M. Beattie, *Crime and the Courts in England, 1660–1800* (Princeton: Princeton University Press, 1986), 352–356; Thomas Green, *Verdict According to Conscience* (Chicago: University of Chicago Press, 1985), 135–136; James S. Cockburn, *The History of the English Assizes, 1578–1714* (London: Cambridge University Press, 1972), 109; Stephen Landsman, "The Rise of the Contentious Spirit: Adversary Procedures in Eighteenth-Century England," *Cornell Law Review* 75 (1990), 498–609; John Langbein, "Shaping the Eighteenth-Century Criminal Trial: A View from the Rider Sources," *University of Chicago Law Review* 50 (1983), 26.

2. Folkes v. Chadd, in Sylvester Douglas, *Reports of Cases Argued and Determined in the Court of the Kings' Bench in the Twenty-Second, Twenty-Third, Twenty-Fourth and Twenty-Fifth Years of the Reign of George III* (London, 1831), 3 vols., 3:160.

3. James B. Thayer, *Select Cases on Evidence at the Common Law* (Boston, 1892), 666; John H. Wigmore, *A Treatise on the Anglo-American System of Evidence in Trials at Common Law* (Boston: Brown, Little, 1923), 2nd ed., 5 vols., 4:103; Anthony Kenny, "The Expert in Court," *Law Quarterly Review* 99 (1983), 199; Stephan Landsman, "Of Witches, Madmen, and Products Liability: A Historical Survey of the Use of Expert Testimony," *Behavioral Sciences and the Law* 13 (1995), 141; idem, "One Hundred Years of Rectitude: Medical Witnesses at the Old Bailey, 1717–1817," *Law and History Review* 16 (1998), 491.

4. Folkes v. Chadd, 157–161, 341–343. See however John Heilbron, "I detriti e la scienza. Leggi della natura e leggi del'uomo nell'Inghilterra del xviii secolo," in Antonio Santucci, ed., *L'età dei lumi: Saggi sulla cultura settecentesca* (Bologna: Il mulino, 1998), 49–67. I thank Prof. Heilbron for a copy of the English version of this paper, "Silt and Science: The Laws of Nature and of Man in Eighteenth Century England," and for other documents cited in the following text.

5. J. A. Steer, "The East Anglian Coast," *Geographical Journal* 69 (1927), 26–33; "Some Notes on the North Norfolk Coast from Hunstanton to Brancaster," *Geographical Journal* 87 (1936), 5–46; "Shoreline Changes of the Marshland Coast of North Norfolk," *Transactions of the Norfolk and Norwich Naturalists' Society* 17 (1953), 322–326.

6. Daniel Defoe, *A Tour through the Whole Island of Great Britain* (London, 1769), 7th ed., 82–84; M. J. Armstrong, *An Essay on the Contour of the Coast of*

Norfolk, but more particularly as it relates to Marum-Banks & Sea-Breaches so Loudly and so Justly complained of? (Norwich, 1789); Tom Williamson, *The Origins of Norfolk* (Manchester: Manchester University Press, 1993), 1–20; Paul W. Blake, *The Norfolk We Live In* (Norwich: Jarrold and Sons, 1964), 3rd ed., 4–7; Arthur W. Purchas, *Some History of Wells-Next-the-Sea and District* (Ipswich: East Anglian Magazine, 1965), 63.

7. Tidal Harbours Commission, *Second Report of the Commissioners with Minutes of Evidence, Appendices, Supplement and Index* (London, 1846), 444–466.

8. Ibid., 451; Nathaniel Kent, "Exported Produce of Norfolk," *Annals of Agriculture* (1784–1815), 46 vols., 22:35; Naomi Riches, *The Agriculture Revolution in Norfolk* (London: Frank Cass, 1967), 15–17, 36–42, 150–152; Blake, *The Norfolk We Live In*, 36

9. Cuchlaine King, *Beaches and Coasts* (London: Arnold, 1959), 274–275; Nathaniel Kent, *General View of the Agriculture of the County of Norfolk* (London, 1794), 9–10; Thomas Badeslade, *The History of the Ancient and Present State of the Navigation of the Port of King's-Lynn, and of Cambridge, and the rest of the trading-towns in those parts: and of the navigable rivers that have their course through the great-level of the Fen* (London, 1725), 13–14.

10. Samuel Smiles, *Lives of the Engineers* (London: John Murray, 1904), 3 vols., 1:35–39; John Armstrong, *Report Concerning the Drainage of Bedford Level and the Port of King's Lynn* (1724).

11. Smiles, *Lives of the Engineers*, 1:vii–viii.

12. Tidal Harbours Commission, *Second Report*, 444–445.

13. John P. McLaren, "Nuisance Law and the Industrial Revolution," *Oxford Journal of Legal Studies* 3 (1983), 155–221.

14. Joel F. Brenner, "Nuisance Law and the Industrial Revolution," *Journal of Legal Studies* 3 (1974) 3:403–434; Tenant v. Goldwin (1705), in Frederick Pollock, ed., *English Reports* (London, 1891), 92: 222.

15. *Case Submitted for Sergeant Grose's opinion* (Jan 1778), in Tidal Harbours Commission, *Second Report*, 444; *Serjeant Grose's Opinion*, ibid., 445. Grose's opinion did not specify its legal authorities. However, Sir Mathew Hale's *De Portibus* governed the relevant late eighteenth-century Common Law. Drawing upon a series of seventeenth-century decisions, Hale specified a series of acts that constituted nuisances to navigation, among them "allowing a port to become silted up or stopped." Hale, *De Portibus Maris*, in Francis Hargrave, *A Collection of Tracts* (1787; London, 1982), 1:85.

16. Anna M. Stirling, *Coke of Norfolk and His Friends: The Life of Thomas William Coke, First Earl of Leicester of Holkham* (London: John Lane, 1912), 2nd ed., 164; Nathaniel Kent, *Hints to gentlemen of landed property* (London, 1775), 8.

17. Quoted in Purchas, *Some History of Wells-next-the-Sea*, 37; Arthur Young, *The Farmer's Tour through the East of England. Being the register of a journey through various counties of this Kingdom, to enquire into the state of agriculture, etc.* (London, 1771), 4 vols., 2:41–59.

18. Robert Parker, *Coke of Norfolk: A Financial and Agricultural Study, 1707–1842* (Oxford: Oxford University Press, 1975), 77. Other estimated the rise in rents to be far higher, from £2,200 in 1776 to £20,000 in 1816. See Stephen Leslie,

The English Utilitarians (London: Duckworth, 1900), 3:76; Baron Ernle, *English Farming Past and Present* (London: Heinemann, 1961), 6th ed., 217–220.

19. Langbein, "Shaping the Eighteenth-Century Criminal Trial," 26.

20. *The King against Sir John, Baronet,* Norfolk Records Office, MC 50 /29/ 2/2. The meeting was described by one of the commissioners, the Reverend Mr. Robinson, in a letter to Sir Martin Folkes of July 10, 1780. Norfolk Records Office, MC 50 /29/8. John Barney, *The Trials of Wells Harbor* (Norwich: Mintaka, 2000), 13.

21. Quoted in *Notes of the Judgment of the 3rd Appeal,* 27/11/1783, Norfolk Records Office, MF/RO 504/2, MS 486, 120; *Notes of the ordering for a 2nd Trial, Folkes, Bart & all agst. Chadd, Esq. & Others,* ibid., 66–67.

22. William S. Holdsworth, *A History of English Law* (Boston: Little, Brown, 1938), 12 vols., 5:242–245; Cecil H. S. Fifoot, *Law and History in the Nineteenth Century* (London: Quaritch, 1956), 151–154.

23. Action for trespass was a writ instituted to recover damages for any unlawful injury to the plaintiff's person, property, or right, usually involving direct force or violence. Steven H. Gifis, *Law Dictionary* (New York: Barron's, 1991), 3rd ed., 499; *Notes of the ordering for a 2nd Trial,* 67; *Notes of the Judgment of the 3rd Appeal,* 120; Folkes v. Chadd, 157.

24. Garth Watson, *The Smeatonians: The Society of Civil Engineers* (London: Telford, 1989), 1–20; Alec W. Skempton, ed., *John Smeaton, FRS* (London: Telford, 1981), 1–5, 22–26; Robert Mylne, Preface, *Reports of the Late John Smeaton, F.R.S., Made on Various Occasions, in the Course of his Employment as a Civil Engineer* (London, 1812), 4 vols., 1:iii–xi; Thomas S. Willan, *River Navigation in England, 1600–1750* (Oxford: Oxford University Press, 1936), 79.

25. Albert E. Richardson, *Robert Mylne: Architect and Engineer, 1733 to 1811* (London: Batsford, 1955); Watson, *The Smeatonians,* 5–6; Skempton, *John Smeaton,* 22–23; W. H. G. Armytage, *A Social History of Engineering* (London: Faber and Faber, 1961), 104–105; James K. Finch, *The Story of Engineering* (New York: Doubleday, 1960), 187, 196; James Oldham, *The Mansfield Manuscripts and the Growth of English Law in the Eighteenth Century* (Chapel Hill: University of North Carolina Press, 1992), 2 vols., Appendix E, 1582, 1588.

26. Robert Mylne, *Report on the Survey of the Harbour, Etc. of Wells, in Norfolk* (Apr. 28, 1781), 4. The dates are based on Mylne's diary given in Richardson, *Robert Mylne,* 114–116.

27. Robert Mylne, *Report,* 6–7.

28. Buckley v. Thomas, in Frederick Pollock, ed., *English Reports,* 75:182, quoted in Wigmore, *A Treatise on the Anglo-American System of Evidence,* 4:102.

29. James Thayer, *A Preliminary Treatise on Evidence at the Common Law* (Boston, 1898), 94; W. Brunner, *The Origin of the Jury Courts* (1827), 452, quoted in John H. Wigmore, *A Treatise on the Anglo-American System of Evidence in Trials at Common Law* (Boston: Little, Brown, 1940), 3rd ed., 10 vols., § 1364.

30. Henry T. Riley, *Memorials of London and London Life, in the XIIIth, XIVth, and XVth Centuries. Being a Series of Extracts, Local, Social, and Political, From the Early Archives of the City of London, 1276–1419* a.d. (London, 1863), 328, 408, 448, 471, 516; Learned Hand, "Historical and Practical Considerations Regarding Expert Testimony," *Harvard Law Review* 15 (1901), 42; James Oldham, "The Ori-

gins of the Special Jury," *University of Chicago Law Review* 50 (1983), 167–168.

31. Learned Hand, "Historical and Practical Considerations," 40–41. The form of the *de ventre inspiciendo* writ given by Bracton was still in practice as late as 1838. See Reg. v. Wycherly, ibid.

32. Thomas R. Forbes, *Surgeons at the Bailey* (New Haven: Yale University Press, 1985), 41; Carol Jones, *Expert Witnesses: Science, Medicine, and the Practice of Law* (Oxford, Oxford University Press, 1994), 29, 35; Landsman, "Of Witches, Madmen, and Products Liability," 135; Learned Hand, "Historical and Practical Considerations," 43, 45.

33. Travers Twiss, *Black Book of the Admiralty* (London, 1876), 463–464; Jacob Beuscher, "Use of Experts by the Courts," *Harvard Law Review* 54 (1941) 1109–1110; Kenneth C. McGuffie, *"Notes on Nautical Assessors"* (unpublished manuscript), quoted in Jones, *Expert Witnesses*, 39; Oldham, *The Mansfield Manuscripts*, 1:146.

34. Cecil H. Fifoot, *Lord Mansfield* (Oxford: Clarendon Press, 1936), 105; Oldham, *The Mansfield Manuscripts*, 1:82–99; Beuscher, "Use of Experts," 1108–1110; Learned Hand, "Historical and Practical Considerations," 43.

35. Wilfrid R. Prest, "The Rise of the Lawyers," in James F. Corkery, ed., *A Career in the Law* (Sydney: Federation Press, 1989), 2nd ed., 20–21.

36. John H. Langbein, *Prosecuting Crime in the Renaissance: England, Germany, France* (Cambridge, Mass.: Harvard University Press, 1974), 118–121; idem, "The Origins of Public Prosecution at Common Law," *American Journal of Legal History* 17 (1973), 263–316; John H. Baker, "Criminal Courts and Procedure at Common Law, 1550–1800," in James. S. Cockburn, ed., *Crime in England, 1550–1800* (Princeton: Princeton University Press, 1977), 20.

37. Eric Powell, "Jury Trial in Gaol Delivery in the Late Middle Ages: The Midland Circuit, 1400–1429," in James. S. Cockburn and Thomas Green, eds., *Twelve Good Men and True: The Criminal Trial Jury in England, 1200–1800* (Princeton: Princeton University Press, 1988); Edmund M. Morgan, "Some Observations Concerning a Model Code of Evidence," *University of Pennsylvania Law Review* 89 (1940), 137.

38. Thomas Smith, *De Republica Anglorum* (1583; London: Alston, 1906), 113.

39. Wigmore, *A Treatise on the Anglo-American System of Evidence*, 2nd ed., 4:101–103.

40. Folkes v. Chadd, 159.

41. Ibid. See Mansfield's decision in Hartley v. Buggin, quoted in Douglas, *Reports*, 40.

42. John Morgan, *Essays upon the Law of Evidence, New Trials, Special Verdicts, Trials at Bar and Repleaders* (London, 1789), vols. 2 and 3; John B. Sellon, *The Practice of the Courts of King's Bench and Common Pleas* (London, 1792), 485–486; Holdsworth, *History of English Law*, 12:493–500; Oldham, *The Mansfield Manuscripts*, 1:157–160. John H. Baker, *English Legal History* (London: Butterworths, 1990), 3rd ed., 116–121.

43. James Oldham, "Eighteenth-Century Judges' Notes: How They

Explain, Correct, and Enhance the Reports," *American Journal of Legal History* 31 (1987), 29–31; idem, *The Mansfield Manuscripts*, 1:131; *Notes of the ordering for a 2nd Trial*, 67; Folkes v. Chadd, 157.

44. *Notes of the ordering for a 2nd Trial*, 66; M. Dixon, "Some Account of the Life, Character, and Works, of Mr. John Smeaton, F.R.S.," in *Reports of the Late John Smeaton*, 1: xv–xxx; Skempton, *John Smeaton*, 7–34.

45. The first report Smeaton signed as a "professional engineer," was his 1768 report on the Forth and Clyde Canal. There is only one known earlier example of this term, in a *London directory* of 1763 where Smeaton was described as a "surveyor and civil engineer." Skempton, *John Smeaton*, 233; Mylne, Preface, *Reports of the Late John Smeaton*, 1:xxi. For an example of a King's Bench trial that Mylne and Smeaton coarbitrated, see *Cook & Others v. Bisson & Others*, in Oldham, *The Mansfield Manuscripts*, Appendix E, 1588.

46. *Reports of the Late John Smeaton*, 1:26. Among the harbors were: Whitehaven, Workington, and Bristol on the west coast; Christchurch, Rye, and Dover on the south; Yarmouth, Lynn, Wisbech, Scarborough, and Sunderland on the east; and Aberdeen, Dundee, and Dunbar in Scotland.

47. Quoted in Skempton, *John Smeaton*, 11.

48. *Reports of the Late John Smeaton*, 3:74–133; Skempton, *John Smeaton*, 206–212.

49. Serjeant Brown, "Minutes of a meeting held betwixt Mr. Milne, Mr Elbstoff and himself in London, the 14th of May, 1781," Sutro Library, San Francisco, H1:75

50. *Norfolk Chronicle*, (Dec. 14, 1782), 3.

51. John Smeaton, *Machine Letters* (London, Institute of Civil Engineering), (Dec. 8, 1781), 1; (Dec. 12, 1781), 8.

52. Alec W. Skempton, "The Engineering Works of John Grundy," *Lincolnshire History and Archaeology* 19 (1984), 65–82; Alec W. Skempton and E. C. Wright, "Early Members of the Smeatonian Society of Civil Engineering," *Transactions of the Newcomen Society* 44 (1971–72), 23–47; Skempton, *John Smeaton*, 24–25; Watson, *The Smeatonians*, 3–21.

53. Holdsworth, *History of English Law*, 12:493–495; Barney, *The Trials of Wells Harbor*, 15–16; Skempton, *John Smeaton*, 28–29; *Reports of the Late John Smeaton*, 3:18.

54. John Smeaton, "Report on Wells Harbour," in *Reports of the Late John Smeaton*, 3: 23–24.

55. Ibid., 25.

56. Ibid., 29.

57. James Hutton, *Theory of the Earth* (London: 1795), 2 vols., 2:90, quoted in Heilbron, "Silt and Science," 7.

58. Bernhard Varenius, *A compleat system of General Geography* (London, 1733), 2 vols.; (London, 1734), 2nd ed., 2 vols.; William Warntz, "Newton, the Newtonians, and the Geographia Generalis Varenii," *Annals of the Association of American Geographer* 79 (1989), 165–191; J. N. L. Baker, "The Geography of Bernhard Varenius," *Transactions of the Institute of British Geographers* 21 (1996), 51–60; *A*

Catalogue of Books, including the Library of J. Smeaton, Esq., F.R.S. (London, 1794).

59. Cf. Heilbron, "Silt and Science"; Kent, *General View of the Agriculture*, 1; John Gascoigne, *Joseph Banks and the English Enlightenment: Useful Knowledge and Polite Culture* (Cambridge: Cambridge University Press, 1994).

60. Cf. Joseph Hodskinson, *Report on the State and Causes of the Decay of Wells Harbor to the Commissioners for the Preservation of the Harbor of Wells* (1782), in Tidal Harbours Commission, *Second Report*, 446–449.

61. Ibid., 446–447.

62. Ibid., 449.

63. Ibid., 449, 450–451.

64. Chief Justice Dudley Ryder's Diary, quoted in Oldham, *The Mansfield Manuscripts*, 1:130.

65. Quoted in Williamson, *The Origins of Norfolk*, 2.

66. Thayer, *Preliminary Treatise on Evidence*, 56.

67. Prest, "The Rise of the Lawyers," 21. Treason trials were an exception. In 1696, Parliament authorized for the first time to employ defense counsel in treason trials.

68. John H. Langbein, "The Criminal Trial Before the Lawyers," *University of Chicago Law Review* 45 (1978), 310; Beattie, *Crime and the Courts in England*, 352–356; Green, *Verdict According to Conscience*, 135–136; Cockburn, *History of the English Assizes*, 109. As many as twenty-five cases could have been heard by the same judge and jury within a single twelve-hour-long session.

69. James F. Stephen, *A History of the Criminal Law of England* (London, 1883), 424; Cf. Landsman, "The Rise of the Contentious Spirit"; Langbein, "Shaping the Eighteenth-Century Criminal Trial"; idem, "The Criminal Trial Before the Lawyers," 312. The prohibition against the defense counsel's addressing the jury in summation was abolished by legislation only in 1836.

70. Cf. Landsman, "The Rise of the Contentious Spirit"; Langbein, "Shaping the Eighteenth-Century Criminal Trial."

71. William M. Best, *A Treatise on the Principles of Evidence* (London, 1849), 133, quoted in Langbein, "Shaping the Eighteenth-Century Criminal Trial," 131–132; James Oldham, "Truth Telling in the Eighteenth Century English Courtroom," *Law and History Review* 12 (1994), 95–121; Stephen Landsman, "From Gilbert to Bentham: The Reconceptualization of Evidence Theory," *Wayne Law Review* 36 (1990), 1149–1186 (1990); Lawrence M. Friedman, *A History of American Law* (New York: Simon and Schuster, 1973), 136; T. P. Gallanis, "The Rise of Modern Evidence Law," *Iowa Law review* 84 (1999), 500–554.

72. George Hardinge, *Dictionary of National Biography*, CD ROM (Oxford: Oxford University Press, 1995); The remarks appeared on the commissioners' copy of Smeaton's report in the British Library, 26.

73. Folkes v. Chadd, 158.

74. The original objection to hearsay evidence was that it was not given under oath. By the end of the seventeenth century, a second type of objection appeared—that such statements could not be tested by cross-examination. Charles T. McCormick, "Some Observations Upon the Opinion Rule and Expert Testi-

mony," *Texas Law Review* 23 (1954), 109–136; Lord Gilbert, *The Law of Evidence* (Dublin, 1795), 4th ed., 152; Morgan, "Some Observations," 52; Wigmore, *A Treatise on the Anglo-American System of Evidence*, 3rd ed., § 1364.

75. Landsman, "The Rise of the Contentious Spirit," 572; Learned Hand, "Historical and Practical Considerations," 37; John H. Wigmore, *"The History of the Hearsay Rule,"* *Harvard Law Review* 17 (1904), 448. Collecting his data from political state trials, Wigmore may have overstated the status of the opinion doctrine in nonpolitical criminal and civil proceedings. There, in 1782, the opinion doctrine was still embryonic and subject to the wide discretion of the individual trial judge.

76. Carter v. Boehm, in James Burrow, ed., *Series of the Decisions of the Court of King's Bench, Upon Settlement-cases; From the Death of Lord Raymond, in March 1732* (London, 1768), 2 vols., 2:1918.

77. Gould Henry, in Edward Foss, *Biographia Juridica. A biographical dictionary of the Judges of England from the Conquest to the present time 1066–1870* (London, 1870), 308.

78. Regular reports of assizes cases were not published until the early nineteenth century. John H. Wigmore, "A General Survey of the History of the Rules of Evidence," in *Selected Essays in Anglo-American Legal History* (Boston: Little, Brown, 1908), 3 vols., 2:696; Holdsworth, *History of English Law*, 12:502–504.

79. Folkes v. Chadd, 159.

80. Ibid., 160.

81. Thayer, *Select Cases on Evidence*, 666.

82. Rex v. Pembroke, in Thomas J. Howell ed., *Cobbett's Complete Collection of State Trials and Proceedings for High Treason and other Crimes and Misdemeanors from the earliest period to the present time* (London: 1809–1826), 33 vols., 7:185–186; A. Rosenberg, "The Sarah Stout Trial: An Early Example of the Doctor as Medical Expert Witness," in Chester. R. Burns, ed., *Legacies in Law and Medicine* (New York: Science History Publications, 1977), 230.

83. R. S. Fitton and A. P. Wadsworth, *The Strutts and the Arkwrights, 1758–1830: A Study of the Early Factory System* (Manchester: Manchester University Press, 1958), 41–46, 81–83.

84. For tax litigation, see the 1781 case of "The King v. Steele and Others," discussed in William Ashworth, "Between the 'Trader and the Public': British Alcohol Standards and the Proof of Good Governance," *Technology and Culture* 42 (2001), 27–50. For nuisance litigation, see the 1770 indictment of the copper works of Charles Roe, discussed in T. C. Barker and J. R. Harris, *A Merseyside Town in the Industrial Revolution, St. Helen's 1750–1900* (London: Cass, 1959), 227.

85. Richard Sorrenson, "Dollond and Son's Pursuit of Achromaticity, 1758–1789," *History of Science* 39 (2001), 31–55; *The case of Richard Arkwright and Co in relation to Mr Arkwright's invention of an engine for spinning cotton etc.* (London, 1782); *The Trial of the cause instituted by Richard Pepper Arden* (London, 1785); *Richard Arkwright Esquire v. Peter Nightingale* (London, 1785); Eric Robinson, "James Watt and the Law of Patents," *Technology and Culture* 13 (1972), 115–139; John Hewish, "From Cromford to Chancery Lane: New Light on the Arkwright Patent Trials," *"Technology and Culture* 28 (1987), 80–86. Arkwright's success was short-lived. His patent was annulled the following year.

86. Like other Common Law judges, Mansfield used to sit with an Admiralty judge in cases arising out of events on the high seas. The Trinity House brethren served as court experts in such cases. On occasion, Mansfield adopted this practice in the Court of the King's Bench.

87. Wigmore, *A Treatise on the Anglo-American System of Evidence*, 2nd ed., 4:101–103, 105.

88. Ibid., 105.

89. Folkes v. Chadd, 160; Oldham, *The Mansfield Manuscripts*, 1:146.

90. On this legal habit, see A. W. Brian Simpson, *Leading Cases in the Common Law* (Oxford: Clarendon Press, 1995); William Twining, "The Rationalist Tradition," in idem, *Rethinking Evidence: Exploratory Essays* (Oxford: Basil Blackwell, 1990), 33–91.

91. Folkes v. Chadd, 160.

92. Hutton, *Theory of the Earth*, 2:90.

93. Folkes v. Chadd, 159.

94. *Norwich Mercury* (Dec. 12, 1782), 1.

95. Smeaton, Letter to Mr. Forster (Jul. 1, 1783), in Smeaton, *Machine Letters*, 180.

96. Mylne, Letter to Hardinge (Jul. 24, 1784), Norfolk Records Office, MF/RO 504/2, MS 486, 133; Folkes v. Chadd, 340–341.

97. Anonymous newspaper clip, "A Summary State of the Trial in 1783," Norfolk Records Office, MF/RO 504/2, MS 486, 131.

98. Mylne to Hardinge, Jul. 24, 1784; Hardinge to Mylne, Jul. 28, 1784; Lord Camelford to Mylne, Jul. 30, 1784, Norfolk Records Office, MF/RO 504/2, MS 486, 133–138.

99. Tidal Harbours Commission, *Second Report*, 462–465; Parker, *Coke of Norfolk*, 91.

100. Tidal Harbours Commission, *Second Report*, 457–466; Purchas, *Some History of Wells-Next-the-Sea*, 62.

101. See, for example, "Point Mugu Soughs as Remembered by Omar Lillevang," *Oral History of Coastal Engineering Activities in Southern California, 1930–1981*, Jan 1986, 25–27.

102. Landsman, "Of Witches, Madmen, and Products Liability," 141.

103. Gilbert, *The Law of Evidence*, 4th ed., 93–135; Oldham, *The Mansfield Manuscripts*, 2:1066–1078.

104. Philip Mason, *The English Gentleman* (London: Deutsch, 1982); Simon Raven, *The English Gentleman* (London: Blond, 1961); Peter Dear, "Totius in Verba: Rhetoric and Authority in the Early Royal Society," *Isis* 76 (1985), 45–161.

105. John Locke, *Some Thoughts Concerning Education* (1690; Oxford: Clarendon, 1989), 126–127, quoted in Steven Shapin, *A Social History of Truth: Civility and Science in Seventeeth-Century England* (Chicago: University of Chicago Press, 1995), 74.

106. Thomas Sprat, *The History of the Royal Society of London for the improving of natural knowledge* (London, 1667), 65; David P. Miller, "The Usefulness of

Natural Philosophy: The Royal Society and the Culture of Practical Utility in the Later Eighteenth Century," *British Journal for the History of Science* 32 (1999), 185–201.

107. Smeaton was elected in 1756, 1759, 1762, and 1784; Mylne in 1781.

108. *Notes of the ordering for a 2nd Trial,* 67; *Norwich Mercury,* Nov. 30 1782.

2. The Common Liar, the Damned Liar, and the Scientific Expert: The Growing Problem of Expert Testimony

1. Geoffrey Gilbert, *The Law of Evidence* (Dublin, 1975), 4th ed., 4.

2. Ibid, 301. "Presumption" translates here as "circumstantial evidence." The test of still birth had been devised on the continent in the seventeenth century and applied frequently in British courts. By the late eighteenth century, it was widely distrusted. William Hunter, "On the Uncertainty of the signs of murder, in the case of Bastard Children," *Medical Observation and Inquiries 6* (1783), 190–266. The test regained medical favor in the early nineteenth century.

3. Thomas Peak, *Compendium of Law on Evidence* (London, 1801), 4.

4. *Notes of the ordering for a 2nd Trial, Folkes, Bart & all agst. Chadd, Esq. & Others,* Norfolk Records Office, MF/RO 504/2, MS 486, 67; Editorial, "Expert Witnesses," *Quarterly Review* 3 (Jan. 11, 1862), 32–33.

5. The historian June Fullmer provided us with a good account of these trials. June Z. Fullmer, "Technology, Chemistry, and the Law in Early Nineteenth-Century England," *Technology and Culture* 21 (1980), 1–28. The London *Times* closely followed the trials. Other detailed reports of the trials include: "Facts Respecting the increased Volatility and Inflammability which Fish Oil and its Vapours acquire by continued or renewed Exposure to certain high Temperatures; elicited by the Examination of Evidence in a late Trial in the Court of Common Pleas, before Lord Chief Justice Dallas and a Special Jury," *Philosophical Magazine* 55 (1820), 252–289; Samual Parkes, "Observation on the Chemical Part of the Evidence, Given Upon the Late Trial of the Action Brought by Messrs. Severn, King, and Co. against the Imperial Insurance Company," *Quarterly Journal of Science and the Arts* 10 (1821), 316–353; J. Bostock, "Some Observation of Whale Oil," *Annals of Philosophy* 1 (Jan 1821), 48–50.

6. *Times* (London), (Dec. 16, 1820), 3.

7. *Times* (London), (Dec. 14, 1820), 3.

8. "Facts Respecting the increased Volatility and Inflammability," *Philosophical Magazine,* 265–266, 281.

9. Ibid., 264–267, 280.

10. John Henry Wigmore, *A Treatise on the Anglo-American System of Evidence in Trials at Common Law* (Boston: Little, Brown, 1923), 2nd ed., 5 vols., 2:667–668.

11. Fullmer, "Technology, Chemistry, and the Law in Early Nineteenth-Century England," 11–12.

12. "Facts Respecting the increased Volatility and Inflammability," *Philosophical Magazine,* 265–268.

13. Ibid., 260–261; *Times* (London), (Apr. 12, 1820), 3.

14. *Times* (London), (Apr. 13, 1820), 3. Dr. Isaac Milner, F.R.S. (1750–1822) was president of Queens College, Cambridge, and the author of the phlogistic text, *A Plan of a Course of Chemical Lectures* (Cambridge, 1788). Smithson Ten[n]ant (1761–1815), was a Cambridge chemist who discovered the elements osmium and iridium in 1803 and occupied the Chair of Chemistry between 1813 and 1815.

15. Cf. Alexander Welsh, *Strong Representations: Narratives and Circumstantial Evidence in England* (Baltimore: John Hopkins University Press, 1992).

16. "Facts Respecting the Increased Volatility and Inflammability," *Philosophical Magazine*, 267, 280–81.

17. *Times* (London), (Apr. 14, 1820), 3.

18. Severn v. Olive, in Frederick Pollock, ed., *English Reports* (London, 1896), 23:565; *Times* (London), (Apr. 12, 1820), 3.

19. *Times* (London), (Dec. 16, 1820), 3.

20. Ibid.

21. Thomson and Accum used the other two sets.

22. Ibid.

23. Ibid.

24. The scorched barrel was presented in the court. *Times* (London), (Dec. 18 1820), 3.

25. Ibid.

26. *Times* (London), (Dec. 19, 1820), 3.

27. Ibid. The sugar industry decided differently. Wilson's process was quickly abandoned and Howard's pricier vacuum process took over. The high costs of its installation were probably responsible for the disappearance of small under-capitalized English sugar refineries. Fullmer, "Technology, Chemistry, and the Law in Early Nineteenth-Century England," 27.

28. Ibid., 24–25.

29. Severn v. Olive, 565–567.

30. Cf. Rosemary O'Day, *The Professions in Early Modern England, 1450–1800: Servants of the Commonweal* (London: Longman, 2000).

31. Severn v. Olive, 565–567; Fullmer, "Technology, Chemistry, and the Law in Early-Nineteenth-Century England," 24–25; Colin A. Russel, *Edward Frankland: Chemistry, Controversy and Conspiracy in Victorian England* (Cambridge, Cambridge University Press, 1996), 445–457.

32. John P. McLaren, "Nuisance Law and the Industrial Revolution," *Oxford Journal of Legal Studies* 3 (1983), 155–221; Joel. F. Brenner, "Nuisance Law and the Industrial Revolution," *The Journal of Legal Studies* 3 (1974), 403–434; Patrick Atiyah, *The Rise and Fall of Freedom of Contract* (Oxford: Clarendon Press. 1979), 238–250.

33. Quoted in Ronald Rees, "The Great Copper Trials," *History Today* 43 (1993), 39; George Grant Francis, *The Smelting of Copper in the Swansea District* (London, 1881).

34. John Henry Vivian, "Statement in Explanation of the Measures taken by

Messrs. Vivian, to do away with the Inconvenience arising from the Copper Smoke," (Nov. 26, 1822), in *Proceedings of the Subscribers to the Fund for Obviating the inconvenience arising from the Smoke produced by Smelting Copper Ores* (Swansea, 1823), 19–20.

35. Ibid., 4–5.

36. Ibid., 17–60; "Results of experiments made at the Hafod Works, by Messrs. Phillips and Faraday, July and August 1922," ibid., 62–67.

37. Ibid., 49–50.

38. Ibid., 60; "The resolution of the Borough of Swansea," ibid., 91; "The Resolution of the meeting of the Subscribers to the Fund," ibid., 89–90.

39. Scarlet made a similar argument in another public nuisance case, R. v. Medley (1834). Frederick Pollock, ed., *English Reports* (London, 1891), 172:1246

40. Rees, "The Great Copper Trials," 40–42.

41. "Report of the Select Committee to Inquire into the Smoke Nuisance," *Parliamentary Papers* (London, 1843), 7:177–178.

42. Brenner, "Nuisance Law and the Industrial Revolution," 403–408; Richard A. Posner, *Economic Analysis of Law* (Boston: Little, Brown, 1986), 3rd ed., 47.

43. Ludwig Fritz Haber, *The Chemical Industry During the Nineteenth Century* (Oxford: Oxford University Press, 1958), 252–254.

44. Quoted in Russel, *Edward Frankland*, 168–169. Frankland later served as the chief member of a Royal Commission on Rivers Pollution that operated between 1868–1873.

45. Edwin Chadwick, *Report to her Majesty's Principal Secretary of State for the Home Department. From the Poor Law Commissioners, On an Inquiry into the Sanitary Conditions of the Labouring Population of Great Britain; with Appendices* (Jul. 1842), 3 vols.

46. 10 & 11 Vict. C. 34, clause 108. Carlos Flick, "The Movement for Smoke Abatement in Nineteenth Century Britain," *Culture and Technology* 21 (1980), 29–50. The major Acts were: Nuisance Removal and Diseases Prevention Act, 1848; Smoke Nuisance Abatement Act, 1853; General Board of Health Act, 1854; Nuisance Removal Act, 1855; Local Government Act, 1858; Public Health Acts 1858 and 1859; Nuisance Removal Act, 1860; Nuisance Removal Act, 1863; Nuisance Removal Act, 1866; Sanitary Acts, 1866 and 1868; Local Government Act, 1871; Public Health Acts, 1872.

47. *Manchester Guardian* (Aug. 24, 1857), 3; (Aug. 26, 1857), 4.

48. Quoted in Robert H. Kargon, *Science in Victorian Manchester* (Baltimore: Johns Hopkins University Press, 1977), 141. Frankland and Spence joined the Manchester Literary and Philosophical Society on the same day, April 29, 1851.

49. *Manchester Guardian* (Aug. 26, 1857), 4.

50. *Chemical News* 14 (1866), 207; For the next important nuisance case see St. Helen's Smelting Co. V. Tipping (1862), Pollock, *English Reports*, 144:348.

51. "Curiosities in Chemical Evidence," *Chemical News* 5 (Mar 29, 1862), 182.

52. Christopher Hamlin, "Scientific Method and Expert Witnessing: Victorian Perspective on a Modern Problem," *Social Studies of Science* 16 (1986), 485–513.

53. "The Thames," *The Field*, (7 Aug. 1869), 121. Quoted in Hamlin, "Scientific Method and Expert Witnessing," 491.

54. Roy MacLeod, *Public science and Public Policy in Victorian England* (Hampshire: Variorum, 1996), Part 1, 85–112; Frederick Filby, *History of Food Adulteration and Analysis* (London: Allen & Unween, 1934), Chap. 8.

55. William H. Brock, "The Spectrum of Science Patronage," in Gerard L. E. Turner, ed., *The Patronage of Science in the Nineteenth Century* (Leyden: Noordhoff International, 1976), 186; a cursory list of Hofmann's famous legal appearances would include the following: the 1854–5 Scottish trial of Gillespie v. Russell, concerning the true identity of Turbanite mineral; the 1856 case of Talbot v. Laroch, concerning processes of photographic developments; the 1860 Medlock patent trial concerning red and purple dyes; the 1861 Gerard's aniline blue patent trial, and the 1864 Young's patent trial concerning paraffin and paraffin oils.

56. This had been required since the first half of the eighteenth century and insisted upon since the 1780s. Allan Gomme, *Patents of Invention* (London, 1946), 25–39.

57. Hamlin, "Scientific Method and Expert Witnessing," 490.

58. August W. Hofmann, "Notes of Research on the Poly-Ammonias—No. V. Action of Bichloride of Carbon on Aniline," *Proceedings of the Royal Society of London* 9 (1858), 284–286; idem, "Recherches pour Servir à l'Histoire des Bases Organiques," *Comptes Rendus* 47 (1858), 492–495; John Beer, *The Emergence of the German Dye Industry* (Urbana: University of Illinois Press, 1959), 26–27. The Battle of Magenta took place in June 4, 1859, between the Austrian and the Franco-Piedmontese armies.

59. *Chemical News* 6 (1862), 302.

60. Henry Medlock, "Improvements in the Preparation of Red and Purple Dyes," *Chemical News* 6 (1862), 299, 302.

61. "Simpson and Others v. Wilson and another, Bill of Complaint In Chancery," no. 153 (Jul. 2, 1861), Public Records Office, London, Chancery Pleadings, C16/47/153, 4–5.

62. Richard Godson, *Practical Treatise on the Law of Patents for Inventions, and of Copyright* (London, 1851), 2nd ed, 230–231; James Watt, "Thoughts upon Patents, or exclusive privileges for new Inventions," unpublished MS (1786), reprinted in Eric Robinson and A. E. Musson, *James Watt and the Steam Revolution* (New York: Kelley, 1969), 213–228. The paper contained a detailed suggestion for a scientific tribunal for patent litigation and was submitted to Lord Chancellor Kenyon in 1890. See Eric Robinson, "James Watt and the Law of Patents," *Technology and Culture* 13 (1972), 115–139; Moureen Coulter, *Property in Ideas: The Patent Question in Mid-Victorian Britain* (Missouri: Thomas Jefferson University Press, 1991), 27–30.

63. "Court of the Queen's Bench. December 10, 1862. Simpson and Others v. Wilson and Another," *Chemical News* 7 (1863), 19.

64. Terminology was as confusing as theory. Each chemist used different

terms to explain this empirical classification. Mechanically combined water was also called "water of constitution" or "water of hydration." Chemically combined water was known also as "basic water." The term "water of combination" was used indifferently for both.

65. Simpson and Co. manufactured Hofmann's famous violet dyes and paid Hofmann consulting fees if not royalties. Beer, *The Emergence of the German Dye Industry*, 28–29.

66. "Court of the Queen's Bench: December 10, 1862, Simpson and Others v. Wilson and Another," *Chemical News* 7 (1863), 4–7, 18–21, 28–30, 40–43, 52–53.

67. Ibid., 19, 20, 41.

68. This is not true for anhydrous arsenic acid, which ceases to be deliquescent.

69. Ibid., 20, 41–42, 53.

70. Ibid., 54.

71. Ibid. Six years later, Dr. Reimann, the author of the first comprehensive treatise on the novel theory and technology of aniline, was in a better position to give an opinion. "This patent specification," Reimann wrote of Medlock's patent, "has the same peculiarity common to a thousand others: it cannot be carried out in practice . . . Only a solution of arsenic acid is able to transform aniline into magenta." M. Reimann, *Aniline and its Derivatives: A Treatise upon the Manufacture of Aniline and Aniline Colours* (New York, 1868), 39.

72. Anthony S. Travis, *The Rainbow Makers: The Origins of the Synthetic Dyestuffs Industry in Western Europe*, (Bethlehem: Lehigh University Press, 1993), 125–137

73. R. M. Jackson, "The Incidence of Jury Trial During the Past Century." *Modern Law Review* 1 (1937), 132–144; W. R. Cornish and G. de N. Clark, *Law and Society in England 1750–1950* (London: Sweet and Maxwell, 1989), 20.

74. *Edinburgh New Philosophical Journal* 16 (1834), 376–384.

75. Quoted in Edmund E. Fournier, *The Life of Sir William Crookes* (New York: Appleton, 1923), 73.

76. *Chemical News* 9 (1864), 274.

77. James Butt, *James 'Paraffine' Young: Founder of the Mineral Oil Industry* (Edinburgh: Scotland's Cultural Heritage, 1983), 5–7.

78. James Young, "Improvements in the Treatment of Certain Bituminous Mineral Substances, and in Obtaining Products Therefrom," *Chemical News* 9 (1864), 249–250; John. A. Hassan, "Relationships Between Coal, Gas, and Oil Production: A Nineteenth Century Case Study," *Indiana Archeology Review* (1978), 281. "Cannel" was derived from the word "candle," as pieces of that coal were used as candles.

79. Russel, *Edward Frankland*, 163; *Chemical News* 4 (1860), 168.

80. Gillespie v. Russel (1854–1855), *Cases Decided in the Court of Session*, (Edinburgh, 1882), 1–15, 533–547.

81. Ibid., 2.

82. Boghead coal, ironically, is neither a coal nor an oil shale, but a geological freak called today "thorbanite."

83. See the text at the end of the "Insurance Litigation: Severn, King and Company v. Imperial Insurance Company" section.

84. Gillespie v. Russel, 12, 15. For a related litigation in North America, in 1851–52, see Paul Lucier, *Scientists and Swindlers: Coal, Oil, and Scientific Consulting in American Industrial Revolution, 1830–1870* (Ph. D. Dissertation: Princeton University, 1994), 2 vols., 1:69–121.

85. Young v. Fernie, in J. W. Gifford, ed., *Report of Cases Adjudged in the High Court of Chancery by Vice Chancellor Sir John Stuart* (London, 1869), 577–613; *Chemical News* 9 (1864), 120, 132, 167, 240, 249–252, 262–264, 273–276; *Young Versus Fernie, Law Times Reports* 10 (1864), 861–865; Paul Lucier, "Court and Controversy: Patenting Science in the Nineteenth Century," *British Journal for the History of Science* 29 (1996), 147–150.

86. What was quoted in court was probably from the 1689 English translation of Glauber's *Opera Chymica*, made by C. Packe. *The Works of the Highly Experienced and Famous Chymist, John Rudolph Glauber: Containing great Variety of choice Secrets in Medicine and Alchymy, in the Working of Metallick Mines, and the Separation of Metals* (London: 1689). *Pharmacopoeia Londinensis* was a series of year books made by the Royal College of London on the various vegetable, animal, and mineral medications available. The other works quoted are: J. F. C. Morand, *L'Art d'Exploiter les Mines de Charbon de Terre* (Paris, 1768–69); M. De Gensanne, *Sur l' Exploitation des Mines d'Alsace & Comte de Bourgogne* (Paris, 1763); R. Kane, *Elements of Chemistry, Including the Most Recent Discoveries and Applications of Science to Medicine and Pharmacy and to the Arts* (Dublin, 1849), 2nd ed.; A. Ure, *A Dictionary of Arts, Manufactures, and Mines, Containing a Clear Exposition of their Principles and Practice* (London, 1839).

87. Young v. Fernie, *Report of Cases Adjudged in the High Court of Chancery*, 602.

88. Ibid., 611–612.

89. Severn v. Olive (1821); Gillespie v. Russel (1854–1855); Dashiell v. Griffith, C. Clark and W. Finnelly, *Reports of Cases heard and decided during the Sessions 1847 and 1848* (London, 1843), 12 vols., 10:190.

90. James F. Stephen, J. F., *A General View of the Criminal Law of England* (London, 1863), 189–90, 199.

91. Roger Smith, *Trial by Medicine: Insanity and Responsibility in Victorian Trials* (Edinburgh: Edinburgh University Press, 1981); Catherine Crawford, *The Emergence of English Forensic Medicine: Medical Evidence in Common-Law Courts, 1730–1830* (Ph.D. Dissertation: Oxford University, 1987); James C. Mohr, Doctors and the Law: Medical Jurisprudence in Nineteenth-Century America (Oxford: Oxford University Press, 1993); Janet E. Tighe, A Question of Responsibility: The Development of American Forensic Psychiatry, 1838-1930 (Ph.D. Dissertation, University of Pennsylvania, 1983).

92. James F. Stephen, *A History of the Criminal Law of England* (London, 1883) 3 vols., 3:464. A panel of three Royal judges was rare. Usually only one judge sat in criminal trials.

93. *Sammlung Kilinischer Vorträge* 69 (1881), 152.

94. G. L. Browne and C. G. Stewart, *Reports of Trials for Murder by Poisoning*

by Prussic Acid, Strychnia, Antimony, Arsenic and Aconitia, with Chemical Introduction and Notes on the Poisons Used (London, 1883), 278.

95. George H. Knott, Trial of William Palmer (Toronto: Canada Law Book Co., 1923), 2nd ed., 262, 307.

96. Browne, Reports of Trials for Murder by Poisoning, 448–475.

97. The copper used had been melted down from halfpennies.

98. Browne, Reports of Trials for Murder by Poisoning, 465, 477.

99. J. E. Rodgers, Letter to the Editor, Times (London), (Aug. 24, 1859), 12. The public bias against the scientific experts for the defense led the secretary of St. George's Hospital to publish a statement a day later in the Times that the hospital had no more ties with Dr. Rodgers. Letter to the Editor, Times (London), (Aug. 25, 1859), 7.

100. William Herapeth, Letter to the Editor, Times (London), (Aug. 26, 1859), 9. Taylor did not find a grain but less than half a grain of arsenic. Herapeth was still right, however, in his calculations. Seven grains of chlorate of potash would dissolve, at the most, 22 grains of copper. Half a grain of arsenic would amount therefore to 2¼ percent in the copper dissolved—an impossible quantity.

101. Henry Letheby, Letter to the Editor, Times (London), (Aug 27, 1859), 8.

102. Stephen, A History of the Criminal Law of England, 3:465; Browne, Reports of Trials for Murder by Poisoning, 478–79.

103. Chemical News 4 (May 19, 1860), 285.

104. Editorial, "Expert Witnesses," Saturday Review (Jan. 11, 1862), 32–33.

105. Ibid., 33.

106. Times (London), (Mar. 25, 1885), 9.

107. Cornelius O'Dowd, "Cornelius O'Dowd upon Men and Women, and Other Things in General," Blackwood's Edinburgh Magazine 96 (Sep. 1864), 284. Beck and Orfila were the authors of two leading text books on medical jurisprudence. O'Dowd's column was widely quoted. See, for example, Chemical News 10 (Dec. 31, 1864), 318; Journal of the Franklin Institute 49 (1865), 110.

3. Who Shall Decide Where Experts Disagree? The Nineteenth-Century Debates

1. Times (London, Apr. 14, 1820), 3; ibid. (Dec. 19, 1820), 3; J. Bostock, "Some Observations on Whale Oil," Annals of Philosophy 17 (Jan. 1821), 50.

2. Bostock, "Some Observations on Whale Oil"; S. Parkes, "Observations on the Chemical Part of the Evidence, Given upon the Late Trial of the Action Brought by Messrs. Severn, King, and Co., against the Imperial Insurance Company," Quarterly Journal of Science 10 (1821), 316–354; J. Kidd, "Observations on Naphthalene," Philosophical Transactions (1821), 209–221; M. Faraday, "On the Mutual Action of Sulfuric Acid and Naphthalene," Philosophical Transactions (1826), 140–162.

3. Voluminous literature exists on this transformation of Early Victorian

science. See for example, J. Morrell and A. Thackray, *Gentlemen of Science: Early Years of the British Association for the Advancement of Science* (Oxford: Clarendon, 1981); R. MacLeod and P. Collins, *The Parliament of Science: The British Association for the Advancement of Science, 1831–1981* (Middlesex: Science Reviews Ltd., 1981); F. M. Turner, *Contesting Cultural Authority: Essays on Victorian Intellectual Life* (Cambridge: Cambridge University Press, 1993).

4. Prince Albert, "'Science and State,' The 1859 Presidential Address of the British Association for the Advancement of Science," in G. Basalla, W. Coleman, and R. Kargon, *Victorian Science: A Self-Portrait from the Presidential Addresses of the British Association for the Advancement of Science* (New York: Doubleday, 1970), 52–53.

5. Voices calling for such a reform had been heard earlier in the century in two specific contexts: medical testimony and patent trials. For medical testimony see C. Crawford, *The Emergence of English Forensic Medicine: Medical Evidence in Common-Law Courts, 1730–1830* (Ph. D. Dissertation: University of Oxford, 1983), ch. 6. For patent litigation see C. MacLeod, *Inventing the Industrial Revolution: The English Patent System 1660–1800* (Cambridge: Cambridge University Press, 1988); M. Coulter, *Property in Ideas: The Patent Question in Mid-Victorian Britain* (Missouri: Thomas Jefferson University Press, 1991).

6. Henry Letheby, Letter to the Editor, *Times* (London), (Aug. 27, 1859), 8; *Transactions of the National Association for the Promotion of Social Science* (1863), xxxix–xl; B. Rodgers, "The Social Science Association, 1857–1886," *The Manchester School of Economic and Social Studies* 20 (1952), 283–310.

7. See Chapter 2, Regina v. Spence..

8. A. Gibson and W. Farrar, "Robert Angus Smith, F.R.S., and 'Sanitary Science,'" *Notes and Records of the Royal Society* 28 (1974), 248; R. A. Smith, "Science and Social Progress, General Principles, especially affecting Courts of Law," *Transactions of the National Association for the Promotion of Social Science* (1859), 517–526; R. A. Smith, "Science in our Courts of Law," *Journal of the Society of Arts* 7 (Jan 20, 1860), 135–147.

9. Smith, "Science in Our Courts of Law," 136–37.

10. Ibid.

11. Ibid., 139, 140, 136, 141.

12. Ibid., 142–143.

13. Ibid., 144. Webster's *Reports and Notes of Cases on Letters Patent for Inventions* (London, 1841) was long the standard text-book on Patent Law.

14. Ibid., 144–145.

15. Ibid., 145.

16. Ibid., 145–146.

17. Ibid., 146.

18. Ibid., 147–49, 220–21.

19. William Odling, "Science in Courts of Law," *Journal of the Society of Arts* 7 (Jan. 27, 1860), 167–168.

20. Ibid.

21. Prince Albert, "'Science and State," 52.

22. *Journal of the Society of Arts* 5 (1860), 141–142; *Chemical News* 6 (1862), 189–190.

23. *Chemical News* 5 (Apr. 5, 1862), 183.

24. Ibid.

25. *Chemical News* 6 (Oct. 11, 1862), 189.

26. Ibid.

27. William Crookes, "Science in Courts of Law," *Chemical News* 10 (1864), 72; A. J. Bernays, "Precautions Which Should Surround Toxicological Investigations in Medico-Legal Cases and On Evidence in Law Courts," *Chemical News* 25 (1872), 97–98; W. Beale, "Scientific Evidence in Law Courts," ibid. (1872), 117; editorial, "Evidence of Experts," ibid. (1872), 196; W. Wahl, "The Evidence of Experts," ibid. (1872); Justice, "The Evidence of Experts," ibid. (1874), 249. For another important discussion on expert testimony see: "Report of the Committee of Scientific Evidence in Courts of Law," *Report of 36th Meeting of the British Association for the Advancement of Science, Nottingham* (London, 1867), 456–457.

28. On this lingering anxiety see: R. Yeo, "Scientific Method and the Image of Science, 1831–1891," in MacLeod, *The Parliament of Science*, 65–88; Morrell, *Gentlemen of Science*, chap. 1.

29. W. J. F. Herschel, *A Preliminary Discourse on the Study of Natural Philosophy* (London, 1831), 114; *Report of the Nineteenth Meeting of the British Association for the Advancement of Science* (London, 1850), xliii–xliv.

30. "Science in 1847,"*Athenaeum* (Jan. 12, 1848), 60, quoted in Yeo, "Scientific Method and the Image of Science," 78; Quoted in W. D. Niven, *The Scientific Papers of James Clerk Maxwell* (New York: Dover, 1965), 2 vols. 2:356.

31. For an excellent analysis of this controversy, see C. Hamlin, "Scientific Method and Expert Witnessing: Victorian Perspective on a Modern Problem," *Social Studies of Science* 16 (1986), 485–513.

32. William Odling, "The Institute of Chemistry: The President's Address," *Chemical News* 52 (1885), 244.

33. Ibid., 244–45.

34. Norman Lockyer, "The Whole Duty of a Chemist," *Nature* 33 (Nov. 26, 1885), 77.

35. Ibid., 74, 75.

36. Edward Frankland, "The Institute of Chemistry," *Chemical News* 52 (Dec. 15, 1885), 305–306.

37. William Crookes, "Science in the Law-Courts," *Chemical News* 52 (Dec. 18 1885), 299.

38. William Odling, "The Whole Duty of a Chemist," *Nature* 33 (Dec. 3, 1885), 99. Odling probably referred to Lockyer's speculation that the atoms, known to the chemist as elements, were themselves groupings of smaller elements. Lockyer's speculation, known as the "dissociation hypothesis," was met with almost universal opposition. H. Dingle, "The Dissociation Hypothesis," in M. T. Lockyer and W. L. Lockyer, *Life and Work of Sir Norman Lockyer* (London: Macmillan, 1928), 292–315.

39. Editorial, "The Institution in the Strand," *Journal of Gas Lighting, Water Supply, & Sanitary Improvement* 46 (Dec. 22, 1885), 1097–1098.

40. Ibid., 1097.

41. Ibid., 1098.

42. William Crookes, "Science in the Law-Courts," *Chemical News* 52 (Jan. 1, 1886), 1–2.

43. Ibid., 2.

44. Editorial, "Science in the Law Courts," *Journal of Gas Lighting, Water Supply, & Sanitary Improvement* 46 (Dec. 29, 1885), 1145.

45. Ibid. "*Caveat emptor*" (let the buyer beware) was a central principle of nineteenth-century commercial law that conceived of commercial transactions as the meeting of free wills and refused to introduce any control elements to ensure their fairness.

46. Ibid., 1146.

47. Editorial, "Science in the Law Courts," *Journal of Gas Lighting, Water Supply, & Sanitary Improvement* 47 (1886), 107. "Science in the Law Courts," *Chemical News* 53 (1886), 39; R. A. Proctor, "The Dignity of Science," *Knowledge* (1886), 93–95; "Scientific Assessors in Courts of Justice," *Nature* 38 (1886), 289–291.

48. In 1863, under the Alkali Act, Smith became the leading scientific figure in the first successful governmental policy directed towards the regulation of the Chemical industry. See Roy MacLeod, *Public Science and Public Policy in Victorian England* (Hampshire: Variorum, 1996), Part 1: 85–112.

49. Emory Washburn., "Testimony of Experts," *American Law Review* 1 (1866), 45–64, on 48, 49.

50. For a long list of scientific expertise that had made appearance in court, see F. Wharton, *A Commentary on the Law of Evidence in Civil Issues* (Philadelphia, 1877), 2 vols., 2:443–444; S. Bell, *The Use and Abuse of Expert Testimony* (Philadelphia, 1879); E. Lewis, *The Law of Expert Testimony* (Philadelphia, 1894).

51. Anonymous, "Expert Testimony," *American Law Review* 5 (1870), 227–246; 428–442, on 228.

52. Morrison Remick Waite, "Testimony of Experts," *The Western Jurist* 8 (Mar. 1874), 129–135, on 134–135. Judicial complaints about scientific expert testimony could be compiled almost at will. For one of the more extensive compilations, see L. G. Kinne, "Expert Testimony—Its Origin, Value, Dangers and Proper Place in Jurisprudence," *American Lawyer* (1896), 201–204.

53. Symposia, "Expert Testimony," *North American Review* 138 (1884), 602–617; Clemens Herschel, *On the Best Manner of Making Use of the Services of Experts in the Conduct of Judicial Inquiries* (Boston, 1886), 39 pp. A shorter version of this pamphlet was published in *American Law Review* 22 (1887), 571–577; J. Trowbridge, "The Imperiled Dignity of Science and the Law," *Atlantic Monthly* (1896), 491–495.

54. Charles. F. Himes, "The Scientific Expert in Forensic Procedure," *Journal of the Franklin Institute* 85 (1893), 407–436, on 411.

55. William Mason, "Expert Testimony," *Science* (Aug. 1897), 243–248, on 244–245.

56. William Best, *A Treatise on the Principles of Evidence and Practices as to Proofs in Courts of Common Law* (London, 1854), 2nd ed., 346; Anonymous, "Expert Testimony," 428–429; Waite, "Testimony of Experts," 131.

57. T. D. Crothers, "The Psychology of a Jury in a Long Trial," *Scientific American* 73 (Aug. 17, 1896), 104.

58. R. M. Jackson, "The Incidence of Jury Trial During the Past Century." *Modern Law Review* 1 (1937), 132–144; W. R. Cornish and G. de N. Clark, *Law and Society in England 1750–1950* (London: Sweet and Maxwell, 1989), 20.

59. J. H. Choate, "Trial By Jury: Annual Address Before the American Bar Association." *American Law Review* 33 (1898), 285–314. The Seventh Amendment to the American Constitution provides as follows: "In suits at common law, where the value in controversy shall exceed twenty dollars, the right of trial by jury shall be preserved, and no fact tried by jury, shall be reexamined in any Court of the United States, other than according to the rules of the common law." U.S. Constitution, Amendment 7.

60. E. R. Sunderland, "The Inefficiency of the American Jury," *Michigan Law Review* 13 (1914), 302–316; Note, "Changes in the Jury," *Virginia Law Review* 17 (1930), 497–501; M. D. Howe, "Juries as Judges of Criminal Law," *Harvard Law Review* 52 (1939), 582–616; Anonymous, "The Changing Role of the Jury in the Nineteenth Century," *Yale Law Journal* 74 (1964), 170–197; I. A. Horowitz, "Changing Views of Jury Power: The Nullification Debate, 1787–1988," *Law and Human Behavior* 15 (1991), 165–182; G. Peters, "Invading the Province of the Jury," *Indiana Law Review* (1926–7), 539–545.

61. Editorial, "Science in the Courts," *Scientific American* (1872), 167.

62. S. G. Kohlstedt, *The Formation of the American Scientific Community: The American Association for the Advancement of Science. 1848–1860* (Urbana: University of Illinois Press, 1976).

63. One of the first reform bills was a joint effort of the American Academy of Arts and Sciences, the Suffolk District Medical Society, the Boston Society for Medical Observation and the Boston Society for Medical Sciences. The bill was written by Judge Emory Washburn who headed the committee of the Academy of Arts and Sciences, see Editorial, *Boston Medical and Surgical Journal* 90 (Apr. 16, 1874), 387–388. For a wider look on the efforts of the medical community, see James Mohr, *Doctors and the Law: Medical Jurisprudence in Nineteenth-Century America* (Oxford: Oxford University Press, 1993).

64. See a long list of such discussions under the entry "Evidence, Expert," in D. Dooley, ed. *Index to State Bar Association Reports and Proceedings* (New York: Voorhis. 1942), 176–177.

65. Gustav Eindlich, "Proposed Changes in the Law of Expert Testimony," *Pennsylvania Bar Proceedings* (1898), 189–221; William Foster, "Expert Testimony—Prevalent Complaints and Proposed Remedies," *Harvard Law Review* 11 (1897–98), 169–186; William Purrington, "The Nature of Expert Testimony,

and the Defects in the Method by which it is now Adduced in Evidence," *Medical Record* (Dec 9, 1899), 849–854; Learned Hand, "Historical and Practical Considerations Regarding Expert Testimony," *Harvard Law Review* 15 (1901), 40–58; Note, "Appointment of Expert Witnesses by the Court," *Harvard Law Review* 24 (1910–11), 483–484.

66. L. H. Korn, "Law, Fact, and Science in the Courts," *Columbia Law Review* 66 (1966), 1080–1116. This is not to say that the English legal system did not use the law of evidence to control the problem of expert testimony. But is seems that the Royal judges who dominated their courts to a degree unacceptable in the U.S. were able to relax this technical corpus to a much greater extent.

67. Charles T. McCormick, "Expert Testimony as an 'Invasion of the Province of the Jury,'" *Iowa Law Review* 26 (1941), 819–840; idem, *Handbook of the Law of Evidence* (St. Paul: West Pub., 1954), 26; "Expert Opinion on Ultimate Facts," in "Notes on Legislation," *Iowa Law Review* (1941), 825–840.

68. M. Ladd, "Expert Testimony," *Vanderbilt Law Review* 5 (1952), 414–431.

69. John H. Wigmore, *A Treatise on the System of Evidence in Trials at Common Law* (Boston: Little, Brown, 1904), § 686; Charles T. McCormick, "Some Observations Upon the Opinion Rule and Expert Testimony," *Texas Law Review* 23 (1954), 128–130.

70. J. Lawson, *The Law of Expert and Opinion Evidence Reduced to Rules: With Illustrations from Adjudged Cases* (St. Louis, 1883), 169–194; H. W. Rogers, *The Law of Expert Testimony* (St. Louis, 1891), 389–414; J. Wigmore, "Scientific Books in Evidence," *American Law Review* 26 (1892), 390–403; Charles Dana, "Admission of Learned Treatises in Evidence," *Wisconsin Law Review* (1945), 414–450; "Expert and Opinion Evidence," *Ruling Case Law* (Rochester, N.Y., 1916), 558–590.

71. James B. Thayer, "The Present and Future of the Law of Evidence," *Harvard Law Review* 12 (1898), 71–74, on 72; Wigmore, *A Treatise on the System of Evidence*; Bond, "Expert Witness," *Chicago Legal News* 42 (1909), 111–116, on 115. Thayer's son later reminisced how a famous judge, J. H. Choate, asked him to tell his father that his famous *Preliminary Treatise on Evidence at the Common Law* (Boston, 1898) was a good book, "but it is a pity he did not publish it while there was still such a thing in existence as the law of evidence." E. R. Thayer, "Observations on the Law of Evidence," *Michigan Law Review* 13 (1915), 355–367, on 364.

72. Much research is still needed on twentieth-century developments in England. Clearly, in spite of the relatively greater judicial control and latitude, the problem of scientific expert testimony has continued to simmer in English courts. A. M. Wood, ed., *Science and Technology in the Eye of the Law* (London: Royal Society, 2000); E. Erzinclioglu, "British Forensic Science in the Dock." *Nature* 32 (1998), 859–860; H. Woolf, *Access to Justice: Final Report* (London: HMSO, 1996).

4. Blood Will Out: Distinguishing Humans from Animals and Scientists from Charlatans

1. Editorial, "Can Human and Animal Blood be Distinguished in Case of Blood Stains?" *Central Law Journal* 10 (1880), 183–187, on 184.

2. Ibid., 184.

3. Ibid., 185.

4. William. D. Sutherland, *Blood-Stains: Their Detection and the Determination of their Sources: A Manual for the Medical and Legal Professions* (New York: W. Wood, 1907), 12–25; Theodore Wormley, *The Micro-Chemistry of Poisons, Appendix on the Detection and Microscopic Determination of Blood,* (Philadelphia, 1885), 2nd ed., 708–714. None of these tests were infallible.

5. Jean-Pierre Barruel, "Mémoire sur l'existence d'un principe propre a caractérizer le sang de l'homme et celui des diverses espèces d'animaux," *Annales d'hygiène publique et de médecine légale* 1 (1829), 267–277.

6. Taddei de Gravina, *British and Foreign Medical Review* 2 (1836) 226; Leuret, "Sur le principe aromatique du sang," *Annales d'hygiène publique et de médecine légale* 2 (1829), 217; Andrew Fleming, "Blood Stains," *American Journal of Medical Sciences* 37 (1859), 84–119, on 105; Matthieu J. B. Orfila, *Traité de médecine légale* (Paris, 1836), 3rd ed., 4 vols., 2:700; Henry Formad, *Comparative Studies of Mammalian Blood, with Special Reference to the Microscopical Diagnosis of Blood Stains in Criminal Cases* (Philadelphia, 1888), 33.

7. Chevallier et J. P. Barruel, "Taches de sang," *Annales d'hygiène publique et de médecine légale* 10 (1833), 160–163; Orfila, Barruel, et Chevallier, "Taches de sang: rapport médico-légale," ibid., 14 (1835), 349; Barruel, Chevallier, et O. Henry, "Affaire Gilbert et Rodolphe: rapport," ibid., 23 (1840), 387–409, on 396.

8. Karl Schmidt, *Die Diagnostik Verdächtiger Flecke in Criminalfällen* (Leipzig, 1848), 19.

9. About the scandal see, Tardieu, Barruel, et Chevallier, "Expériences sur l'odeur du sang," *Annales d'hygiène publique et de médecine légale* 49 (1853), 413–417. For other unfavorable reports see, Chevallier, "Examen du sang desséché," *Journal de Chimie Médicale* 5 (1853), 2nd ser., 490–498, on 493; Alfred S. Taylor, "Remarks on Death from Strangulation," *Guy's Hospital Reports* 5 (1851), 371–425, on 413.

10. Fleming, "Blood Stains," 87–101; Sutherland, *Blood-Stains,* 12–25; Wormley, *Micro-Chemistry of Poisons,* 708–714.

11. Lane Butler, "On the Detection of Blood Stains," *Medical Times* (London, 1850), 647–649.

12. Alfred S. Taylor, *Principles and Practice of Medical Jurisprudence,* Fred J. Smith, ed., (London: Churchill, 1905), 5th ed., 127; Thomas Price, "Examination of Blood Stains," *Pacific Medical and Surgical Journal* (Dec. 1867), 289–293.

13. The German physiologist Bernhard Ritter was probably the first to suggest, in an elaborate essay that received a prize from the Prussian government, using size measurements of red blood corpuscles for the identification of human blood. Bernhard Ritter, *Über die Ermittelung von Blutflecken an Mettallischen Instrumenten* (Berlin, 1846), 22.

14. Gulliver's work fills the pages of the *Zoological Society of London, Proceedings.* For example: "Measurements of the Red Corpuscles of the Blood of Vertebra," (1845), 93; "Additional Measurements of the Red Corpuscles of the Blood of Vertebra," (1848), 36–38; "On the Red Corpuscles of the Blood of Vertebrata, and on the Zoological Import of the Nucleus, with Plans of their Structure, Form and Size (on a uniform scale), in Many Different Orders," (1862), 91–93.

Numerous observations on the size, shape, and structure of the blood corpuscles can also be found in the notes to *The Works of William Hewson*, ed. George Gulliver (London, 1846).

15. A detailed list of such moisteners, with formulas for their preparation, can be found in Formad, *Comparative Studies of Mammalian Blood*, 289; Fleming, "Blood Stains," 109–110; Joseph G. Richardson, *Handbook of Medical Microscopy* (Philadelphia, 1871), 289–290.

16. Schmidt, *Die Diagnostik Verdächtiger Flecke*, 5.

17. Hermann Welcker, "Grösse, Zahl, Volum, Oberfläche und Farbe der Blutkörperchen bei Menschen und bei Theirenen," *Zeitschrift für Rationelle Medicin* 20 (1863), 257–307, on 257; Rudolf Virchow, "Über die forensiche Untersuchung von trockenen Blutflecken," *Virchows' Archiv* 12 (1857), 334; Ernst Brücke, "Über die gerichtsärztliche Untersuchung von Blutflecken," *Wiener medizinische Wochenschrift* (1857), 425; *Lancet* 1 (1852), 321; Fleming, "Blood Stains," 119.

18. Hermann Friedberg, *Histologie des Blutes mit besonderer Rücksicht auf forensiche Diagnostik* (Berlin, 1852), 57, quoted in Fleming, "Blood Stains," 119; Bernhard Ritter, "*Über die Ermittelung von Blut, Samen, und Exkrementenflecken in Kriminalfällen: Ein spezieller beiträge zur gerichtlichen Arzneikunde*" (Würzburg, 1854), 139. Virchow, "Über die forensiche Untersuchung von trockenen Blutflecken," 336.

19. Taylor, *Principles and Practice of Medical Jurisprudence*, 127.

20. See for example, J. Towler, "Analysis of Blood Stains as a Means of Detecting Crime," *Transactions of the New York State Medical Society* (1864), 115–120.

21. Joseph G. Richardson, "On the Detection of Red and White Corpuscles in Blood Stains," *American Journal of Medical Sciences* 58 (1869), 50–58, on 54–55.

22. John C. Dalton, *Treatise on Human Physiology* (London, 1882), 214; George. F. Dowdeswell, "On Some Appearances of the Red Blood corpuscles of Man and other Vertebrata," *Quarterly Journal of Microscopical Science* 21 (1881), 154–161; Edward. S. Wood, "Examination of Blood and other Stains," in Rudolph A. Witthaus and Tracy C. Becker, eds., *Medical Jurisprudence: Forensic Medicine and Toxicology* (New York, 1894), 2 vols., 2:39–41.

23. Richardson, "On the Detection of Red and White Corpuscles," 53, 54, 56.

24. Gulliver, "On the Red Corpuscles," 93; Fleming, "Blood Stains," 109.

25. Richardson, "On the Detection of Red and White Corpuscles," 53; idem, "On the Cellular Structure of the Red Blood corpuscles," *Transactions of the American Medical Association* (1870), 17–25. Menobranchus is a reptile.

26. Alfred S. Taylor, *A Manual of Medical Jurisprudence*, J. J. Reese, ed. (Philadelphia, 1873), 7th ed., 307. On Richardson's frustration see "Proceedings of the Biological and Microscopical Section of the Academy of Natural Sciences," *Medical Times* (Philadelphia), (Jul. 18, 1874), 663.

27. Joseph G. Richardson, "On the Value of High Powers in the Diagnosis of Blood Stains," The *Monthly Microscopical Journal* 12 (Sep. 1874), 130–141, on 134–136, 141, also published in *American Journal of The Medical Sciences* 68 (Jul. 1874), 102–110.

28. "Proceedings of the Biological and Microscopical Section of the Academy of Natural Sciences," *Medical Times* (Philadelphia), (Jul. 18, 1874), 663–665; (Sep. 26, 1874), 826–827; (Sep. 30, 1876), 638–639; Joseph G. Richardson, "Notes on the Performance of Two One-Fiftieth Objectives," *Medical Times* (Philadelphia), (Nov. 28, 1874), 132–133.

29. Editorial, "The Diagnosis of Blood-Stains," *Lancet* (Aug. 1874), 210–211. For other responses, pro and con, see *Medical Times and Gazette* (Aug. 1874), 151; *Medical Record* (London, Sept. 1874), 560–562; "Proceedings of the Biological and Microscopical Section of the Academy of Natural Sciences," *Medical Times* (Philadelphia), (Sep. 30, 1876), 638–639; *American Journal of Microscopy and Popular Science* 2 (Jan. 1877), 1–3.

30. Anonymous, "Expert Testimony," *American Law Journal* 5 (1870), 227–246, 428–442; *Scientific American* (1872), 167; James C. Mohr, *Doctors and the Law: Medical Jurisprudence in Nineteenth-Century America* (Oxford: Oxford University Press, 1993), 197–213; Janet A. Tighe, *A Question of Responsibility: The Development of American Forensic Psychiatry, 1838–1930* (Ph.D. dissertation, University of Pennsylvania, 1983).

31. *The Baltimore American and Commercial Advertiser* (Jan. 26, 1872), quoted in Mohr, *Doctors and the Law*, 192; "Science in the Court-Room," *Industrial Monthly* 5 (Feb. 1874), 44, and 7 (Mar. 1876), 53.

32. Joseph J. Woodward, "On the Similarity Between the Red Blood Corpuscles of Man and Those of Certain Other Mammals, Especially the Dog; Considered in Connection with the Diagnosis of Blood-Stains in Criminal Cases," *American Journal of the Medical Sciences* 137 (Jan. 1875), 152–162, on 152–153, also published in *Monthly Microscopical Journal* (London), 13 (1875), 65–76; *Bulletin of the Philosophical Society* (Washington), 2 (1874), 78.

33. Joseph J. Woodward, "The Application of Photography to Micrometry, with Special Reference to the Micrometry of Blood in Criminal Cases," *American Medical Association, Transactions* 27 (1876), 303–314, on 31; "discussion," ibid., 296.

34. Woodward, "On the Similarity Between the Red Blood Corpuscles," 155.

35. Ibid., 159–160.

36. Ibid., 160–162.

37. Joseph G. Richardson, "Explanatory Note in Regard to the Diagnosis of Blood Stains," *American Journal of the Medical Sciences* (Apr. 1875), 575–576, also published in the *Monthly Microscopical Journal* 13 (1875), 213–217. See also Richardson's comments in "Report on the Meeting of the Biological and Microscopical Section of the Academy of Natural Sciences," *Medical Times* (Philadelphia), (Oct. 28, 1876), 43–44, and his retrospective evaluation, "Upon Measurements of Red Blood corpuscles and the Diagnosis of Blood Stains," *Gaillard's Medical Journal* 31 (1881), 44–52.

38. Richardson, "Explanatory Note," 576.

39. Woodward, "On the Similarity Between the Red Blood Corpuscles," 152–153.

40. George Gulliver, *Zoological Society of London, Proceedings* (Jun. 15 1875), 484; idem, "Comparative Photographs of Blood-Disks," *Monthly Microscopical*

Journal 16 (1876), 240; Lionel S. Beale, President of the Royal Microscopical Society: "We must admit that we are unable to decide with sufficient certainty to justify us giving our evidence in a court of law." L. S. Beale, *The Microscope in Medicine*, (Philadelphia, 1878), 207.

41. "Proceedings of the Biological and Microscopical Section of the Academy of Natural Sciences," *Medical Times* (Philadelphia), (Sep. 30 1876), 638.

42. Editorial, "Marvels of Microscopy," *Medical Times* (Philadelphia), (Aug. 18, 1878), 564; Joseph Jones, "Medico-legal Evidence Relating to the Detection of Human Blood Presenting the Alterations Characteristic of Malarial Fever, on the Clothing of a Man Accused of Murder," *New Orleans Medical and Surgical Journal* 6 (1878), 139–156.

43. Herbert H. Hayden, *Poor Mary Stannard! A Full and Thrilling Story of the Circumstances Connected with her Murder. The Only True and Reliable Account of the Most Mysterious of all the Cases Which Have Baptized Connecticut in Blood* (New Haven, 1879), 18–20.

44. Ibid., 35.

45. *New York Times* (Jan. 20, 1880), 4; *New York Sun* (Jan. 14, 1880), 3.

46. Herbert H. Hayden, *An Autobiography: The Mary Stannard Murder, Tried on Circumstantial Evidence, Including Records of the Trial* (Hartford, 1880), 11–12.

47. *New York Sun* (Jan 14, 1880), 3.

48. *New York Sun* (Jan. 15, 1880), 3.

49. Ibid.; *New York Sun* (Jan. 21, 1880), 3.

50. *New York Times* (Jan. 20, 1880), 4.

51. Editorial, "The Buffalo Meeting of the American Society of Microscopists: Address by President R. H. Ward," *American Monthly Microscopical Journal* (1880), 29–34, 47–51, on 47–48; Editorial, *American Monthly Microscopical Journal* (Jan. 1880), 17; editorial, "The Expert Testimony in the Hayden Trial," *Medical Record* (New York), (Nov. 29, 1879), 27, 517–518.

52. Editorial, "Some 'Expert' Evidence," *American Monthly Microscopical Journal* (Mar. 1880), 55–57, published also in *American Journal of Microscopy* 5 (1880), 65–67, and as "The Size of the Blood Corpuscles," *Medical Record* (New York), (Jan. 31. 1880), 131–132. See also "Further Expert Testimony at the Hayden Trial," *Medical Record* (New York), 28 (Jan. 10, 1880), 39–40; "The Size of the Blood Corpuscles," *American Journal of Microscopy* 5 (Jan. 1880), 65–68.

53. *New York Times*, (Dec. 28, 1879), 6.

54. See the start of this chapter.

55. *Central Law Journal*, "Can Human and Animal Blood be Distinguished?" 183–184, 187.

56. Ibid., 183; *New York Times* (Mar. 16, 1880), 4.

57. Charles. O. Curtman, "Blood Stains as Evidence in Criminal Cases," *American Monthly Microscopical Journal* (Oct. 1880), 184–186, reprinted from *St. Louis Courier of Medicine and Collateral Sciences* 4 (1880), 229–231; Editorial, "Microscopical Evidence Concerning Blood Corpuscles," *American Monthly Microscopical Journal* (Sep. 1883), 175–176.

58. Richard A. Lemmon, "Blood Corpuscles in Medical Jurisprudence,"

American Monthly Microscopical Journal (Sep. 1880), 207–208; S. H. Gage, "Blood Corpuscles in Medical Jurisprudence," ibid. (Oct. 1880), 235–236; D. S. Kellicott, "Easy Methods of Detecting Blood-Stains," *Buffalo Medical and Surgical Journal* 20 (1880–81), 150–154.

59. Richard U. Piper, "Expert Testimony and the Microscopical Examination of Blood," *American Law Register* 16 (1880), 529–541, on 534, 537; 17 (1880), 593–614, on 599–600.

60. Richard U. Piper, "Microscopical Examination of Blood in Its Relation to Criminal Trials," *American Law Register* 15 (1876), 561–569; 16 (1877), 257–267; 17 (1878), 554–561, on 555.

61. Ibid., 563. For another series of Piper's articles in another leading legal journal see, "Use of the Microscope in Medico-Legal Cases, Pt. I, II, & III," *Chicago Legal News,* 10 (1877), 110–111, 136–137, 150; "Examination of Alleged Blood Spots," ibid. 16 (1883), 56–57.

62. A brief reference in a scientific paper given at the microscopical section of the Iowa Medical Society is the only exception found. *American Monthly Microscopical Journal* (Jul. 1881), 140. At least in one case, Piper's overenthusiastic testimony led to a successful appeal. "We think," the Wisconsin Supreme Court decided, "it [Piper's testimony] was clearly incompetent, and must work a reversal of judgment." Knoll v. State, *Wisconsin Reports* 55 (May 1882), 547–551, on 549.

63. Wormley, *Micro-Chemistry of Poisons,* 736; *American Monthly Microscopical Journal* 5 (Oct. 1884), 181–183.

64. Marshall D., Ewell, "Measurement of Blood Corpuscles," *American Monthly Microscopical Journal* (Aug. 1885), 150–151, also published in *Chicago Legal News* 17 (1885), 387–388.

65. T. J. Gallaher, "On Blood and Blood Stains," *Pittsburgh Medical Journal* 1 (1881), 163–172; Charles. M. Vorce, "The Microscopical Discrimination of Blood," *American Monthly Microscopical Journal* (Dec. 1883), 223–224; idem, "The Measurements of Blood Corpuscles," ibid. (Jan. 1884), 223–224; T. Up de Graff, "Measuring Blood Corpuscles," ibid. (Feb. 1884), 26–27; S. G. Shanks, "A Contribution to Blood Measurements," ibid. (Feb. 1886), 25–26; idem, "Measuring Blood corpuscles," ibid. (Jul 1886), 138–139; Marshall D. Ewell, "On Fine Measurements," ibid. (Jun. 1886), 119–120; F. Detmers, "The Comparative Size of Blood Corpuscles of Man and Domestic Animals," *American Society of Microscopists, Proceedings* 9 (1887), 216–224; W. J. Lewis, "Forensic Microscopy, or the Microscope in its Legal Relations," *American Monthly Microscopical Journal* (Sep. 1889), 197–207, on 198, 207.

66. C. Seiler, "High Powers in Micro-Photography," *Medical Times* (Philadelphia), (Feb. 19, 1876), 247–251; idem, "Photographic Enlargements of Microscopical Objects," ibid. (1876), 563–576; Joseph J. Woodward, "The Application of Photography to Micrometry, with Special Reference to the Micrometry of Blood in Criminal Cases," *American Medical Association, Transactions* (1876), 303–314; G. M. Sternberg, *Photo-Micrographs and How to Make Them* (Boston, 1884); Formad, *Comparative Studies of Mammalian Blood,* 31, 43. In one of Formad's court appearances (Killer v. Com, *Pennsylvania Reports* 124 (1888), 92) he testified that blood found on a handkerchief was human and not beef blood as the defense argued.

67. Georges Hayem, *Recherches sur l'Anatomie normale et pathologique du sang*, (Paris, 1878).

68. H. C. Hyde, "The Microscope in Medical Jurisprudence," *American Journal of Microscopy*, 4 (1879), 11–13; Vorce, "The Microscopical Discrimination of Blood"; Patricia Gossel, "A Need for Standard Methods: The Case of American Bacteriology," in Adele Clarke, ed., *The Right Tools for the Job* (Princeton: Princeton University Press, 1992).

69. Marshall D. Ewell, "The Relations of the Microscope to the Administration of Justice," *American Microscopical Society, Proceedings*, 14 (1892), 1–11, on 8–9.

70. E. Boring, *History of Experimental Psychology* (New York: Century, 1950), 134–156; E. C. Sanford, "Personal Equation," *American Journal of Psychology* 2 (1888), 3–38, 271–298, 403–430; I. Kirsch, "The Impetus to Scientific Psychology: A Recurrent Pattern," *Journal of the History of the Behavioral Sciences* 12 (1976), 12–29.

71. Marshall D. Ewell, "A Micrometric Study of Four-Thousand Red Blood Corpuscles in Health and Disease," *Medico-Legal Journal* (New York), 10 (1892), 175–201, on 184; idem, "On the Accuracy of Measurement of Blood Corpuscles in Criminal Cases," *American Law Register* 26 (1898), 20–21.

72. Ewell, "The Relations of the Microscope to the Administration of Justice," 9. Ewell repeated the same conclusion before the Medico-Legal Society of New York, Ewell, "A Micrometric Study of Four-Thousand Red Blood Corpuscles," 189.

73. Editorial, "The Charlatanry of Microscopy," *Chicago Evening Journal* (Sep. 9, 1891), quoted in Ewell, "The Relations of the Microscope to the Administration of Justice," 2–3.

74. Moses White, "The Red Blood Corpuscle in Legal Medicine," *American Microscopical Society, Transactions* 18 (1896), 201–219, on 203, 217–218; idem, "The Blood Corpuscle in Legal Medicine," *Yale Medical Journal* 1 (1895), 95–100; Wormley, *Micro-Chemistry of Poisons*, 728. Clark Bell, President of the American International Congress on Medical Jurisprudence, entertained a similar opinion. Like Ewell and White, Bell argued his view before both the New York Medico-Legal Society and the American Microscopical Society, Bell, "Blood and Blood Stains in Medical Jurisprudence," *American Microscopical Society, Proceedings* 14 (1892), 91–120, and *Medico-Legal Journal* (New York), 10 (1892), 129–174.

75. Henry. L. Tolman, "On the Means of Distinguishing Human Blood," *American Monthly Microscopical Journal* 15 (Apr. 1894), 97–104, on 97, 103; William. S. Thorne, "The Red Blood Corpuscle its Value in Judicial Investigations," *Pacific Medical Journal* 39 (Nov. 1896), 673–680, on 673; James F. Babcock, "Blood and Other Stains," in Allan M. Hamilton and Lawrence Godkin, eds., *A System of Legal Medicine* (New York, 1894), 2 vols., 1:167–184; Joseph Jones, "Micro-chemical Examination of Blood Stains," *Medico-Legal Journal* (New York), 10 (Sep. 1892), 249–252.

76. Lewis, "Forensic Microscopy," 207.

77. Richardson, "Explanatory Note," 575–576; White, "The Red Blood Corpuscle in Legal Medicine," 218; Woodward, "On the Similarity Between the

Red Blood Corpuscles," 162; Ewell, "The Relations of the Microscope to the Administration of Justice," 9.

78. P. Edwards, "Chemical Experts: A Trio of Important Factors in the Detection of Crime," *Central Law Journal* 42 (1896–97), 323–326, on 324.

79. Ibid.

80. See *The trial of Joseph LaPage the French monster, for the murder of the beautiful school girl Miss Josie Langmaid. Also, the account of the murder of Miss Marietta Ball, the school teacher, in the woods, in Vermont* (Philadelphia, 1876). In this trial three eminent doctors—Horace Chase, Joshua Treadwell, and Dana Hayes—testified for the state that blood corpuscles may be restored to perfect shape even after the lapse of ten years, and that it could then be infallibly determined whether the dried blood was human or not. Two equally eminent doctors—John B. Edwards and Gilbert R. Guilder—swore for the defense, however, that after lapse of only two weeks dried corpuscles cannot be restored to their original size.

81. F. J. Parker, "Micrometry of Human Red Blood Corpuscle," *Transactions of the American Microscopical Society* 20 (1898), 201–219; J. N. Jenne, "Blood Stains," *Vermont Medical Monthly* 2 (1896), 127–137; John H. Linsley, "Some Suggestions Concerning the Examination of Blood," *Medical Record* (Philadelphia), 48 (Nov. 1895), 685–688; "Detecting Human Blood," *Green Bag* 7 (1895), 61–63; E. R. Axtell, "The Medico-Legal Examination of the Red Stains Found on the Clothes of Charles Ford, with a Plea for the Use of the Camera Lucida in the Microscopic Examination of Blood Stains," *Journal of the American Medical Association* (Jul. 17, 1895), 139–144; J. H. Linsley, "Some Suggestions Concerning the Examination of Blood," *Medical Record* (Philadelphia), 48 (Nov. 1895), 685–688.

82. A. M. Bleile, "The Detection and Recognition of Blood: the Annual Address of the President," *Transactions of the American Microscopical Society* 22 (1900), 4.

83. Paul Uhlenhuth, "Neuer Beitrag zum spezifischen Nachweis von Eiereiweiss auf biologischem Wege," *Deutsche medizinische Wochenschrift* 46 (Nov. 15, 1900), 1–5; idem, "Eine Methode zur Unterscheidung der verschiedenen Blutarten, im besonderen zum differentialdiagnostischen Nachweise des Menschenblutes," *Deutsche medizinische Wochenschrift* 6 (Feb. 7, 1901), 1–3, on 3.

84. A. J. Ferreira da Silva and Alberto d'Aguiar, *L'examen médico légale des taches de sang, et spécialement la méthod d'Hulenhuth*, (Porto, 1906); Harry T. Marshall, "Blood Spots as Evidence in Criminal Trials," *Virginia Law Review* 2 (Apr. 1915), 481–492; Jurgen Thorwald, *The Century of the Detective* (New York: Harcourt, 1965), 148–152; M. Ainsworth, "Science and the Detective," in William T. Shore, ed., *Crime and its Detection* (London: Gresham Pub. Co., 1932), 70–71. The serological test is also not infallible. On the doubts that the late twentieth-century DNA fingerprinting test cast on the reliability of serology, see Andre A. Moenssens, "Novel Scientific Evidence in Criminal Cases: Some Words of Caution," *Journal of Criminal Law and Criminology* 84 (1993), 1–21, on 12–15.

85. G. L. E. Turner, "The Microscope as a Technical Frontier in Science," in L. E. T. Gerard, ed., *Essays on the History of the Microscope* (Oxford: Senecia, 1980), 159–183; Arleen Tuchman, *Science, Medicine, and the State in Germany: The Case of Baden, 1815–1871* (New York: Oxford University Press, 1993); R. L. Kremer,

"Building Institutes for Physiology in Prussia, 1836–1846," in A. Cunningham and P. Williams, eds., *The Laboratory Revolution in Medicine* (Cambridge: Cambridge University Press, 1992), 72–109.

5. The Authority of Shadows: The Law and X-Rays

1. R. Brecher, and E. Brecher, *The Rays: A History of Radiology in the United States and Canada* (Baltimore: Williams and Wilkins, 1969), 104.

2. R. Arns, "The High-Vacuum X-ray Tube: Technological Change in Social Context," *Technology and Culture* (1997), 38:852-590.

3. J. Howell, *Technology in the Hospital: Transforming the Patient Care in the Early Twentieth century* (Baltimore: John Hopkins University Press, 1995); S. Reiser, *Medicine and the Reign of Technology* (Cambridge: Cambridge University Press, 1979), 45–68.

4. B. Pasveer, "Knowledge of Shadows: The Introduction of X-ray Images in Medicine," *Sociology of Health and Illness* (1989), 3:360-81; H. Lerner, "The Perils of 'X-ray Vision': How Radiographic Images Have Historically Influenced Perception," *Perspectives in Biology and Medicine* (1992), 35:382-97.

5. Editorial, "X-ray in Evidence," *Rocky Mountain Daily News* (Dec. 2, 1896), 5; S. Withers, "The Story of the First Evidence," *Radiology* 17 (1934), 99–100; Editorial, "Indemnity Bond for Plaintiffs in Damage Suits," *Colorado Medical Journal* 2 (1896), 396; Editorial, "Novel Idea in Court," *Denver Evening Post* (Dec. 2, 1896), 2; R. H. Shikes, *Rocky Mountain Medicine: Doctors, Drugs and Disease in Early Colorado* (Boulder: Johnson, 1986), 89; W. W. Grant, "History of St. Joseph's Hospital," *Transactions of the Colorado State Medical Society* 31 (1901), 519–520.

6. Cf. L. A. Stimson, *A Practical Treatise on Fractures and Dislocations* (Philadelphia, 1899); L. A. Stimson and J. Rogers, *Manual of Operative Surgery* (New York, 1895); H. O. Thomas, *The Principles of the Treatment of Fracture and Dislocations* (London, 1886).

7. Editorial, "Reliability of X-Rays," *Rocky Mountain Daily News* (Dec. 3, 1896), 7.

8. K. A. De Ville, *Medical Malpractice in Nineteenth-Century America: Origins and Legacy* (New York: New York University Press, 1990). For statistics, see H. W. Smith, "Legal Responsibility for Medical Malpractice," *Journal of the American Medical Association* 118 (1941), 2149–2159, 2670–2679; A. A. Sandor, "Medicine and Law: The History of Professional Liability Suits in the United States," ibid., 163 (1957), 459–466.

9. G. Law, "Ethics in Malpractice," *Denver Medical Times* 11 (1892), 582; idem, "Malpractice Suits," *Denver Medical Times* 16 (1896), 2–11; E. F. Sanger, "Report on Malpractice," *Boston Medical and Surgical Journal* 100 (1879), 46; J. Mohr, *Doctors and the Law: Medical Jurisprudence in Nineteenth-Century America* (Oxford: Oxford University Press, 1993), 113–115.

10. W. W. Grant, "Appendicitis," *Transactions of the Colorado State Medical Society* 22 (1892), 140–148; L. Freeman, "In Memoriam: William W. Grant,

1846–1934," *Transactions of the Western Surgical Association* 44 (1934), 522–523; M. Price, "Remarks on Suits for Malpractice," *New York Medical Journal* 65 (1897), 676–678.

11. *Rocky Mountain Daily News* (Dec. 3, 1896), 7; Withers, "The Story of the First Evidence," 100.

12. Early x-ray images carried various names. The more popular terms were Roentgengraphs, radiographs, shadowgraphs, skiagraphs, and x-ray pictures or photographs. I use the last term but my quotations contain other terms as well. For the origins and reasoning behind some of these terms see A. W. Goodspeed, "Roentgen's Discovery" *Medical News* 68 (1896), 169.

13. O. Glasser, *Wilhelm Conrad Röntgen and the Early History of the Roentgen Rays* (London: Bale, 1933), 29–46; C. T. Holland, "X-Rays in 1896," in A. Bruwer, ed., *Classic Descriptions in Diagnostic Roentgenology* (Springfield, Ill.: Thomas, 1964), 70–84; A. E. Barclay, "The Old Order Changes," *British Journal of Radiology* 22 (1949), 300–308; N. Knight, "The 'New Light': X-rays and Medical Futurism," in J. J. Corn, ed., *Imagining Tomorrow: History, Technology and the American Future* (Cambridge: MIT Press, 1986), 10–34; J. Howell, *Technology in the Hospital: Transforming the Patient Care in the Early Twentieth century* (Baltimore: John Hopkins University Press, 1995), Ch. 5: "The X-ray Image: Meaning, Gender, and Power," 133–168.

14. H. W. Cattell, "Roentgen's Discovery—Its Application in Medicine," *Medical News* 68 (1896), 169–170.

15. B. H. Lerner, "The Perils of 'X-ray Vision': How Radiographic Images Have Historically Influenced Perception," *Perspectives in Biology and Medicine* 35 (1992), 382–397; B. Pasveer, "Knowledge of Shadows: The Introduction of X-ray Images in Medicine," *Sociology of Health and Illness* 3 (1989), 360–381.

16. A. E. Barclay, "The Old Order Changes," *British Journal of Radiology* 22 (1949), 300; Editorial, "A New Kind of Light," *Journal of the American Medical Association* (Feb. 15, 1896). Reprinted in *Radiology* 45 (1945), 436.

17. *Colorado Medical Journal* 2 (1896), 88, quoted in Shikes, *Rocky Mountain Medicine*, 136.

18. Ibid., 152.

19. Ibid., 154–155.

20. R. Brecher, and E. Brecher, *The Rays: A History of Radiology in the United States and Canada* (Baltimore: Williams and Wilkins, 1969), 64–65.

21. Shikes, *Rocky Mountain Medicine*, 136.

22. R. Arns, "The High-Vacuum X-ray Tube: Technological Change in Social Context," *Technology and Culture* 38 (1997), 582–590.

23. Prof. Röntgen had already advanced the tripartite classification of the x-rays to soft, medium, and hard in his original lecture. See W. Röntgen, "On a New Kind of Ray," *Nature* 53 (Jan. 23, 1896), 274.

24. Sixteen years later, this kind of discourse was still dominating the field, see A. J. Quimby, "Laboratory Notes on Radiography," *Post Graduate* 27 (1912), 103–115, 174–189.

25. W. J. Morton, *The X-ray, or Photography of the Invisible and its Value to Surgery* (New York, 1896), 79–133.

26. P. C. Hodges, "Development of Diagnostic X-ray Apparatus During the First Fifty Years," *Radiology* 45 (1945), 439.

27. Withers, "The Story of the First Evidence," 99.

28. W. D. Coolidge and E. E. Charlton, "Roentgen-Ray Tubes," *Radiology* 45 (1945), 449–466.

29. *Rocky Mountain Daily News* (Dec. 9, 1896), 8.

30. L. Daston and P. Galison, "The Image of Objectivity," *Representation* 40(1992), 81-128, on 81.

31. J. Mnookin, "The Image of Truth: Photographic Evidence and the Power of Analogy," *Yale Journal of Law and the Humanities* (1998), 10:1–74.

32. Anonymous, "The Legal Relations of Photographs," *American Law Register* (Jan. 1869), 6; Eborn v. Zimpleman, *American Reports* 26 (1877), 315; Rulloff v. People, *New York Reports* 45 (1771), 224.

33. Anonymous, "The Photograph as a False Witness," *Virginia Law Review* 10 (1886), 644.

34. Mnookin, "The Image of Truth," 41–42.

35. The 1880s saw the spread of the dry plate, which simplified further the process of photographing and allowed for the flourishing of amateur photography.

36. Cowley v. the People, *New York Reports* 83 (1881), 478.

37. "The Rule is that a witness may use a plate, diagram, or map, made in any way to explain or make himself intelligible to a jury, though it cannot go to them as evidence." Quoted in John H. Wigmore, *A Treatise on the System of Evidence in Trials at Common Law* (Boston: Little, Brown, 1904), §791; Campbell v. State, *Alabama Reports* 23 (1853), 83.

38. Cowley v. the People, 478.

39. Editorial, "Photographs as Evidence," *Minnesota Law Review* 2 (1894), 91–96; G. Lawyer, "Photographs as Evidence," *Central Law Journal* 41 (1895), 52–56; S. Kenner, "Photographs as Evidence," *Central Law Journal* 60 (1905), 406–410.

40. Mnookin, "The Image of Truth." 45–50.

41. In most states, the trial judge was not allowed to comment on the weight of the evidence to the jury. E. Sunderland, "The Inefficiency of the American Jury." *Michigan Law Review* 13 (1914), 307–309.

42. Wigmore, *A Treatise on the System of Evidence*, §790.

43. Mnookin, "The Image of Truth," 45–50. Mnookin also argues that instead of diminishing the value of photographic evidence, thinking of the photograph in terms of illustrative evidence had raised the evidentiary status of the whole category. Mnookin, "The Image of Truth," 58-62.

44. M. J. Stern, "Report of Work with Roentgen Rays at the Polyclinic Hospital," *Medical and Surgical Reporter* (Philadelphia), 75 (1896), 676–79, 678–79; A. W. Fuchs, "Evolution of Roentgen Film," *American Journal of Radiology* 75 (1956), 30–47.

45. Quoted in Withers, "The Story of the First Evidence," 100.

46. *Canadian Medical Review* 3 (1896), 102; *British Journal of Photography* 43 (Mar. 20, 1896), 179; Editorial, "New Photography in the Courts," *Literary Digest* 12 (Apr. 11, 1896), 707; Anonymous, "A County Court Judge and the Roentgen Rays," *British Journal of Photography* 43 (Jul. 17, 1896), 461, 683.

47. M. D. Howe, "Juries as Judges of Criminal Law," *Harvard Law Review* 52 (1939), 582–616; I. A. Horowitz, "Changing Views of Jury Power: The Nullification Debate, 1787–1988," *Law and Human Behavior* 15 (1991), 165–182.

48. *Rocky Mountain Daily News* (Dec. 3, 1896), 7; Withers, "The Story of the First Evidence," 99–100.

49. *Rocky Mountain Daily News* (Dec 3, 1896), 7.

50. Withers, "The Story of the First Evidence," 100; *Rocky Mountain Daily News* (Dec 3, 1896), 7.

51. Smith v. Grant, *Chicago Legal News* 29 (1896), 145.

52. Cowley v. the People, 478; Franklin v. State.

53. Smith v. Grant.

54. Late nineteenth-century malpractice law bound the medical practitioner to the ordinary degree of care, skill, and diligence as exercised by those in the same line of work in his or her locality. C. Bell, "Malpractice," *New Jersey Law Journal* 16 (1893), 103–105; eight jurors were in favor of Smith, four for Grant. See Editorial, "Grant Jury Discharged: Was Unable to Reach a Verdict," *Daily News* (Dec. 15, 1896), 5.

55. Editorial, "How Shall Physicians Protect Themselves," *Colorado Medical Journal* 3 (1897), 34–35; editorial, "Indemnity Bond for Plaintiffs in Damage Suits," ibid., 35–36, 115, 122.

56. Tal Golan, "The Authority of Shadows: The Legal Embrace of the X-ray." *Historical Reflections* 22 (1998), 437–458; Bruce v. Beall; *Southwestern Reports* 41 (1897), 445, reported also in *Central Law Journal* 45 (1897), 185.

57. Anonymous, "First Radiograph in Evidence," *American X-ray Journal* 2 (1898), 155; W. W. Goodrich, "The Legal Status of the X-ray," *Brooklyn Medical Journal* 17 (1903), 515–517; O. F. Scott, "Röntgenograms and their Chronological Legal Recognition," *Illinois Law Review* 24 (1929), 674–679; V. P. Collins, "Origins of Medico-legal and Forensic Roentgenology," in A. Bruwer, ed., *Classic Descriptions in Diagnostic Roentgenology* (Springfield, Ill.: Thomas, 1964), 2 vols., 2:1578–1604; E. C. Halperin, "X-rays at the Bar, 1896–1910," *Investigative Radiology* 23 (1988), 639–646; Wigmore, *A Treatise on the System of Evidence*, §795, note 3.

58. Bruce v. Beall, 445.

59. Miller v. Dumon, *Pacific Reporter* 64 (1901), 804.

60. Elzig v. Bales, *Iowa Reports* 135 (1907), 208.

61. De Ville, *Medical Malpractice*, 138–155; E. A. Tracy, "The Fallacies of X-ray Pictures," *Journal of the American Medical Association* 29 (1897), 949–951.

62. Editorial, "Discussing the X-ray: Judge Lefevre's Decision Is Said to Be Far Reaching," *Rocky Mountain Daily News* (Dec. 9, 1896), 8.

63. W. White, "Röntgen Rays in Surgery," *Transactions of the American Surgical Association* (1897), 70. For similar discussions at the 1897 annual meeting of the American Medical Association, see C. L. Leonard, "The Application of the Roent-

gen Rays to Medical Diagnosis," *Journal of the American Medical Association* 29 (1897), 1157–1158, and Deforest Willard, "Röntgen Ray Skiagraphy," Ibid., 30 (1898), 1016.

64. White, "Röntgen Rays in Surgery," 70.

65. Ibid., 83–84,88.

66. *Third Annual Report of the Hospital Funds of the Denver and Rio-Grande Railroad Company* (Mar. 31, 1887), quoted in Shikes, *Rocky Mountain Medicine*, 202.

67. Fairbrother, "Legal Railway Surgery," *Railway Surgeon* 8 (1902), 254.

68. R. H. Reed, "The X-ray from a Medico-Legal Standpoint," *Journal of the American Medical Association* 30 (1898), 1013–1019, on 1016.

69. Dr. Grant was talking from personal experience. He was the first American physician to be sued successfully on the basis of x-ray photographs. See previous "Malpractice" section.

70. Reed (ref. 102), 1017–1019.

71. H. P. Pratt, "X-ray Essentials—The Value of X-ray from a Diagnostic and Therapeutic Standpoint," *American X-ray Journal* 4 (1899), 546.

72. F. W. Ross, "The X-ray in Forensic Medicine," *American X-ray Journal* 4 (1899), 502–503. Dr. Ross repeated the same warning before the prestigious New York Medico-Legal Society. See *Medico-Legal Journal* 16 (1898), 142–150.

73. "Report of the Committee of the American Surgical Association on the Medico-Legal Relations of the X-rays," *American Journal of the Medical Sciences* 120 (1900), 7–36.

74. W. W. Grant, "Elbow Fractures and the X-ray," *Journal of the American Medical Association* 36 (1901), 777–780.

75. C. Beck, "Errors Caused by False Interpretation of the Roentgen Rays, and their Medico-Legal Aspects," *Medical Record* 58 (1900), 285.

76. Carl Beck, *Roentgen Ray Diagnosis and Therapy* (New York: Appleton, 1904), 339. The surgeons, Beck suggested elsewhere, following the 1900 resolution of the American Surgical Association, "are afraid that the X-rays will disclose their errors, and hence they regard the rays as an enemy instead of availing themselves to them early." C. L. Leonard, "What Reliance Can be Placed Upon the Image Produced by the X-ray from a Medico-Legal Standpoint," *Proceedings of the Medical Society of New-Jersey* (1901), 277.

77. C. L. Leonard, "What Reliance Can be Placed Upon the Image Produced by the X-ray," 268–269, 275. See also Leonard's 1905 presidential address to the American Röntgen Ray Society: "The Past, Present, and Future of the Röntgen Ray," *American Medicine* 10 (1905), 1082–1085.

78. M. Kassabian, "The Roentgen Rays in Forensic Medicine," *Medico-Legal Journal* 1 (1901), 407, 416. See also Kassabian's 1904 address before to the American Electro-Therapeutical Association: "The Value of the Roentgen Rays in the Diagnosis of Fractures," *Archives of the Roentgen Ray* 9 (1904), 142–146.

79. Cf. R. Brecher and E. Brecher, *The Rays: A History of Radiology in the United States and Canada* (Baltimore: Williams and Wilkins, 1969). For a foreign perspective on the rising American specialty, see the report of a German doctor on his American tour. Paul Krause, "Zur Kenntnis der Röntgenologie in den Verein-

igten Staaten von Nord Amerika," *Fortschritte auf dem Gebiete der Röngenstrahlen* 13 (1909), 326–333.

80. S. Monell, *Studies in X-ray Diagnosis* (New York: Pelton, 1902), 170–171; S. Lange, "The Present Status of the Roentgen Ray," *Lancet-Clinic* (Jan. 26, 1907), 79. Compare this with Wigmore comment on regular photography that heads this chapter.

81. Lange, "The Present Status of the Roentgen Ray," 82. Variations of this slogan could be found in many articles of the period. For example, "The Roentgen rays never lie, but that it is entirely our own imperfections which induce us to err under peculiar circumstances," in Beck, "Errors Caused by False Interpretation of the Roentgen Rays," 285.

82. Kassabian, "The Roentgen Rays in Forensic Medicine," 408.

83. Quoted in R. Carman, "Medical Roentgenology as a Specialty," *Proceedings of the Missouri State Medical Association* 7 (1910–11), 122.

84. G. H. Stover, "The Professional Position of the Röntgenonolgist," *New York Medical Journal* 91 (1910), 17. See also Lange's advice: "Under no circumstances should the plate or print be put in the hands of the patient, because of the readiness with which such evidence lends itself to unscrupulous criticism and manipulation." Lange, "The Present Status of the Roentgen Ray," 82. See also H. Albers-Schoenberg, "The Roentgenologist is a Medical Specialist, and All Roentgen Plates, Prints, Tracing, and other Documents are his Sole Property," *Archives of the Roentgen Ray* 19 (1914), 94–97.

85. E. H. Skinner, "The Ownership of X-ray Plate," *The Modern Hospital* 1 (1913), 31.

86. E. Eliot, "The Legal Responsibility to the Surgeon and Practitioner Which the Use of the X-ray Involves," *Annals of Surgery* (Philadelphia), 3 (1916), 483.

87. For an extensive list of cases, see L. P. Wilson, "The X-ray in Court," *Cornell Law quarterly* 7 (1922), 215, note 39.

88. For an extensive list of cases, see ibid., note 40.

89. Lang v. Marshalltown L. & R. Co., *Iowa Reports* 185 (1919), 940. See also Daniels v. Iowa City, *Iowa Reports* 191 (1921), 811.

90. Lange, "The Present Status of the Roentgen Ray," 79; Stover, "The Professional Position of the Röntgenonolgist," 16; idem, "Medico-Legal Value of the X-ray," *Medical Journal* (Philadelphia), (1898), 801–802.

91. Marion v. Coon Construction Co., *New York Reports* 216 (1915), 178.

92. C. Scott, *Photographic Evidence: Preparation and Presentation* (Kansas City, MO: Vernon Law Book Co. 1942), §269.

93. The paradigmatic case was People v. Doggett, *California Appeals* 83 (1948) 2nd ed., 405. In that case a husband and wife were convicted of oral sex perversion. The only evidence introduced at the trial was a photograph taken by the defendants in *flagrante delicto*. No verifying witness was available of course, but the photograph was admitted anyway. See also Hartley v. A. I. Rudd Lumber Co., *Michigan Reports* 282 (1937), 652; Carner v. St. Louis-San Francisco Ry. Co., *Southwestern Reports* 89 (1935), 947; Watkins v. Reinhardt, *Alabama Reports* 293 (1942), 243. Lohman v. Wabash, *Southwestern Reports* 269 (1954), 885.

94. The legal compilation, *Corpus Juris*, suggested a new evidentiary category—best secondary evidence—in order to describe the status of photographs in court. W. Mack, ed., *Corpus Juris*, (New York: American Law Book Co., 1914–1922), 72 vols., 22:992. See also J. Anderson, "The Admissibility of Photographs as Evidence," *North Carolina Law Review* 7 (1929), 443–449; Scott, *Photographic Evidence*, §1; J. Mouser, "Photographic Evidence—Is There a Recognized Base for Admissibility," *Hastings Law Journal* 8 (1956), 310–314.

95. J. McKelvey, *Handbook on the Law of Evidence* (St. Paul, Minn.: West Pub., 1944), 5th ed., §380, §669.

96. D. S. Gardner, "The Camera Goes to Court," *North Carolina Law Review* 24 (1946), 244, 246.

97. At first some courts tended to exclude motion pictures. See Gibson v. Gunn, *New York State Reports* 202 (1923), 19. The court remarked that a certain film, which showed a vaudeville actor perform prior to his injury, "tended to make farce of the trial."

98. The leading cases were: United States v. Hobbs, *Federal Reports* 403 (1968) 2 ed., 977; United States v. Taylor, Ibid. 530 (1976), 639; United States v. Calyton, Ibid. 643 (1981), 1071. See also cases where the incriminating photographs were taken by the defendants: Bergner v. State, *Northeastern Reports* 397 (1979), 2 ed., 1012; United States v. Stearns, *Federal Reports* 550 (1977), 1167; State v. Holderness, *Northwestern Reports* 293 (1980), 226.

99. E. Olson, "Case Note: Evidence—Adoption of the 'Silent Witness Theory'—Bergner v. State," *Indiana Law Review* (1980), 1025–1053; J. McNeal, "Silent Witness Evidence in Relation to the Illustrative Evidence Foundation," *Oklahoma Law Review* 37 (1984), 219–244; B. Madison, "Seeing Can Be Deceiving: Photographic Evidence in a Visual Age—How Much Weight Does it Deserve?" *William and Mary Law Review* 25 (1984), 705–742.

100. Mnookin, "The Image of Truth;" J. Dumit, "Objective Brains, Prejudicial Images," *Science in Context* 12 (1999), 173–202; R. Brain and D. Broderick, "The Derivative Relevance of Demonstrative Evidence: Charting its Proper Evidentiary Status," *University of California at Davis Law Review* 25 (1992), 957–1012.

6. Science Unwanted: The Law and Psychology

1. By 1909, according to one account, 60 percent of the cases brought before the Massachusetts Superior Court used expert testimony. H. W. Smith, "Scientific Proof," *Southern California Law Review* 16 (1943), 122.

2. H. N. Scheiber, "The Impact of Technology on American Legal Development, 1790–1985," in J. Colton and S. Bruchey, eds., *Technology, the Economy, and Society: The American Experience* (New York: Columbia University Press, 1987), 83–124. On Pollution, see F. Quivik, *Smoke and Tailings: An Environmental History of Copper Smelting Technologies in Montana, 1880–1930* (Ph.D. dissertation, University of Pennsylvania, 1998); On mining, C. Spence, *Mining Engineers and the American West* (New Haven: Yale University Press, 1970), Ch. 6: "We Must Overwhelm

Them with Testimony," 195–230. On food and drug adulteration, Mitchell Okun, *Fair Play in the Marketplace: The First Battle for Pure Food and Drugs* (Dekalb, Ill.: Northern Illinois University Press, 1986). On malpractice, K. A. De Ville, *Medical Malpractice in Nineteenth-Century America: Origins and Legacy* (New York: New York University Press, 1990). On steamboats explosions, R. Andrist, *Steamboats on the Mississippi* (New York, American Heritage Pub. Co, 1962). On agriculture, J. Lawson, *The Law of Expert and Opinion Evidence Reduced to Rules: With Illustrations from Adjudged Cases* (St. Louis, 1883), Ch. 2: "Farmers and Agriculturalists," 13–25; On insurance, ibid., Ch. 3: "Insurers and Insurance," 26–49; John H. Wigmore, "Expert Opinions as to Insurance Risk," *Columbia Law Review* 2 (1902), 67–78. On handwriting, A. Hayward, *The Evidence of Handwriting, with preface and collateral evidence by Hon. Edward Twisleton* (Cambridge, 1874); Evan B. Lewis, *The Law of Expert Testimony* (Philadelphia, 1894), Ch. 5 "Handwriting," 28–34. Voluminous literature exits on patents, but see the personal account by A. Browne, "Patent Litigation from the Expert's Standpoint," *Proceedings of the American Bar Association* 25 (1902), 670–674.

3. These are but some of the headings of Hans Gross' popular handbook *Criminal Investigation: A Practical Handbook for Magistrates, Police Officers, and Lawyers* (1897; Madras: Krishnamachari, 1906). Anthropometry was the science of identification by the measurements of the size and proportions of the human body.

4. Cf. Gross, *Criminal Investigation*. On the methods of reconstructing corpses see K. Pearson, "On the Reconstruction of the Stature of Prehistoric Races," *Philosophical Transactions of the Royal Society* 192 (1898), 169. On forensic chemistry, mainly in poisoning cases, J. Mohr, *Doctors and the Law: Medical Jurisprudence in Nineteenth-Century America* (Oxford: Oxford University Press, 1993). On insanity pleas in murder trials, J. A. Tighe, *A Question of Responsibility: The Development of American Forensic Psychiatry, 1838–1930* (Ph.D. dissertation, University of Pennsylvania, 1983). On microscopy and radiology, see Chapters 4 and 5 of this book. On fingerprints and anthropometry, see S. Cole, *Suspect Identities: A History of Fingerprinting and Criminal Identification* (Cambridge, MA: Harvard Univeristy Press, 2002).

5. T. L. Haskell, *The Authority of Experts: Studies in History and Theory* (Bloomington: Indiana University Press, 1984), part 3, 180–241.

6. O. W. Holmes, "The Path of the Law," *Harvard Law Review* 10 (1897), 469; The judicial willingness to rely on nondoctrinal sources of information about the real world was first legitimized by the readiness of the U.S. Supreme Court in 1908 to rely on the "Brandeis brief," a party brief that employed, among other things, economic, sociological and statistical analyses. See, Muller v. Oregon, *United States Supreme Court Reports* 204 (1908), 412.

7. J. Christison, *Tragedy of Chicago: How an Innocent Young Man was Hypnotized to the Gallows* (Chicago: 1906), 95 pp.

8. *Chicago Tribune* (Jan. 13–16, 1906).

9. Ibid. (Jan. 15–16, 1906).

10. Quoted in the trial records, *Tragedy of Chicago*, 16.

11. Ibid., 14.

12. H. Goddard, *Feeble Mindedness: Its Cause and Consequences* (New York: Macmillan, 1914); F. Galton, *Inquiries Into Human Faculty and Its Development* (London, 1883); C. Lombroso, *L'Homme Criminel* (Paris, 1895); H. Maudsley, *Responsibility in Mental Disease* (London, 1874).

13. L. M. Terman, *The Measurement of Intelligence* (Boston: Houghton, 1916), 11; H. Goddard, *The Kallikak Family: A Study in the Heredity of Feeble Mindedness* (New York: Macmillan, 1912), 101–102.

14. *Chicago Tribune* (Jan. 14, 1906), 1; Christison, *Tragedy of Chicago*, 43, 76.

15. *Chicago Chronicle* (Jun. 23, 1906), 4; *The Public* (Jun. 30, 1906), 1; Christison, *Tragedy of Chicago*, 60.

16. A. Binet, *Alterations in Personality* (New York, 1896); B. Sidis, *The Psychology of Suggestion* (New York, 1898), with a preface by William James; J. M. Bramwell, *Hypnotism* (London: Richards, 1903); J. Grasset, *L'hypnotisme at la suggestion* (Paris, Doin, 1903); W. D. Scott, "Personal Differences in Suggestibility," *Psychology Review* 15 (1908), 147–154.

17. Quoted in Christison, *Tragedy of Chicago*, 51; *Chicago Tribune* (Jun. 13, 1906), 4.

18. *Chicago Evening Journal* (Jun. 14, 1906); *Chicago Tribune* (Jun. 21 and 23, 1906); H. Münsterberg, *Science on the Witness Stand* (New York: McClure's, 1908), 141–142.

19. *Chicago Tribune* (Jun 23, 1906), 4; Editorial, "The 'Third-Degree' and the Position of the Trial Judge in Illinois," *Illinois Law Review* 7 (1912), 303–310.

20. C. Klein, *The Third Degree: A Play in Four Acts* (New York: French, 1908); H. Münsterberg, "Untrue Confessions," *Times Magazine* (Jan. 1907).

21. M. Münsterberg, *Hugo Münsterberg: His Life and Work* (New York: Appleton, 1922); M. Hale., *Human Science and Social Order: Hugo Münsterberg and the Origins of Applied Psychology* (Philadelphia, Temple University Press, 1980); P. Keller, *States of Belongings: German-American Intellectuals and the First World War*, (Cambridge, MA: Harvard University Press, 1979).

22. A. Tuchman, A., *Science, Medicine and the State in Germany: The Case of Baden, 1815–1871*, (Oxford: Oxford University Press, 1993); T. Lenoir, "Science for the Clinic: Science Policy and the Formation of Carl Ludwig's Institute in Leipzig," in W. Coleman and F. Holmes, eds., *The Investigative Enterprise: Experimental Physiology in Nineteenth-Century Medicine* (Berkeley: University of California Press, 1988), 100–138.

23. W. Wundt, "Philosophy in Germany," *Mind* 2 (1877), 493–518.

24. R. Kremer, *The Thermodynamics of Life and Experimental Physiology, 1770–1880* (New York: Garland, 1990); T. Lenoir, *The Strategy of Life: Teleology and Mechanics in Nineteenth-Century German Biology* (Dordrecht: Reidel, 1982); P. Cranefield, "The Organic Physics of 1847 and the Biophysics of Today," *Journal of the History of Medicine and Allied Sciences* 12 (1957), 407–423; D. Galaty, "The Philosophical Basis of Mid-Nineteenth Century German Reductionism," *Journal for the History of Medicine and Allied Sciences* 29 (1974) 295–316.

25. Approximately one-fourth of Johannes Müller's *Handbuch der Physiologie des Menschen für Vorlesungen* (Coblenz: Holscher, 1833–37), which systematically

summarized the physiological knowledge of the day, was dedicated to the physiology of the senses and nervous system.

26. A good entry point for these developments, including an extensive bibliography of primary sources, can be found in E. G. Boring, *A History of Experimental Psychology* (New York: Appleton, 1950), 2nd ed., 27–116.

27. E. Cassirer, *The Philosophy of the Enlightenment* (Princeton: Princeton University Press, 1979), Ch. 3, "Psychology and Epistemology," 93–133.

28. D. Robinson, *William Wundt and the Establishment of Experimental Psychology, 1875–1914: The Context of a New Field of Scientific Research* (Ph. D. dissertation, University of California at Berkeley, 1987), 61. Wundt himself attributed his appointment to personal connections.

29. Wundt's *Grundzüge der physiologischen Psychologie* (Leipzig, 1873–1874), which was published just before his appointment in Leipzig, went through six editions, grew in size from one to three volumes, and served as the definitive textbook for the first generation of experimental psychologists.

30. W. G. Bringmann, and G. A. Ungerer, "The Foundation of the Institute of Experimental Psychology at Leipzig University," *Psychological Researcher* 42 (1980), 5–18; K. Danziger, "Wundt's Psychological Experiment in Light of his Philosophy of Science," ibid., 109–122; T. Mischel, "Wundt and the Conceptual Foundations of Psychology," *Philosophy and Phenomenological Research* 31 (1970), 1–26.

31. Münsterberg's laboratory was the fourth created in Germany, after Leipzig (1879), Göttingen(1887), and Breslau (1888). Münsterberg's major publications in these years were: *Die Willenshandlung: Ein Beiträg zur Physiologischen Psychologie* (Freiburg, 1888) and *Beiträge zur experimentellen Psychologie* (Freiburg: 1889–1892), 4 vols.

32. F. Ringer, *The Decline of the German Mandarins: The German Academic Community 1890–1933* (Cambridge, MA: Harvard University Press, 1969), 117, 295.

33. This rudimentary description cannot do justice to Münsterberg's complex philosophy. For the most accessible accounts see H. Münsterberg, "Psychology and Life," in idem, *Psychology and Life* (Boston, 1899), 1–34; Hale, *Human Science and Social Order,* 31–44; B. Kuklick, *The Rise of American Philosophy: Cambridge, Massachusetts 1860–1930* (New Haven: Yale University Press, 1977), 196–214.

34. T. J. Merz, *A History of European Thought in the Nineteenth Century* (London: Blackwood, 1904), 2 vols., 1:518, 521. One contemporary commentator argued that it was Münsterberg's experiments more than anything else that contributed to the decline of Wundt's program. W. Heinrich, *Die moderne physiologische Psychologie* (Zürich, 1895), 154.

35. E. Tichener, "Dr. Münsterberg and Experimental Psychology," *Mind* 16 (1891), 534.

36. For the source of this tradition see J. Lock, *Essay Concerning Human Understanding* (London, 1721). For leading nineteenth-century representatives see A. Bain, *The Emotions and the Will* (London, 1859) and G. F. Stout, *A Manual of*

Psychology (London, 1899). Münsterberg, of course, did not subscribe to utilitarian ethics.

37. C. Robertson, "Dr. Münsterberg on Apperception," *Mind* 15 (1890), 243–244; idem, "Münsterberg on 'Muscular Sense' and 'Time Sense'," ibid., 524–536.

38. W. James, "What Is an Emotion?" *Mind* 9 (1884), 188–205; idem, *Principles of Psychology* (New York, 1890), 2 vols., 2:442–485; idem, "A Plea for Psychology as a Natural Science," *The Philosophical Review* (1892), 146–153.

39. James, *Principles of Psychology*, 2:486–492, 505n. James sent some of his advanced students to complete their training in Münsterberg's laboratory, for example, E. D. DeLabrre, who later founded the Department of Psychology at Brown.

40. Quoted in Keller, *States of Belongings*, 26.

41. Hale, *Human Science and Social Order*, 47.

42. K. Danziger, "The Social Origin of Modern Psychology", in A. R. Buss, ed., *Psychology in Social Context* (New York: Irvington, 1979), 18.

43. M. A. Tinker, "Wundt's Doctorate Students and their Theses," *American Journal of Psychology* 44 (1932), 630–637; Boring, *A History of Experimental Psychology*, 344; L. Veysey, *The Emergence of the American University* (Chicago: University of Chicago Press, 1965).

44. E. W. Scripture, *Thinking, Feeling, Doing* (New York, 1895), 24–25; On the warm reaction of the American philosophical orthodoxy to psychology see J. M. O'donnell, *The Origins of Behaviorism* (New York: New York University Press, 1985), 52–65.

45. W. O. Krohn, "Facilities in Experimental Psychology in Colleges of the United States," *in Reports of the Commission on Education* (1891). Compare with idem, "Facilities in Experimental Psychology at the various German Universities," *Journal of American Psychology* 4 (1892), 585–594.

46. Hale, *Human Science and Social Order*, 9, 49; H. Nichols, "The Psychological Laboratory at Harvard," *McClure's Magazine* 1 (1895), 399–409.

47. H. Münsterberg, "The Teacher and the Laboratory: A Reply," *Atlantic Monthly* 81 (1898), 824; idem, *Grundzüge der Psychologie* (Leipzig, 1900), Band I: Allgemeiner Teil: Die Prinzipen der Psychologie.

48. Cf. Hale, Human Science and Social Order; M. G. Ash, "Academic Politics in the History of Science: Experimental Psychology in Germany, 1879–1941," *Central European History* 13 (1981), 255–286; D. Coon, *Inventory of Historical Apparatus from the Psychological Laboratory of Harvard University (1890–1927)*, unpublished manuscript, Harvard Library; J. Spillmann and L. Spillmann, "The Rise and Fall of Hugo Münsterberg," *Journal of the History of the Behavioral Sciences* 29 (1993), 322–338, on 326

49. H. Münsterberg, *American Traits from the Point of View of a German* (Boston: Houghton, 1901), v–vi; H. Münsterberg, *Die Amerikaner* (Berlin: 1904), translated by his student E. Holt as *The Americans* (New York: McClure, 1904), vii–ix.

50. H. Münsterberg, "Untrue Confessions," *Times Magazine* (Jan. 1907);

"On the Witness Stand," ibid. (Mar. 1907); "Suggestion in Court," *Reader Magazine* (Apr. 1907).

51. I. Kant, *Metaphysical Foundations of Natural Science* (1786; Indianapolis: Bobbs-Merrill, 1970), 6–8; A. Comte, *Positive Philosophy* (1830–42; London, 1853), 1:458–462; James, *Principles of Psychology*, 1:187–192; J. F. Herbart, *A Text Book in Psychology* (1824; New York, 1897), 3. Wundt preferred to call the procedure used in his laboratory *internal perception(innere Wahrnehmung)* to differentiate it from the more "primitive" procedure of *introspection(Selbstbeobachtung)*. W. Wundt, "Selbstbeobachtung und innere Wahnehmung," *Philosophische Studien* 4 (1888), 1–12.

52. G. T. Ladd, *Psychology: Descriptive and Explanatory* (New York, 1894), 23; E. B. Titchener, "Prolegomena to a Study of Introspection," *American Journal of Psychology* 23 (1912), 427–508; James, *Principles of Psychology*, 1:182, 185.

53. D. Coon, "Standardizing the Subject: Experimental Psychologists, Introspection, and the Quest for a Technoscientific Ideal," *Technology and Culture* 34 (1993), 757–783.

54. For a mature expression of this interest see E. H. Weber, *Tastsinn und Gemeingefühl*, (Leipzig, 1846).

55. C. Jungnickel, "Teaching and Research in the Physical Sciences and Mathematics in Saxony, 1820–50," *Historical Studies in the Physical Sciences* 9 (1979), 37; C. Stumpf, "Hermann von Helmholtz and the New Psychology," *Psychological Review* 2 (1895), 1–13; S. R. Turner, "Hermann von Helmholtz and the Empiricist Vision," *Journal of the History of the Behavioral Sciences* 13 (1977), 48–58; idem, "Helmholtz, Sensory Physiology, and the Disciplinary Development of German Psychology," in W. Woodward and M. G. Ash, eds., *The Problematic Science: Psychology in Nineteenth Century Thought* (New York: Fraeger, 1982), 147–166.

56. E. C. Sanford, "Personal equation," *American Journal of Psychology* 2 (1888), 3–38, 271–298, 403–430; I. Kirsch, "The Impetus to Scientific Psychology: A Recurrent Pattern," *Journal of the History of the Behavioral Sciences* 12 (1976), 12–29; G. Zilboorg and G. Henry, *A History of Medical Psychology* (New York: Norton, 1941), Ch. 9: "the Discovery of Neuroses," 342–378.

57. W. Stern, "Psychology of Testimony," *Journal of Abnormal and Social Psychology* 34 (1939), 4; M. K. Matsuda, *The Memory of the Modern* (Oxford: Oxford University Press, 1996), Ch. 5: "Testimonies: Deserving of Faith," 101–141; S. Felman and D. Laub, *Testimony: Crises of Witnessing in Literature, Psychoanalysis, and History* (New York: Routledge, 1992); G. Gangèbe, *Du faux témoignage* (Paris, 1900).

58. A. Schrenck-Notzing, *Über Suggestion und Erinnerungsfälschung im Berchthold-Prozess* (Leipzig, 1897). The quote is from Hale, *Human Science and Social Order*, 113.

59. H. Gross, *Criminal Psychology: A Manual for Judges, Practitioners, and Students* (1897; Boston: Little and Brown, 1911), 3, 492.

60. A. Binet, *La Suggestibilité* (Paris, 1900); W. Stern, "Zur Psychologie der Aussage: Experimentelle Untersuchungen über Erinnerungstreue," *Zeitschrift für die gesammte Strafrechtswissenschaft* 22 (1902), 1–14. "Aussage Studium," *Beiträge zur Psychology der Aussage* 1 (1903–4), 46–78. Many followed Stern's paradigmatic experiments, for example, A. Wreschner, "Zur Psychologie der Aussage," *Archiv für*

gesammte Psychologie 1 (1902), 148–166; O. Lipmann, "Experimentelle Aussagen über einen Vorgang und eine Lokalität," *Beiträge zur Psychology der Aussage* 1 (1903–4), 90–115.

61. F. Galton, "Psychometric Experiments," *Brain* 2 (1879), 149–162; M. Wertheimer and J. Klein, "Psychologische Tatbestanddiagnostik," *Archiv für Kriminal-Anthroplogie und Kriminalistik* 15 (1904), 72–76; "Experimentelle Untersuchungen zur Tatbestanddiagnostik," *Archiv für gesammte Psychologie* 6 (1905), 72; K. Jung. "Die psychologische Diagnose des Tatbestandes," *Zeitschrift für Schweizeriche Straftrecht* 18 (1905), 369–394. For the connection to psychoanalysis see S. Freud, "Psychoanalysis and the Ascertaining of Truth in Courts of Law," first published in the *Archiv für Kriminalanthropologie und Kriminalistik* 26 (1906), and reprinted in P. Rieff, ed., *Collected Papers of Sigmund Freud* (New York: Collier, 1963), 1:115–125; M. Wertheimer et al., "Carl Jung and Max Wertheimer on a Priority Issue," *Journal of the History of the Behavioral Sciences* 28 (1992), 45–56.

62. After 1907 the journal continued under the name *Zeitschrift für angewandte Psychologie und psychologische Sammelfurschung*. Gross' older *Archiv für Kriminal-Anthropologie und Kriminalistik* and the *Zeitschrift für Pädagogische Psychologie* also served as important organs.

63. A. Binet, "La science du témoignage," *L'année Psychologique* 11 (1904), 128–136, and 12 (1906) 230–274; É., Claparède, "La fidélité et l'éducabilité du témoignage," *Archive des Sciences Physiques et Naturalles* (Apr. 7, 1904), 44–61; "La psychologie judiciare", *L'année Psychologique* 11 (1905), 128–136 and 12 (1906), 275–302; idem, "Expériences collectives sur le témoignage," *Archives de Psychologie* 5 (1906), 344–387.

64. See Stern's summary of his own work for the American audience at the famous Clark Conference, where Sigmund Freud also spoke. W. Stern, "Lectures on the Psychology of Testimony," *American Journal of Psychology* 21 (1910), 271–282.

65. G. M. Beard, "The Scientific Study of Human Testimony," *Popular Science Monthly* 13 (1878), 53–64, 173–183, 328–338; J. Cattell, "Measurements of Accuracy of Recollection," *Science* 2 (1895), 765–766; G. F. Arnold, *Psychology Applied to Legal Evidence* (Calcutta and New York; Thacker and Spink, 1906), 6; G. M. Whipple, "The Observer as a Reporter: A Survey of the Psychology of Testimony," *Psychological Bulletin* 6 (1909), 154.

66. D. H. Grover, *Debaters and Dynamiters: The Story of the Haywood Trial* (Corvallis: Oregon State University Press, 1964); M. Münsterberg, *Hugo Münsterberg*, 144; H. Münsterberg, "Experiments with Orchard," Unpublished manuscript, quoted in Hale, *Human Science and Social Order*, 117.

67. Hale, *Human Science and Social Order*, 117.

68. M. Münsterberg *Hugo Münsterberg*, 148–149.

69. Editorial, "Precipitate Psychology," *New York Times* (Jul. 5, 1907); letter to the *New York Evening Post* (Jul. 12, 1907), Münsterberg Collection, Mss. Acc. 2374, Boston Public Library; Justice Grier in *Livingston vs. Jones, Federal Reports*, 15:668.

70. Letter to Clarence Darrow (Jul. 14, 1907), Münsterberg Collection, Mss. Acc. 2311; Letter to *McClure's Magazine* (Jul. 14, 1907), Münsterberg Collection,

Mss. Acc. 2358; H. Münsterberg, "Letter to the Editor," *Nation* 85 (18 Jul., 1907), 55.

71. H. Münsterberg, "Nothing but the Truth," *McClure's Magazine* (Sep. 1907), 533.

72. Charles C. Moore, "Yellow Psychology," *Law Notes* (Oct. 1907), 125; Charles C. Moore, *A Treatise on Facts or the Weight and Value of Evidence* (New York: Thompson, 1908), 2 vols.

73. Moore, "Yellow Psychology," 125.

74. H. Münsterberg, "Yellow Psychology," *Law Notes* (Nov. 1907), 145.

75. The four articles are: "The Third Degree," "Hypnotism and Crime," and "Prevention of Crime," all in *McClure's Magazine*, and "Traces of Emotions" in the *Cosmopolitan*. The quotes are from the later collection of these articles: Münsterberg, *Science on the Witness Stand*, 108, 117, 45, 9–10.

76. Editorial, "Invents Machines for 'Cure of Liars,'" *New York Times* (Sep. 11, 1907). For a taste of Münsterberg's lectures, see "Courts Need Psychologists," *Cornell Daily Sun* (Dec. 6, 1908), 1. For Moore's continuing attacks, see Charles C. Moore, "Psychology in the Courts," *Law Notes* (Jan. 1908), 185–187.

77. J. Jastrow, "Science on the Witness Stand," *Dial* 45 (Jul. 16, 1908), 38; S. Baldwin, "On the Witness Stand," *Science* 29 (Feb 19, 1909), 301–302; *Nation* 86 (1908), 472; P. E. Winter, "Psychological Literature," *American Journal of Psychology* 20 (1909), 136.

78. R. H. Gault, "Memories," Wigmore Collection, Northwestern University Library; Letter to Münsterberg (Nov 11, 1908), Münsterberg Collection, Mss. Acc. 2244.

79. Chief Justice Edward Coke had been a symbol of the legal defiance of external intrusion ever since he had been committed to the Tower of London in 1616 for defying the King. William Blackstone was famous for making Common Law intellectually respectable through his *Commentaries On the Law of England* of 1765–1769.

80. J. H. Wigmore, "Professor Münsterberg and the Psychology of Testimony," *Illinois Law Review* 3 (Feb. 1909), 399–445.

81. Editorial, "Other Experts to Bite 'Em, etc.," *Law Notes* (Jun 1910), 42.

82. S. Greenleaf, *A Treatise on the Law of Evidence* (Boston: Little, Brown, 1899), 16th ed., 3 vols., 1:22.

83. Stern, "Psychology of Testimony," 4.

84. "Professor Langdell's Speech at the 'Quarter Millennial' Celebration of Harvard University," *Law Quarterly Review* 3 (1887), 123–125, on 124; R. A. Cosgrove, *Our Lady the Common Law: An Anglo-American Legal Community, 1870–1930* (New York: New York University Press, 1987), 25–53; M. H. Hoeflich, "Law and Geometry: Legal Science from Leibniz to Langdell," *American Journal of Legal History* 30 (1986), 95–121; M. Speziale, "Langdell's Concept of Law as Science: The Beginning of Anti-Formalism in American Legal Theory," *Vermont Law Review* 5 (1980), 3–37.

85. H. Main, *Ancient Law, its Connection with the Early History of Society and its Relation to Modern Ideas* (London: J. Murray 1906); J. Wigmore, "Planetary Theory

of the Law's Evolution," in A. Kocourek and J. Wigmore, eds., *Formative Influences of Legal Development* (Boston: Little, Brown, 1918), 3 vols., 3:531–541.

86. C. Haney, "Criminal Justice and the Nineteenth-Century Paradigm," *Law and Human Behavior* 6 (1982), 191–235.

87. J. Morse, "The Value of Psychology to the Lawyer," *Case and Comment* 19 (May 1913), 795–799, on 796.

88. J. H. Wigmore, "Science of Criminology," *Proceedings of the Iowa State Bar Association* (1909), 113–123, on 114. The same year Wigmore initiated a national conference on Criminology, which resulted in the formation of a national institute and a journal.

89. J. H. Wigmore, *Principles of Judicial Proof As Given by Logic, Psychology, And General Experience and Illustrated in Judicial Trials* (Boston: Little, Brown, 1913), 1–2.

90. Wigmore, *Principles of Judicial Proof,* 4. The method, presented in §376, captured many of the essential ingredients of modern Data Relational Analysis, whose significance begins only now to be fully appreciated by legal theorists. See P. Tillers and D. Schum, "Charting New Territory in Judicial Proof: Beyond Wigmore," *Cardozo Law Review* 9 (1988), 907–966.

91. W. Twining, *Theories of Evidence: Bentham and Wigmore* (London: Weidenfeld and Nicolson, 1985).

92. Hutchins, "The Law and the Psychologists," *Yale Review* 16 (1927), 678–690, on 678; H. Cairns, *Law and the Social Sciences* (New York: Harcourt, 1935), 169.

93. J. A. Larson, *Lying and Its Detection* (Chicago: University of Chicago Press, 1932), 65–94; L. Clendening, "The History of Certain Medical Instruments," *Annals of Internal Medicine* 4 (1931), 176–189; S. W. Mitchell, "The Early History of Instrumental Precision in Medicine," *Transactions of the Congress of American Physicians and Surgeons* 2 (1891), 159–198.

94. P. Trovillo, "A History of Lie Detection," *Journal of the American Institute of Criminal Law and Criminology* 29 (1938–39), 849–881 and 30 (1940), 105–119; Lombroso, *L'Homme Criminel,* 336–346; J. Tarchanoff, "Über die galvanisch Erscheinungen," *Pflüger's Archives* 46 (1890), 46; F. Peterson and C. G. Jung, "Psychophysical Investigations with the Galvanometer and Pneumograph in Normal and Insane Individuals," *Brain* 30 (1907), 153–218; *Studies in Word-association: Experiments in the Diagnosis of Psychopathological Conditions Carried out at the Psychiatric Clinic of the University of Zurich, under the Direction of C. G. Jung* (London: W. Heinemann, 1918), 448–449; "Discovery Made by a Swiss Doctor May Play an Important Part in Criminal Trials," *New York Times* (Jun. 9, 1907).

95. C. Morgan, "A Study in the Psychology of Testimony," *Journal of American Institute of Criminal Law and Criminology* 8 (1917), 232; "Applied Psychology and its Possibilities," *New York Times* (Sep. 22, 1907); "The Soul Machine," *Harper's Weekly* 52 (1907), 12–13, 32–33.

96. V. Benussi, "Die Atmungsymptome der Lüge," *Archiv für gesammte Psychologie* 31 (1914), 244–273. For an English translation, see "The Respiratory Symptoms of Lying," *Polygraph* 4 (1975), 52–76; H. Crane, "A Study in Association Reaction and Reaction Time," *Psychological Monographs* 18 (1915), 1–61; W. Mar-

ston, "Systolic Blood Pressure Symptoms of Deception," *Journal of Experimental Psychology* 2 (1917), 117–163, on 162; Marvin Bowman, "New Machine Detects Lies," *Boston Sunday Advertiser* (May 8, 1921), B3; W. Marston, *Lie Detector Test* (New York: Smith, 1938), 46–48; "Marston, William Moulton," *Encyclopedia of American Biography*, (New York: American Historical Society, 1937), 7:23.

97. Marston, *Lie Detector Test*, 59–68; J. A. Matte, *The Art and Science of the Polygraph Technique* (Springfield, Illinois: Thomas, 1980), 729.

98. W. Marston, "Psychological Possibilities in the Deception Tests," *Journal of the American Institute of Criminal Law and Criminology* 11 (1921), 551–570; H. Burtt, "The Inspiration-Expiration Ratio During Truth and Falsehood," *Journal of Experimental Psychology* 4 (1921), 1–21; J. Larson, "Modification of the Marston Deception Test," *Journal of the American Institute of Criminal Law and Criminology* 12 (1921), 390–399.

99. Marston, *Lie Detector Test*, 70–72.

100. Quoted in J. E. Starrs, "A Still-Life Watercolor: Frye v. United States," *Journal of Forensic Sciences* 27 (Jul 1982), 684–694, on 689.

101. Marston, *Lie Detector Test*, 71. One should treat Marston's comments with a grain of salt. He was heavily invested in the outcome of the trial. His description is noticeably biased and he cited no verifiable references. On the other hand, Marston's quotes from the trial do correspond to the trial records and to the description found in the newspapers. The available trial records are at File 3968, retired files, National Record Center, Suitland, MD. See also Pardon File RG 204, Box 1583, Pardon Case Files 1853–1946, Record 56, pp. 384–412, ibid.

102. Marston, *Lie Detector Test*, 71.

103. Marston, *Lie Detector Test*, 72; Starrs, "A Still-Life Watercolor," 691; *The Evening Star* (Jul. 20, 1922), 1.

104. Marston, *Lie Detector Test*, 72; Starrs, "A Still-Life Watercolor," 691. For McCoy's legal career, see J. C. Proctor, *Washington: Past and Present* (New York: Lewis Historical Publishing Co., 1930), 237.

105. Anonymous, *History of the United States Court of Appeal for the District of Columbia Circuit in the Country Bicentennial Year* (Washington: U.S. Government Printing Office, 1977), 15; O. Richard, "Lie Detector," *Family Circle* 7 (Nov. 1, 1935), 14–22, on 21; Starrs, "A Still-Life Watercolor," 691.

106. Burtt, "The Inspiration-Expiration Ratio." Marston himself claimed that in the last ten years he had obtained accuracy of over 95 percent. Marston, "Sex Characteristics of Systolic Blood Pressure Behavior," *Journal of Experimental Psychology* 6 (1923), 387–419.

107. For the notorious reputation of handwriting and medical evidence see Anonymous, "Expert Testimony," *American Law Journal* 5 (1870), 228. The difficulties with ballistics evidence were clearly displayed a year earlier, in the notorious 1921 trial of Sacco and Vanzetti. See O. K. Fraenkel, *The Sacco-Vanzetti Case* (New York: Knopf, 1931), 331–405; L. Joughin and E. Morgan, *The Legacy of Sacco and Vanzetti* (New York: Harcourt, 1948), 126–131.

108. Quoted in Starrs, "A Still-Life Watercolor," 692.

109. J. R. Richardson, *Modern Scientific Evidence: Civil and Criminal* (Cincin-

nati: Anderson, 1974), 2nd ed., 315; S. Landsman, "Of Witches, Madmen, and Products Liability: An Historical Survey of the Use of Expert Testimony," *Behavioral Sciences and the Law* 13 (1995), 153.

110. C. T. McCormick, "Expert Testimony as an Invasion of the Province of the Jury," *Iowa Law Review* 26 (1941), 839–840; Austin, "Some Rules Governing the Examination of Expert Witnesses in Illinois," *Illinois Law Review* 19 (1924), 6.

111. Z. Chafee, "The Progress of the Law, 1919–1921: Evidence" *Harvard Law Review* 35 (1921–22), 302–317, on 309.

112. Editorial, "Electrical Machines to Tell Guilt of Criminals," *New York Times* (Sep. 10, 1911), part 6.

113. C. T. McCormick, *McCormick's Handbook of the Law of Evidence* (St. Paul: West Publishing Co., 1972), 2nd ed., 489.

114. W. Best, *A Treatise on the Principles of Evidence and Practices as to Proofs in Courts of Common Law* (London, 1854), 2nd ed., 346; J. Thayer, *A Preliminary Treatise on Evidence at the Common Law* (Boston: Little, Brown, 1940), 3rd ed., 269; H. Trautman, "Logical or Legal Relevancy," *Vanderbilt Law Review* 5 (1952), 388–395; M. McCormick, "Scientific Evidence: Defining a New Approach to Admissibility," *Iowa Law Review* 67 (1982), 880–882.

115. Frye v. United States, *Federal Reports* 293 (1923), 1013–1014.

116. Note, "The Use of Psychological Tests to Determine the Credibility of Witnesses," *Yale Law Journal* 33 (1923–24), 771–774, on 774; Note, "Evidence—Expert Testimony—Admissibility of Deception Tests," *Harvard Law Review* 37 (1924), 1138; Editorial, "Psychology in Court," *Law Notes* (July 1924), 28; Note, "Admissibility of Deception Tests," *Columbia Law Review* (1924), 429–430.

117. C. T. McCormick, "Deception-Tests and the Law of Evidence," *California Law Review* 15 (1926–27), 484–504.

118. See Chapters 4 (on microscopy) and 5 (on radiology).

119. A. Johnston, "The Magic Lie Detector," in V. A. Leonard, ed., *Academy Lectures on Lie Detection* (Illinois: Police Science Series, 1958), 25.

120. A. A. Moenssens and F. E. Inbau, *Scientific Evidence in Criminal Cases* (Mineola, NY: Foundation Press, 1978), 2nd ed., 5–6.

Epilogue

1. Editorial, "Science," *Scientific American* 71 (1894), 243.

2. W. L. Foster, "Expert Testimony—Prevalent Complaints and Proposed Remedies," *Harvard Law Review* 11 (1897–98), 169–186, on 169.

3. Gustav Eindlich, *Expert Testimony: What is to be Done with it?* (Philadelphia, 1896), 9–11.

4. Ibid., 12–13.

5. Quoted in A. Kidd, "The Proposed Expert Evidence Bill," *California Law Review* 3 (1914–1915), 216–226, on 218.

6. *People v. Dickerson* (1910), *Northwestern Reports* 129:198; Note, "Appoint

ment of Expert Witnesses by the Court," *Harvard Law Review* 24 (1910–1911), 483–484.

7. T. L. Haskell, *The Authority of Experts: Studies in History and Theory* (Bloomington: Indiana University Press, 1984), part 3, 180–225.

8. L. Friedman, "Expert testimony: Its Abuse and Reformation." *Yale Law Journal* 29 (1910), 247–257, on 252; Kidd, "The Proposed Expert Evidence Bill," 218, 222.

9. Wigmore offered a similar but less-demanding formulation in a new section: "Scientific Experimental Tests by Psychologists," which he added to the 1923 edition of his famous treatise on evidence. "All that should be required as a condition is the preliminary testimony of a scientist that the proposed test is an accepted one in his profession." J. Wigmore, *A Treatise on the Anglo-American System of Evidence in Trials at Common Law* (Boston: Little, Brown, and Co., 1923), 2nd ed., 5 vols., 2:419.

10. See the section "Winnowing the Wheat from the Chaff in *Folkes v. Chadd*," Chapter 1.

11. D. Faigman, E. Porter and M. Saks, "Check Your Crystal Ball at the Courthouse Door, Please: Exploring the Past, Understanding the Present and Worrying about the Future of Scientific Evidence," *Cardozo Law Review* 15 (1994), 1803–1807.

12. A. O. Lovejoy, "The Thirteen Pragmatisms." *Journal of Philosophy* (1908), 1–12, 29–39; H. S. Thayer, *Meaning and Action: A Critical History of Pragmatism* (Indianapolis: Hackett, 1968).

13. R. Pound, "The Scope and Purpose of Sociological Jurisprudence," *Harvard Law Review* 24 (1911–1912), 591–619 and 25 (1913), 140–168, 489–516.

14. C. L. Smallwood, "Evidence: Lie-Detector: Discussion and Proposals." *Cornell Law Quarterly* 29 (1944), 535–545; "The Lie Detector in the Courts," *Annual Report of the New York State Judicial Council* 14 (1948), 264–266.

15. P. Giannelli, "The Admissibility of Novel Scientific Evidence: Frye v. United States, A Half-Century Later, "*Columbia Law Review* 80 (1980), 1197–1250.

16. F. A. Rowell, "Comments: Admissibility of Evidence Obtained by Scientific Devices and Analysis," *Arkansas Law Review* 6 (1952), 181–198; A. Moenssens, "Admissibility of Scientific Evidence—An Alternative to Frye," *William and Mary Law Journal* 25 (1984), 5–6; Daubert v. Merrell Dow Pharmaceutical, Inc., *Federal Reports* 43 (1995), 1311–1317.

17. M. Ladd, "Expert Testimony," *Vanderbilt Law Review* 5 (1952), 414–431; H. Trautman, "Logical or Legal Relevancy," *Vanderbilt Law Review* 5 (1952), 385–413; C. T. McCormick, "Some Observations Upon the Opinion Rule and Expert Testimony." *Texas Law Review* 23 (1954), 109–137.

18. Giannelli, "The Admissibility of Novel Scientific Evidence." Taking Frye to the extreme, some courts considered the general acceptance of suggested evidence not only by scientists but also by other courts. Moenssens "Admissibility of Scientific Evidence," 25.

19. *Federal Rules of Evidence* (New York: Federal Judicial Center, 1975), Rule 702.

20. P. H. Schuck, *Agent Orange on Trial: Mass Toxic Disasters in the Courts* (Cambridge, MA: Belknap Press of Harvard University Press, 1987), 33; M. Saks, "Do we Really know Anything about the Behavior of the Tort Litigation System, and Why Not?" *University of Pennsylvania Law Review* (1992), 140; L. Brickman, "On the Relevance of the Admissibility of Scientific Evidence: Tort System Outcomes are Principally Determined by Lawyers' Rates of Return." *Cardozo Law Review* 15 (1994), 1755.

21. M. Angell, *Science on Trial: The Clash of Medical Evidence and the Law in the Breast Implant Case* (New York: Norton, 1996); President's Council on Competitiveness, *Agenda for Civil Justice Reform in America: a Report from the President's Council on Competitiveness* (Washington, D.C.:U.S. G.P.O, 1991); P. Huber, *Galileo's Revenge: Junk Science in the Courtroom* (New York: Basic Books, 1991).

22. M. McCormick, "Scientific Evidence: Defining a New Approach to Admissibility," *Iowa Law Review* 67 (1982), 879–916.

23. Daubert v. Merrell Dow Pharmaceutical, Inc., *Federal Reports* 951 (1991), 1128.

24. Daubert v. Merrell Dow Pharmaceutical, Inc., *United States Law Week* 61 (1993), 4805–4811. The petitioners won the battle in the Supreme Court but lost the war. The court reconsidered their evidence and excluded it again, this time under the new Daubert criteria. See Daubert v. Merrell Dow (1995). For a detailed analysis of the Daubert decision and its criteria see K. R. Foster and P. Huber, *Judging Science: Scientific Knowledge and the Federal Courts* (Cambridge, MA: MIT Press, 1997); Joseph Sanders, *Bendectin on Trial: A Study of Mass Tort Litigation* (Ann Arbor: University of Michigan Press, 1998).

25. Cf. Symposium, "Scientific Evidence After the Death of Frye," *Cardozo Law Review* 15 (1994); B. Black, J. F. Ayala and C. Saffran-Brinks, "Science and Law in the Wake of Daubert: A New Search for Scientific Knowledge." *Texas Law Review* 72 (1994), 715–802; Angell, *Science on Trial*, 127.

Index